Performance Evaluation of Industrial Systems

Discrete Event Simulation in Using Excel/VBA

Second Edition

Performance Evaluation of Industrial Systems

Discrete Event Simulation in Using Excel/VBA

Second Edition

David Elizandro and Hamdy Taha

CRC Press
Taylor & Francis Group
Boca Raton London New York

CRC Press is an imprint of the
Taylor & Francis Group, an **informa** business

CRC Press
Taylor & Francis Group
6000 Broken Sound Parkway NW, Suite 300
Boca Raton, FL 33487-2742

© 2012 by Taylor & Francis Group, LLC
CRC Press is an imprint of Taylor & Francis Group, an Informa business

No claim to original U.S. Government works

Printed in the United States of America on acid-free paper
Version Date: 20120215

International Standard Book Number: 978-1-4398-7134-8 (Hardback)

Visit the Taylor & Francis Web site at
http://www.taylorandfrancis.com

and the CRC Press Web site at
http://www.crcpress.com

Contents

PART III   APPLICATIONS

Preface

The focus of this book is on systems modeling and analysis to assess the performance of manufacturing, logistics, and other production systems that range from health care to computers and data communications. The objective of systems modeling is to develop an understanding of trade-offs between system performance and utilization of resources. System performance is either quantitative system-oriented characteristics, which refer to system productivity, reliability, flexibility, and security, or qualitative measures of responsiveness and perceived fairness by the user. Most systems are used by humans, whose behavior dramatically affects system performance. Because of the inherent randomness in systems, including the effects of human behavior, system performance is expressed in statistical terminology.

Performance analysis of production systems has become a paradigm across many industries. In a broader sense, systems modeling is the basis for coordinating the physical workplace with the people and work of a particular organization to support organizational objectives. Improvements in these production systems may be described in terms of reduced cycle time, reduction in cost, increase in capacity utilization, and the elimination of bottlenecks. In any production environment, discrete event simulation is a powerful tool for the analysis, planning, design, and operation of those facilities.

This book is written for the novice who wants to learn the basics of discrete simulation as well as for professionals who wish to use discrete event simulation to model systems described above. The book assumes that the reader has a fundamental familiarity with modeling concepts and Excel; however, it does not assume any prior programming experience. The book is organized into three parts. Part I presents concepts of discrete simulation, Part II covers the design environment for event-driven simulation (DEEDS), and Part III presents a variety of applications using DEEDS. The DEEDS environment is itself an Excel/VBA add-in.

Background

Basic approaches to discrete simulation have been process simulation languages (e.g., General Purpose Simulation System (GPSS)) and event-scheduling type (e.g., SIMSCRIPT). The trade-offs are that event-scheduling languages offer more modeling flexibility, and process-oriented languages are more intuitive to the user.

Process-oriented languages are based on blocks (or nodes) that perform functions in serial fashion. For example, suppose that a transaction leaving a queue must dispose of a previously acquired resource before it enters one of several available facilities for processing. The leaving transaction must pass through a special block/node to dispose of the resource and then through another to select the desired facility.

In a process-oriented language, special blocks/nodes must be designed to respond to distinct modeling needs. The result is that the language is not as user-friendly because of the high level of model abstractness and the large number of blocks/nodes in the language. Also, the fact that transactions must pass through these blocks/nodes serially does indeed reduce the language flexibility and may create the need for external (FORTRAN, Visual Basic, or Java) program inserts. A third disadvantage is that each special block/node is designed to respond to specific modeling needs, an approach that inevitably leads to a degree of redundancy, and hence inefficiency, in the language.

Objectives of the Book

With these considerations in mind, we embarked on the development of a new discrete simulation environment. The design objectives are to

- Achieve the modeling flexibility of an event-driven simulation language
- Achieve the intuitive nature of a process-oriented language
- Develop a user-friendly implementation environment
- Provide language visualization capability for testing models and resource allocation decisions

In essence, the goal was to design a development environment that is easy to use, yet flexible enough to model complex production systems.

Approach

Using pioneering ideas of Q-GERT's network simulation (nodes connected by branches), Hamdy Taha developed the simulation language SIMNET. The basis of SIMNET was that the majority of discrete simulations may be viewed in some form or other as a complex combination of simple queuing systems. SIMNET has four basic types of nodes: a source from which transactions arrive, a queue where transactions may wait when necessary, a facility where service is performed, and a "Delay" node, which is an infinite-capacity generalization of a facility node.

DEEDS implements the basic tenets of SIMNET in the Excel/VBA environment. Each node is designed to be self-contained, in the sense that user-developed event handlers written in Visual Basic for Applications (VBA) manage the transaction before it enters and as it leaves the node. In essence, such a definition implements the sequential nature of a process-oriented language, without the many blocks, and provides for the flexibility of event-driven simulation languages. This approach has proven to be effective and convenient in handling very complex simulation models.

DEEDS has been used by undergraduate industrial engineering students in the first simulation course at Tennessee Technological University. These students have completed an information systems course based on Excel and VBA. It is gratifying to report that these students were able to use DEEDS effectively to program fairly complex models after only four weeks of instruction. Some of these models are included in this book.

The students indicated the following advantages of DEEDS:

■ Excel facilitates model development and presentation of results.
■ With a VBA background, DEEDS is easy to learn.
■ A user interface facilitates model development and program management.
■ An interactive VBA debugger and "like English" trace report reduce the effort to validate a model.
■ VBA class definitions facilitate managing the four network nodes.
■ Explicit simulator messages detail unusual events that occur during the simulation.

Contents

The first part (Chapters 1 through 6) focuses on the fundamentals of simulation, and the second part (Chapters 7 through 16) covers DEEDS. The third part, Chapters 17 through 25, presents examples of production systems models and techniques for using DEEDS to improve decision making.

In Part I, the subject of simulation is treated in a generic sense to provide the reader with an understanding of the capabilities and limitations of this important tool. An overview of systems modeling and analysis to assess system performance described in Chapter 1 is followed by an introduction to the theory of queues in Chapter 2, which presents tools for analyzing the effects of various resource contentions on such performance measures as throughput, utilization, and cycle time. Chapter 3 provides an overview of simulation modeling with emphasis on the statistical nature of simulation. Chapter 4 gives a summary of the role of statistics in the simulation experiment. Chapter 5 is devoted to introducing the elements of discrete simulation, including types of simulation, methods for sampling from distributions, and methods for collecting data in simulation runs. The material in Chapter 6 deals with the statistical peculiarities of the simulation experiment and ways to circumvent the difficulties arising from these peculiarities.

Although the material in Part I treats simulation in a general sense, it ties to DEEDS by introducing and defining the terminology and the concepts used in the development of the language. The reader is encouraged to review Part I before beginning Part II.

In Part II, Chapter 7 provides a concepts overview of the DEEDS environment. Chapter 8 presents the basic design of the DEEDS environment. Presented in Chapter 9 are introductory VBA language constructs necessary to develop DEEDS models. Chapter 10 describes *ProgramManager*, the user interface for developing the simulation model. Also in Chapter 10 is a description of the *EventsManager* module where the simulation program resides and how to develop and execute a program.

Chapter 11 presents details of VBA and DEEDS class procedures used to develop simulation models. Output features of DEEDS and model validation tools such as simulation reports and the VBA Interactive Debugger are presented in Chapter 12. Chapter 13 details how to conduct simulation experiments in DEEDS. Chapter 14, a new chapter in this edition, describes model visualization features of DEEDS that enable an analyst to present an interactive view of model activity. Chapter 15 presents a variety of programs that demonstrate the flexibility of DEEDS and Chapter 16 presents advanced features of DEEDS that assist the modeler with complex decision-making schemes.

In Part III, Chapter 17 presents concepts that facilitate the design and development of large and complex simulation models. Also presented in Chapter 17 is an overview of management issues related to simulation projects. In Chapters 18 through 24, DEEDS simulation programs are designed and developed for a spectrum of traditional application areas that include facilities layout, material handling, inventory control, scheduling, maintenance, quality control, and supply chain logistics. For the practitioner, these models have been organized into chapters to address common operations management topics. However, several models could easily fit into two or more chapters.

All of the models were developed by senior students in an advanced simulation course at Tennessee Tech University. Their models have been edited to emphasize certain DEEDS features. Solutions may be regarded as examples of how these problems may be modeled in DEEDS. To conserve space, only selected output results based on a single run for each problem are presented. Most of these problems have been widely circulated in the literature as the basis for comparing simulation languages.

In Chapter 25, the capstone chapter, the research features of DEEDS are demonstrated. In contrast to the traditional "what if" approach to simulation, design of experiments and genetic algorithms techniques are integrated into the DEEDS environment to produce a powerful tool that can be used for "optimizing" the models in Chapters 18 through 24.

Use of This Book

The flexibility of DEEDS makes it a great tool for students or novices to learn the concepts of discrete simulation. Therefore, it can be the basis for an undergraduate course in introduction to simulation. By extending the depth of coverage, it can be the basis for a graduate simulation course. It can also be a reference book for practitioners engaged in simulation projects. It may also be used as a research tool by faculty and graduate students who are interested in "optimizing" production systems.

Acknowledgments

Several semesters of simulation classes at Tennessee Tech University have demonstrated that not only is design environment for event-driven simulation (DEEDS) an effective tool for teaching concepts of discrete event simulation, it is a viable tool for modeling real production systems in a variety of application areas. Without assistance from several students in particular, the development of the DEEDS software and production of this book would not have been possible. They are Clinton Thomas, Julie Braden, Chad Watson, Jacob Manahan, Chris Potts, Cody Allen, and Ben Reily. Each of these students made unique contributions to the design of DEEDS.

Diane Knight and my wife Marcia were very generous to contribute many hours proofreading the manuscripts of the first and second editions. Randy Smith at Microsoft provided valuable advice on VB programming in Visual Studio.

Finally, a very special thanks to Brandon Malone, who did a great job as the systems programmer by helping to port the simulator Addin from VBA to VB.net.

Authors

David W. Elizandro is a professor of industrial engineering at Tennessee Tech University, where he teaches operations research and simulation. He earned a BS in chemical engineering and an MBA, and a PhD in industrial engineering. Professor Elizandro has served in a variety of administrative and leadership roles in science and engineering education.

Dr. Elizandro has written publications and made presentations in areas such as expert systems, data communications, distributed simulation, adaptive control systems, digital signal processing, and integrating technology into engineering education. He has also been an industry consultant on projects that include developing computer models to assess manpower production requirements and resource utilization, evaluation of an overnight freight delivery network, performance of digital signal processing boards, and response time for a flight controller communications system.

Professor Elizandro received the University Distinguished Faculty Award, Texas A & M, Commerce and College of Engineering Brown–Henderson Award at Tennessee Tech University. He served on the National Highway Safety Advisory Commission and is a member of Tau Beta Pi, Alpha Pi Mu, and Upsilon Pi Epsilon.

Hamdy A. Taha is a university professor emeritus of industrial engineering with the University of Arkansas, where he taught and conducted research in operations research and simulation. He is the author of four other books on integer programming and simulation, and his work has been translated into Chinese, Korean, Spanish, Japanese, Malay, Russian, Turkish, and Indonesian. He is also the author of several book chapters, and his technical articles have appeared in the *European Journal of Operations Research*, *IEEE Transactions on Reliability*, *IIE Transactions*, *Interfaces*, *Management Science*, *Naval Research Logistics Quarterly*, *Operations Research*, and *Simulation*.

Professor Taha was the recipient of the Alumni Award for excellence in research and the university-wide Nadine Baum Award for excellence in teaching, both from the University of Arkansas, and numerous other research and teaching awards from the College of Engineering, University of Arkansas. He was also named a Senior Fulbright Scholar to Carlos III University, Madrid, Spain. He is fluent in three languages and has held teaching and consulting positions in Europe, Mexico, and the Middle East.

MODELING
FUNDAMENTALS

Chapter 1

Introduction to Modeling

1.1 Introduction

The focus of this book is on using systems modeling and analysis to assess the performance of manufacturing, logistics, and other production systems that range from health care to computers and data communications systems. The purpose of systems modeling and analysis is to develop an understanding of trade-offs between system performance and utilization of resources. System performance is either quantitative system-oriented characteristics, which refer to system productivity, reliability, flexibility, and security, or qualitative measures of responsiveness and perceived fairness by the user. Most systems are used by humans, whose behavior dramatically affects system performance. Because of the inherent randomness in systems that include the effects of human behavior, system performance is expressed in statistical terminology.

A classic example is a telephone exchange, in which system performance, as perceived by the user, is affected by the behavior of others. The scope of performance evaluation has expanded a great deal since then, but a 1909 publication by A. K. Erlang [1], the Danish mathematician and engineer, is viewed as the original system performance evaluation. Because performance is affected by statistical data and performance metrics are expressed using probabilities Erlang's publication has the basic features of system performance evaluation.

1.2 Model Design

The engineering design process for performance evaluation is presented in Figure 1.1. As shown, the system is designed to meet the system requirements. These requirements are the basis for performance metrics used to evaluate the system. Some systems have very small tolerances between specifications and actual system performance. Others, especially those with extensive human interaction, will have a very large variability in performance metrics. Understanding the concept of variability is important in the design phase when alternative components of the system are selected.

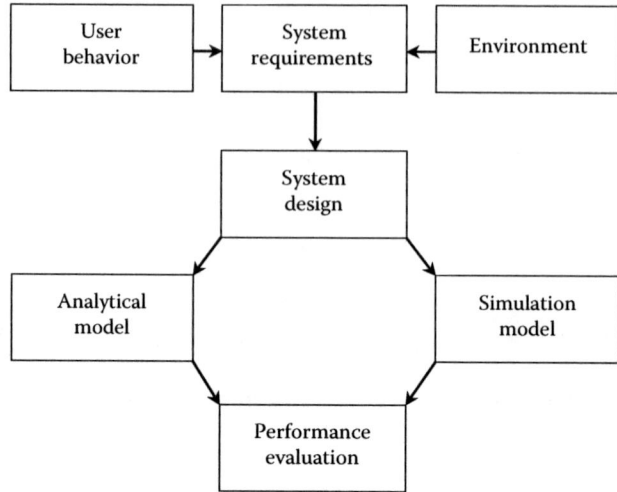

Figure 1.1 Performance evaluation structure.

1.3 Hierarchical Modeling

The practice of designing complex systems includes decomposing the system into manageable subsystems and analyzing these individual components to determine the respective performance levels necessary to meet the system requirements. Similarly, a model of the system should reflect the system's modular structure. The model of the system is also modularized using a top-down procedure. The functionality of each layer reflects the functionality of its underlying layers. This iterative procedure is the basis for successive model layers until the basic modeling components are implemented in primitive operations. Modeling, prediction analysis, measurement, and evaluation are performed at each layer.

Clearly, the most important step in any performance study is the decomposition of the system model into a set of submodels of manageable complexity and well-defined interfaces. Much of the difficulty in systems modeling is a result of the failure to properly design a hierarchical model. Figure 1.2 presents a schematic of a hierarchical model structure. Micro-level modules are at the lowest layer. Analysis results of these modules should present results in a usable form to the intermediate-level modules. Similarly, results from intermediate-level modules are incorporated into macro-level modules. In some situations a time scaling factor must be applied to information shared between layers in order to synchronize the layers with respect to time. Micro-level module activities may be measured in seconds, while intermediate-level activities are expressed in minutes.

For example, in a production line with multiple workstations, workstation cycle times may be measured in minutes while individual tasks within the workstation are expressed in seconds. The line may be sufficiently long that production line cycle time is expressed in fractional hours. Time scales for modules within a system model are application specific. In a computer system the range could be from microseconds to minutes. The first pass at decomposition is based on the configuration of the actual system. Thereafter, the criteria include grouping modules with frequent interactions and modules that require synchronization with respect to time. A basic decomposability requirement is that time scaling for a given layer is sufficiently smaller than for activity times of predecessor layers.

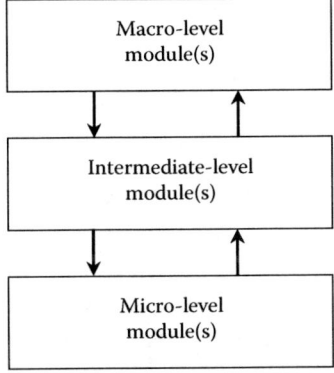

Figure 1.2   The hierarchical model structure.

1.4 Analytic Models

System performance models may be classified into analytic and simulation models. A nuance of simulation and analytic models is empirical models that are the basis for design of experiment techniques described in Section 25.3. Each of these modeling techniques provides insight into the system behavior. Analytic models are functional relations between the parameters of a system and the system performance expressed in equations that are solved using analytic or numeric techniques. Queuing theory is an important analytic tool because most performance problems are related to queuing delays caused by contention for resources. The operations research discipline focuses on formulating and solving such models. The origin of queuing theory dates to Erlang's work. He showed that the origin of random telephone calls can be represented by a Poisson distribution. Since the pioneering work by Erlang and others, queuing theory has become an important branch of applied probability and a basic tool for the management science discipline.

Although analytic models are a very useful tool, some systems are sufficiently complex that formulating an analytic model for performance evaluation is not a viable option. When analytic models are mathematically intractable, simulation becomes the only feasible approach to performance estimation. However, it is important that some attempt at analysis, using even a simplified model, be made wherever possible to validate results of the simulation model.

Many of the assumptions made in simple queuing models may seem oversimplified. However, the prediction results obtained from a simplified analytic model will often agree surprisingly well with the actual system observations as well as estimates obtained from a simulation model.

1.5 Simulation Models

Although simulation has a number of connotations, in this book simulation is a numeric technique for conducting a computer based experiment on a time dependent system. Therefore, the concept of time is explicit in a simulation. Even though the analyst may be interested in only one steady-state performance measure (e.g., resource utilization, response time, or network capacity) a simulation model must include the dynamic behavior of the entire system.

A computer simulation overcomes many of the problems associated with the performance evaluation of systems. For example, relatively inexpensive computer capability provides a platform for estimating system performance without the expense of constructing a system and finding that it does meet performance specifications. Although characterizing user behavior remains a problem, it is possible to use the actual measured distributions rather than simplified distributions that facilitate analytic modeling.

When using computer simulation to model a system, it is important to ensure that the simulation is adequately representative of the system. What are perceived as minor features of a system may be ignored and the model will be unable to predict performance of the system accurately. The other caution in using simulation is to validate the accuracy of the results. It is possible to conduct a valid simulation and, because of the randomness inherent in the system, the performance estimates are biased and therefore unrepresentative.

The decomposition modeling technique described in Section 1.4 facilitates a hybrid modeling technique that is a combination of analytic and simulation modeling. The integration of these two techniques to conduct performance evaluation should be considered because it combines the efficiency of analytic modeling and advantages of simulation modeling.

1.5.1 Simulation Model Complexity

The structure and complexity of a simulation model depends on the scope of the simulation experiment. Although the motivation is now different, there are at least two ancillary benefits derived from the hierarchical structure as an important modeling technique. A hierarchical structure allows modularization of a simulation program into a set of subprograms. Also, a modular design provides flexibility in the program structure that enables model extensions and changes to be easily incorporated into the program.

A hierarchical model structure may also substantially reduce the simulation run time. In general, the simulation run length is determined by the required accuracy of simulation estimates and the transient time necessary for the system to achieve steady-state conditions. Assume that in the model structure shown in Figure 1.2 the basic level of activity in micro-level modules is 10 seconds and the basic level of activity in intermediate-level modules is 10 minutes. In eight hours of simulation time approximately 2440 micro-level module activities will have been performed and it is very likely these modules will have achieved steady state.

During the same time period approximately 48 activities will have been performed in intermediate-level modules. Even if these modules have achieved steady state, the sample size will likely be insufficient for an unbiased estimate of performance measures for these modules. A similar analysis can be made for macro-level modules. However, clearly there are an insufficient number of observations to achieve steady state in these modules. As implied by this situation, it is important to note that most simulation programs specify lapsed time to conduct the experiment. However, the length of time to conduct the simulation is actually a proxy measure for the total number of events processed.

As indicated in Figure 1.1, a model's steady-state performance measures are dependent on the environment. Because a simulation model processes each change in the state of the system caused by the individual events, an unsophisticated simulation model often requires an unnecessarily long period of time. A more efficient approach to modeling is to run module layers separately. Separate runs of the model are needed for different sets of parameters that characterize different environmental conditions. For example, the intermediate-level modules of Figure 1.2 can be modeled separately over a range of intermediate-level environmental conditions. Then macro-level input parameters can be developed for each of the environmental conditions identified. Using this approach, interaction between lower-level and higher-level modules can be achieved by using summary statistics as a basis for resource requests.

1.5.2 Simulation Scripts

A probabilistic simulation model uses random-number generators to map random variables with known distributions to a sequence activity and resource demands. An advantage of a probabilistic simulation model is that the event stream is generated artificially and therefore workload parameters are easily adjustable.

After the system is operational, the model and associated measurement data can then be used to improve system performance. An alternative to a probabilistic simulation model to improve system performance is a script-based simulation model. A script is a time-ordered list of events that occurred in an operational system. A script is therefore a representative segment of the actual workload of the system. Building scripts requires considerable effort because it involves extracting data from historical date. However, the obvious advantage over a probabilistic model is that scripts are based on actual workloads.

With a simulation model of the system, and to a lesser extent when there is an operational system, an empirical model can be constructed. An empirical model is a statistical characterization of system performance based on data from a real system or a simulation model that has been reconfigured. Of course, using an actual production system to conduct experiments is always problematic because of opportunity costs associated with lost or inefficient production capacity while the system is being reconfigured. Of course, the logical structure of the system should direct the effort to design empirical models.

Often, because of the complexities of interactions among various system components and user programs, it is difficult to identify which system and algorithmic variables are significant, and in what functional form these variables relate to a particular performance measure. An empirical model can be used to capture the subtleties of such system behavior. Empirical characterization may also reveal unexpected system behaviors, which may lead to detecting deficiencies in the system design or implementation errors. Empirical observations of an actual system or a simulation model can help to resolve these issues.

1.5.3 System Performance Measures

The accuracy of predicting system performance is determined primarily by the accuracies of the system design parameters and system workloads. Therefore, the selection of model parameters is as important as the modeling technique. Model verification based on actual system measurements is also extremely important. The accuracy of a model can be validated only when modeling results are compared with performance data from the real environment. For a new system, domain experts who developed the system design can provide insights on comparisons between model predictions and system specifications. However, reconciliation of significant differences between specifications and model predictions becomes problematic.

Statistical methods provide a systematic procedure to assess the degree of uncertainty inherent in a statistical inference. A measurement experiment minimizes the uncertainty in performance measures. Such procedures are an experimental design or a design of experiments. Experimental design techniques enable the analyst to evaluate an environment in which changes in the system workload during a measurement period may introduce bias as well as variability in performance measures.

Experimental design is also useful for assessing the effects on performance measures by changes to the system. For example, the system may have multiple factors, such as system parameters and scheduling or resource allocation policies, which can vary. Factors may be quantitative (numeric) or qualitative (categorical). The effects of levels of these factors on a performance measure are not generally cumulative. Also, when factors interact, the significance of each factor cannot be evaluated in isolation. Experiments are conducted using combinations of values that the factors may assume. Exhausting all possible combinations of the experimental arrangements becomes impractical as the number of factors and the range of their level increases.

1.6 Organization of This Book

This book encompasses three major parts. In Part I, an overview of systems modeling and analysis to assess system performance characteristics in Chapter 1 is followed by an introduction to the theory of queues in Chapter 2 that includes tools for analyzing the effects of various resource contentions on such performance measures as throughput, utilization, and cycle time. Chapters 3 through 6 cover prominent theoretical aspects of discrete event simulation. Included in these four chapters are the mechanics of using the concept of events to construct a discrete simulation model and how to prepare input data for the model. Methods for internal data collection by a simulation language processor are also detailed in these chapters. Finally, the statistical aspects of the simulation experiment are explained, and its impact on simulation output is emphasized. Although Chapters 5 and 6 treat simulation in a general sense, they introduce and define the terminology and concepts used in the development of design environment for event-driven simulation (DEEDS). The reader is encouraged to at least review Part I before beginning Part II.

In Part II, Chapter 7 provides a concepts overview of the DEEDS environment. Chapter 8 presents the basic design of the DEEDS environment. Presented in Chapter 9 are introductory Visual Basic for Applications (VBA) language constructs necessary to develop DEEDS models. Chapter 10 describes *ProgramManager*, the user interface for developing the simulation model. Also in Chapter 10 is a description of the *EventsManager* module where the simulation program resides and how to develop and execute a program. Chapter 11 presents details of VBA and DEEDS class procedures used to develop simulation models. Output features of DEEDS and model validation tools such as simulation reports and the VBA Interactive Debugger are presented in Chapter 12. Chapter 13 details how to conduct simulation experiments in DEEDS. Chapter 14, a new chapter in this edition, describes model visualization features of DEEDS that enable an analyst to present an interactive view of model activity. Chapter 15 presents a variety of programs that demonstrate the flexibility of DEEDS and Chapter 16 presents advanced features of DEEDS that assist the modeler with complex decision-making schemes.

In Part III, Chapter 17 presents concepts that facilitate the design and development of large and complex simulation models. Also presented in Chapter 17 is an overview of management issues related to simulation projects. In Chapters 18 through 24, DEEDS simulation programs are designed and developed for a spectrum of traditional application areas that include facilities layout, material handling, inventory control, scheduling, maintenance, quality control, and supply chain logistics. In Chapter 25, the research features of DEEDS are demonstrated. In contrast to the traditional "what if" approach to simulation, design of experiments and genetic algorithms techniques are integrated into the DEEDS environment to produce a powerful tool that can be used for "optimizing" models.

Reference

1. Erlang, A. K., Solution of some problems in the theory of probabilities of significance in automatic telephone exchange, *Post Office Electrical Engineers Journal*, 1918, 8, pp. 33–45.

Suggested Reading

1. Ghanbari, M. et al., *Principles of Performance Engineering for Telecommunications and Information Systems*, IEE Telecommunications Series 35, 1997.
2. Kobayashi, H., and Brian, M., *System Modeling and Analysis: Foundations of System Performance Evaluation*. Upper Saddle River, NJ: Pearson Prentice Hall, 2009.

Chapter 2

Basic Queuing Models

2.1 Introduction

Waiting for service is a part of life. Customers wait in line at restaurants, grocery store checkout counters, and the post office. Jobs wait to be processed on a machine, airplanes wait in holding patterns for permission to land at airports, and cars wait at traffic lights. Although often expensive to achieve, waiting time can usually be eliminated. The design criterion is the cost associated with reducing waiting time to a "tolerable" level.

Queuing systems analysis includes quantifying waiting by using measures of performance, such as average queue length, average waiting time in queue, and average server utilization. Example 2.1 demonstrates how these measures are used to design a service system.

Example 2.1

McBurger is a fast-food restaurant with three service counters. A study to investigate quality of service reveals the following relationship between the number of service counters and waiting time for service:

Number of cashiers	1	2	3	4	5	6	7
Average waiting time (mins.)	16.2	10.3	6.9	4.8	2.9	1.9	1.3

From the data, there is a seven-minute average waiting time for the present three-counter situation and five counters will reduce waiting time to just under three minutes.

Unlike other operations research techniques, queuing theory is not an optimization tool. However, queuing analysis can be used in the context of optimization. As shown in Chapter 25, the total handling cost of arriving freight may be reduced by strategically assigning dock stripping doors to arriving trailers and loading doors to specific destinations. A queuing system model is used by the genetic algorithm to evaluate the handling cost of various dock configurations. In general, the major limitation of queuing system models that evaluate service cost is obtaining reliable cost estimates of waiting, especially when people are an integral part of the system.

2.2 Elements of a Queuing Model

Principals in a queuing model are arriving customers and the server. Upon arrival a customer may immediately begin service or wait in a queue while the server is busy. When a server completes serving a customer, a waiting customer, if any, is automatically "pulled" from the queue and "placed" into service. If the queue is empty, the server becomes idle and remains idle until a new customer arrives.

The arrival of customers is characterized by the interarrival time between successive customers, and service is the service time per customer. Generally, the interarrival and service times are probabilistic, as in a post office environment, or deterministic, as in the arrival of applicants for job interviews with a fixed time for each interview.

Queue capacity has an important role in the analysis of queues. Capacity may be finite, as in a buffer area between two successive machines, or it may be infinite, as in a mail order warehouse. Queue discipline refers to the order in which customers are selected from a queue. The discipline is also an important factor in queuing analysis. The most common discipline is first come, first served (FCFS). Others include last come, first served (LCFS) and service in random order (SIRO). Customers may also be selected from the queue based on priority. For example, rush jobs at a shop are processed ahead of regular jobs.

The queuing behavior of customers also affects waiting-line analysis. Customers may jockey from one queue to another, hoping to reduce waiting time. They may also balk the system because of anticipated long delay, or renege from a queue because of excessive waiting time in the queue. The service system may have parallel servers (e.g., post office or bank operation). Servers may also be in series (e.g., jobs are processed on successive machines), or networked (e.g., routers in a computer network). The source of customers may be finite or infinite. A finite source limits the number of customers arriving for service (e.g., machines needing service by a repairperson). Customers will continue to arrive from an infinite source (e.g., calls arriving at a telephone exchange). Variations in elements of a queuing environment are the basis for a variety of queuing models. A convenient notation for characterizing a queuing system is

$$(a/b/c):(d/e/f)$$

where
a = arrivals distribution
b = departures (service time) distribution
c = number of parallel servers (= 1, 2,..., ∞)
d = queue discipline
e = maximum number (finite or infinite) allowed in the system (in-queue plus in-service)
f = size of the source (finite or infinite)

The standard notation for arrivals and departures (symbols a and b) is

M = Markovian (Poisson) arrivals or departures distribution (exponential interarrival or service time distribution)
D = constant (deterministic) time
E_k = Erlang or gamma distribution of time (or, equivalently, the sum of independent exponential distributions)

GI = general (generic) distribution of interarrival time
G = general (generic) distribution of service time

Queue discipline notation (symbol *d*) includes

FCFS = first come, first served
LCFS = last come, first served
SIRO = service in random order
GD = general discipline (i.e., any type of discipline)

To illustrate the notation, the model (*M/D/10*):(*GD/20/∞*) has Poisson arrivals (or exponential interarrival time), constant service time, and 10 parallel servers. The queue discipline is *GD*, and there is a limit of 20 customers on the entire system. The source limit from which customers arrive is infinite. The first three elements of the notation (*a/b/c*) were devised by D. G. Kendall in 1953 and are in the literature as the Kendall notation. In 1966, A. M. Lee added the symbols *d* and *e* to the notation. H. Taha added the last element, symbol *f*, in 1968.

Figure 2.1 depicts a Poisson distribution based queuing environment with *c* parallel servers. A waiting customer is selected from the queue to begin service by the first available server. Customer arrival rate is λ customers per unit time. Because the servers are identical the service rate for each server is μ customers per unit time. The number of customers in the system includes those in service and those waiting in queue.

Common queuing system performance measures are

L_s = expected number of customers in system
L_q = expected number of customers in queue
W_s = expected waiting time in system
W_q = expected waiting time in queue
$\bar{c}$ = expected number of busy servers

where the system, as shown in Figure 2.1, includes the queuing and server environment.

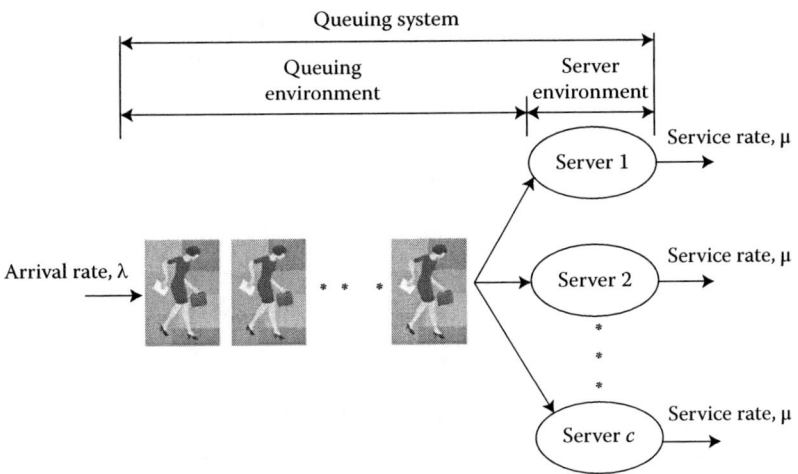

Figure 2.1   Schematic representation of a queuing system with *c* parallel servers.

2.3 Role of the Exponential Distribution

To appreciate the most basic queuing models a fundamental background in the theory of queues is necessary. Therefore this chapter begins with a presentation of the "total randomness" property of two important distributions: the Poisson and the exponential. These distributions are important for identifying situations where queuing results may be applied in practice. The remainder of this chapter emphasizes the interpretation of basic queuing model analyses that require computationally difficult formulas. To avoid tedious computational details, *PoissonQueuingModel.xls* is recommended for solving problems and conveniently testing various scenarios.

In most queuing situations customer arrivals occur in a totally random fashion. Randomness assumes that the occurrence of an event (e.g., arrival of a customer or completion of a service) is not influenced by the elapsed time since the last event occurred. Random interarrival and service times are described by the exponential distribution, defined as

$$f(t) = \lambda e^{-\lambda t}, \quad t \geq 0$$

The expected value of the exponential distribution is

$$E\{t\} = \frac{1}{\lambda}$$

And the probability that t is less than some value T is

$$P\{t \leq T\} = \int_0^T e^{-\lambda t} dt = 1 - e^{-\lambda T}$$

In the above equations λ is the rate per unit time for arrivals or departures. The randomness of the exponential distribution is illustrated in the following example: If the time is 8:20 a.m. and the last arrival occurred at 8:02 a.m., the probability that the next arrival will occur by 8:29 is a function of the time interval from 8:20 to 8:29 and is independent of the elapsed time since the last event (8:02 to 8:20). This is the forgetfulness or lack of memory property of the exponential distribution.

If the exponential distribution, $f(t)$, that represent the time, t, between successive events and S is the time interval since the last event occurred, then the forgetfulness property implies that

$$P\{t > T + S \mid t > S\} = P\{t > T\}$$

For the exponential with mean $\dfrac{1}{\lambda}$

$$P\{t > Y\} = 1 - P\{t < Y\} = e^{-\lambda Y}$$

therefore

$$P\{t > T + S \mid t > S\} = \frac{P\{t > T + S, t > S\}}{P\{t > S\}} = \frac{P\{t > T + S\}}{P\{t > S\}}$$

$$= \frac{e^{-\lambda(T + S)}}{e^{-\lambda S}} = e^{-\lambda T} = P\{t > T\}$$

Example 2.2

A machine has a standby unit for immediate replacement upon failure. The machine's (and its standby unit's) time to failure is exponential and occurs on average every five hours. The operator claims that the machine breaks down every evening around 8:30 p.m. Evaluate the operator's claim.

The average failure rate is $\lambda = 0.2$ failures per hour and the exponential distribution for time to failure is

$$f(t) = 0.2\,e^{-0.2t} \quad t > 0$$

Because the time between breakdowns is exponentially distributed and therefore random, the operator's claim cannot be true. The probability that a failure will occur by 8:30 p.m. cannot be used to support or refute the claim because the probability depends on the time (relative to 8:30 p.m.). For example, if the time is 8:20 p.m., the probability of the operator's claim is

$$p\left\{t < \frac{10}{60}\right\} = 1 - e^{-0.2*(10/60)} = 0.03278$$

If the time now is 1:00 p.m., the probability that a failure will occur by 8:30 p.m. increases to approximately 0.777. These extreme values indicate the operator's claim cannot be supported.

2.4 Pure Arrival and Departure Models

In pure arrival models only arrivals occur. In pure departure models only departures are allowed. An example of a pure arrival model is the creation of birth certificates for newly born babies. The pure departure model is demonstrated by random sales of an item in a store. Interarrival times in the pure birth model and the interdeparture times in the pure departure model are described by the exponential distribution. Demonstrated in this section is the relationship between the exponential and the Poisson distributions, in the sense that one distribution defines the other.

2.4.1 Pure Arrival Model

For the pure arrival model define

$$p_0(t) = (\text{Probability that no arrivals occur during time } t)$$

If the interarrival time is exponential and arrival rate is λ customers per unit time, then

$$p_0(t) = P\{\text{interarrival time} \geq t\} = 1 - P\{\text{interarrival time} \leq t\} = 1 - (1 - e^{-\lambda t}) = e^{-\lambda t}$$

For sufficiently small time interval $h > 0$,

$$p_0(h) = e^{-\lambda h} = 1 - \lambda h + \frac{(\lambda h)^2}{2!} - \cdots = 1 - \lambda h + 0(h)^2$$

The exponential distribution is based on the assumption that during $h > 0$, at most one arrival can occur. Therefore, as $h \to 0$

$$p_1(h) = 1 - p_0(h) \approx \lambda h$$

As shown, the probability of an arrival during time interval h is directly proportional to h, with proportionality constant λ, the arrival rate. For the distribution of the number of arrivals during time interval t when the interarrival time is exponential with mean $1/\lambda$, define

$$p_n(t) = (\text{Probability of } n \text{ arrivals during } t)$$

For sufficiently small $h > 0$,

$$p_n(t + h) \approx p_n(t)(1 - \lambda h) + p_{n-1}(t)\lambda h, \quad n > 0$$
$$p_0(t + h) \approx p_0(t)(1 - \lambda h), \quad n = 0$$

In the first equation n arrivals occur during $t + h$ when there are n arrivals during t and no arrivals during h or when there are $n - 1$ arrivals during t and one arrival during h. According to the exponential distribution all other combinations are not allowed because at most only one arrival can occur during a very small period h. The product law of probability is applicable to the right-hand side of the equation because arrivals are independent. For the second equation, zero arrivals during $t + h$ occur only if no arrivals occurred during t and h.

By rearranging terms and taking limits as $h \to 0$, the result is

$$p_n'(t) = \lim_{h \to 0} \frac{p_n(t + h) - p_n(t)}{h} = -\lambda p_n(t) + \lambda p_{n-1}(t), \quad n > 0$$
$$p_0'(t) = \lim_{h \to 0} \frac{p_0(t + h) - p_0(t)}{h} = -\lambda p_0(t), \quad n = 0$$

$p_n'(t)$ is the derivative of $p_n(t)$ with respect to t. The solution of the preceding difference-differential equations yields

$$p_n(t) = \frac{(\lambda t)^n e^{-\lambda t}}{n!}, \quad n = 0, 1, 2, \ldots$$

This is a Poisson distribution with mean $E\{n|t\} = \lambda t$ arrivals during t. The preceding shows that when the time between arrivals is exponential with mean $\frac{1}{\lambda}$ the number of arrivals during elapsed timed t is Poisson with mean λt. The converse is true also. The following table summarizes exponential and Poisson distribution relationships for λ arrivals per unit time.

	Exponential	Poisson
Random variable	*Time* between successive arrivals, t	*Number* of arrivals, n, during specified period T
Range	$t \geq 0$	$n = 0, 1, 2, \ldots$
Density function	$f(t) = \lambda e^{-\lambda t}, \quad t \geq 0$	$p_n(T) = \dfrac{(\lambda T)^n e^{-\lambda T}}{n!}, \quad n = 0, 1, 2, \ldots$
Mean value	$\dfrac{1}{\lambda}$ time units	λT arrivals during T
Cumulative probability	$P\{t \leq A\} = 1 - e^{-\lambda A}$	$p_{n \leq N}(T) = p_0(T) + p_1(T) + \cdots + p_N(T)$
P{no arrivals during period A}	$P\{t > A\} = e^{-\lambda A}$	$p_0(A) = e^{-\lambda A}$

Example 2.3

Babies are born at a rate of one birth every 12 minutes and the time between births is exponentially distributed. Find

    a. The average number of births per year.
    b. The probability that no births will occur in any one day.
    c. The probability of issuing 50 birth certificates in three hours given that 40 certificates were issued during the first two hours of the three-hour period.

The birth rate per day is $\lambda = \dfrac{24 \times 60}{12} = 120$ births/day. The number of births per year is $\lambda t = 120 \times 365 = 43,800$ births/year. The probability of no births on any day, using the Poisson distribution, is

$$p_0(1) = \frac{(120 * 1)^0 \, e^{-120 \times 1}}{0!} = e^{-120} = 0$$

No births in one day is equivalent to the time between successive births exceeding one day. From the exponential distribution the probability is

$$P\{t > 1\} = e^{-120} = 0$$

Because the distribution of the number of births is Poisson, the probability of issuing 50 certificates by the end of three hours given that 40 certificates were issued during the first two hours is equivalent to having 10 (= 50 – 40) births in one (= 3 – 2) hour. For $\lambda = \dfrac{60}{12} = 5$ births per hour,

$$p_{10}(1) = \frac{(5 * 1)^{10} \, e^{-5*1})}{10!} = 0.01813$$

Queuing formulas are tedious and require some programming skills to achieve reasonable computational accuracy. For calculations associated with the Poisson distribution, Excel *POISSON*, *POISSONDIST*, and *EXPONDIST* functions may be used to compute individual and cumulative probabilities for Poisson and exponential distributions. These functions are automated in PoissonTables.xls. For example, with a birth rate of 5 babies per hour, the probability of exactly 10 births in 0.5 hours is computed by entering 2.5 in cell F16, 10 in cell J16, and obtaining the answer 0.000216 in cell M16. The cumulative probability of at most 10 births is in cell O16 (= 0.999938). For the probability of time between births being less than or equal to 18 minutes, use the exponential distribution by entering 2.5 in cell F9 and 0.3 in cell J9. The answer 0.527633 is in cell O9. Use PoissonQueue.Model.xls to determine all significant (10^{-7}) Poisson probabilities. Under Lambda is $\lambda t = 5 \times 1 = 5$ births per day. For the pure birth model of Example 2.3 data are entered as follows:

(M/M/c):(GD/N/K) Queueing Model (includes N and/or K = infinity)			
Input Data (to enter an infinite value, type i or infinity):			
λ =	5	μ =	0
c =	0		
System Limit N=	infinity	Source limit K =	infinity

2.4.2 *Pure Departure Model*

In the pure departure model no arrivals are allowed and the system is initialized with N customers at time 0. Departures occur at the rate of μ customers per unit time. To develop the difference-differential equations for the probability $p_n(t)$ of n customers in the system after t time units, the same approach as used with the pure arrival model and is as follows:

$$p_N(t+h) = p_N(t)(1-\mu h)$$
$$p_n(t+h) = p_n(t)(1-\mu h) + p_{n+1}(t)\mu h, \quad 0 < n < N$$
$$p_0(t+h) = p_0(t) + p_1(t)\mu h, \quad n = 0$$

As $h \to 0$,

$$p_N'(t) = -\mu p_N(t)$$
$$p_n'(t) = -\mu p_n(t) + \mu p_{n+1}(t), \quad 0 < n < N$$
$$p_0'(t) = \mu p_1(t), \quad n = 0$$

These equations yield the truncated Poisson distribution:

$$p_n(t) = \frac{(\mu t)^{N-n} e^{-\mu t}}{(N-n)!}, \quad n = 1, 2, \ldots N$$

$$p_0(t) = 1 - \sum_{n=1}^{N} p_n(t)$$

Example 2.4

At the beginning of each week a flower shop orders 18 dozen roses. On average, 3 dozen are sold (one dozen at a time) each day, and the demand follows a Poisson distribution. When the stock level reaches 5 dozen a new order of 18 dozen is scheduled for delivery at the beginning of the following week. At the end of the week all remaining roses are disposed. Determine

    a. The probability of placing an order on any one day of the week.
    b. The average number of dozen roses discarded at the end of the week.

Because sales occur at the rate of $\mu = 3$ dozen per day, the probability of placing an order by the end of day t is given as

$$p_{n \leq 5}(t) = p_0(t) + p_1(t) + \cdots + p_5(t)$$

$$= p_0(t) + \sum_{n=1}^{5} \frac{(3t)^{18-n} e^{-3t}}{(18-n)!}, \quad t = 1, 2, \ldots, 7$$

Calculations for $p_{n \leq 5}(t)$ shown are from PoissonQueuingModel.xls. Input data for the pure departure model corresponding to $t = 1, 2, \ldots, 7$ are

$$\lambda = 0, \ \mu = 3t, \ c = 1, \ \text{system } \lim(N) = 18, \ \text{and source } \lim(K) = 18$$

Output is summarized as

t (days)	1	2	3	4	5	6	7
μt	3	6	9	12	15	18	21
$p_{n \leq 5}(t)$	0.0000	0.0088	0.1242	0.4240	0.7324	0.9083	0.9755

The average number of dozen roses discarded at the end of the week ($t = 7$) is $E(n|t = 7)$. Values of $p_n(7)$, $n = 0,1,2,\ldots,18$, can be determined using PoissonQueuingModel.xls.

$$E\{n \mid t = 7\} = \sum_{(n=0)}^{18} np_n(7) = 0.664 \approx 1 \text{ dozen}$$

2.5 General Poisson Queuing Model

This section develops a general queuing model that combines both arrivals and departures based on the Poisson assumptions that interarrival and service times are exponentially distributed. The model is the basis for the derivation of the specialized Poisson models, also in this section.

The generalized model is based on long-run or steady-state behavior, which is achieved after the system has been in operation for a sufficiently long time. This type of analysis contrasts with the transient behavior during the early operation of the system. One reason for not discussing the transient behavior in this chapter is the analytic complexity of transient behavior. Another is that the study of most systems reflects steady-state conditions.

The generalized model assumes that arrival and departure rates are dependent on the number of customers in the service facility. For example, at a highway toll booth, attendants tend to speed up toll collection during rush hours. Another example occurs in a machine shop where the rate of breakdown decreases as the number of broken machines increases (only working machines are capable of breakdowns).

2.5.1 Steady-State Systems

The generalized model derives p_n as a function of λ_n and μ_n. These probabilities are used to determine the system's performance measures such as the average queue length, average waiting time, and average server utilization. Define

n = number of customers in the system (in-queue plus in-service)
λ = arrival rate given n customers in the system
μ = departure rate given n customers in the system
p_n = steady-state probability of n customers in the system

The probabilities p_n are determined from the transition-rate diagram in Figure 2.2. The system is in state n when n customers are in the system. As explained in Sections 2.4.1 and 2.4.2, the probability of more than one event occurring during a small interval h tends to zero

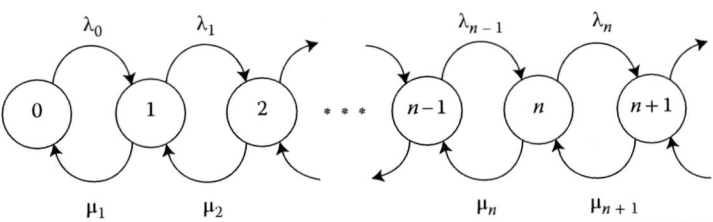

Figure 2.2 Poisson queues transition diagram.

as $h \to 0$. For $n > 0$, state n can change only to two possible states: $n - 1$ when a departure occurs at rate μ_n and $n + 1$ when an arrival occurs at rate λ_n. State 0 can only change to state 1 when an arrival occurs. The rate $\lambda_0 p_{-1}$ is undefined because no departure can occur when the system is empty.

The governing equations are

$$(\text{Expected flow rate into state } n) = \lambda_{n-1} p_{n-1} + \mu_{n+1} p_{n+l}$$

$$(\text{Expected flow rate out of state } n) = \lambda_n p_n + \mu_n p_n$$

Under steady-state conditions, for $n > 0$, expected flow rates into and out of state n must be equal.

$$\lambda_{n-1} p_{n-1} + \mu_{n+1} p_{n+l} = (\lambda_n + \mu_n) p_n, \quad n = 1, 2, \ldots$$

From Figure 2.2, the steady-state equation associated with $n = 0$ is

$$\lambda_0 p_0 + \mu_1 p_1$$

These equations are solved recursively in terms of p_0 as follows: For $n = 0$,

$$p_1 = \left(\frac{\lambda_0}{\mu_1} \right) p_0$$

For $n = 1$,

$$\lambda_0 p_0 + \mu_2 p_2 = (\lambda_1 + \mu_1) p_1$$

Substitute $p_1 = \left(\dfrac{\lambda_0}{\mu_1} \right) p_0$ and simplify to

$$p_2 = \left(\frac{\lambda_1 \lambda_0}{\mu_2 \mu_1} \right) p_0$$

By induction

$$p_n = \left(\frac{\lambda_{n-1} \lambda_{n-2} \ldots \lambda_0}{\mu_n \mu_{n-1} \ldots \mu_1} \right) p_0$$

p_0 is determined from $= 1 \displaystyle\sum_{n=0}^{\infty} p_n = 1$.

2.5.2 Steady-State Performance Measures

Before presenting details of the specialized Poisson queuing systems, steady-state measures of performance for the generalized queuing system are derived (directly or indirectly) from the steady-state probability of n in the system, p_n as

$$L_s = \sum_{n=0}^{\infty} n p_n$$

$$L_q = \sum_{n=c+1}^{\infty} (n-c) p_n$$

The well-known Little's formula expresses the relationship between L_s and W_s (as well as L_q and W_q):

$$L_s = \lambda_{eff} W_s$$

$$L_q = \lambda_{eff} W_q$$

These relationships are valid for rather general conditions. The parameter λ_{eff} is the effective arrival rate at the system and equals λ when all arriving customers join the system. When arriving customers are denied access (e.g., a full parking lot), then $\lambda_{eff} < \lambda$. Development of λ_{eff} is in Section 2.5.4.1.

The relationship between W_s and W_q is

(Expected waiting time in the system) = (Expected waiting time in the queue) + (Expected service time)

Expressed algebraically,

$$W_s = W_q + \frac{1}{\mu}$$

The relationship between L_s and L_q is obtained by multiplying by λ_{eff}, and applying Little's formula:

$$L_s = L_q + \frac{\lambda_{eff}}{\mu}$$

By definition, the difference between the average number in the system, L_s, and the average number in the queue, L_q, is the average number of busy servers, $\bar{c}$:

$$\bar{c} = L_s - L_q = \frac{\lambda_{eff}}{\mu}$$

It follows that

$$(Server\ Utilization) = \frac{\bar{c}}{c}$$

2.5.3 Single-Server Model

This section presents the single-server case ($c = 1$) with no limit on the number in the system and an infinite-capacity source. Respective arrival rate and service rate are λ and μ customers per unit time. Because the derivations of p_n in Section 2.5 and performance measures in Section 2.5.2 are independent of the queue discipline, *GD* (general discipline) will be used in the Kendall notation. The notation that summarizes this environment is $(M/M/1):(GD/\infty/\infty)$. From the generalized model:

$$\left.\begin{array}{l} \lambda_n = \lambda \\ \mu_n = \mu \end{array}\right\} n = 0,1,2,\dots$$

Also, $\lambda_{eff} = \lambda$ because all arriving customers join the system. With $\rho = \dfrac{\lambda}{\mu}$, the expression for p_n in the generalized model reduces to

$$p_n = \rho^n p_0, \quad n = 0,1,2,\dots$$

The value of p_0 is obtained from the identity

$$p_0(1 + \rho + \rho^2 + \cdots) = 1$$

Assuming $\rho < 1$, the geometric series will have a finite sum equal to $\dfrac{1}{1-\rho}$ and $p_0 = (1-\rho)$
The general formula for p_n is the following geometric distribution:

$$p_n = (1-\rho)\rho^n, \quad n = 1,2,\dots$$

The derivation of p_n assumes $\rho < 1$, or $\lambda < \mu$. When $\lambda \ge \mu$, the geometric series does not converge and steady-state probabilities, p_n, are undefined. Unless the service rate exceeds the arrival rate, the queue length will continue to increase and steady state will not be achieved.
The performance measure L_s is derived as follows:

$$L_s = \sum_{n=0}^{\infty} n p_n = \sum_{n=0}^{\infty} n(1-\rho)\rho^n$$

$$= (1-\rho)\rho \frac{d}{d\rho} \sum_{n=0}^{\infty} \rho^n$$

$$= (1-\rho)\rho \frac{d}{d\rho}\left(\frac{1}{1-\rho}\right) = \frac{\rho}{1-\rho}$$

Because $\lambda_{eff} = \lambda$ other performance measures are from relationships in Section 2.5.2.

$$W_s = \frac{L_s}{\lambda} = \frac{1}{\mu(1-\rho)} = \frac{1}{\mu - \lambda}$$

$$W_q = W_s - \frac{1}{\mu} = \frac{\rho}{\mu(1-\rho)}$$

$$L_q = \lambda W_q = \frac{\rho^2}{1-\rho}$$

$$\bar{c} = L_s - L_q = \rho$$

Example 2.5

Automata car wash has only one bay. Cars arrive according to a Poisson distribution with a mean of four cars per hour, and may wait in the parking lot if the bay is busy. The time for washing and cleaning a car is exponential with a mean of 10 minutes. Cars that cannot park in the lot can wait in the street bordering the facility. For all practical purposes there is no limit on the size of the system. The facility manager wants to determine the size of the parking lot so that an arriving car will find a parking space at least 90% of the time. Since $\lambda = 4$ cars per hour, and $\mu = \dfrac{60}{10} = 6$ cars per hour, then $\rho \leq 1$, and the system will operate under steady-state conditions. *PoissonQueuingModel.xls* input for this model is shown in Figure 2.3.

The output is shown in Figure 2.4. The average number of cars waiting, L_q, is 1.33333. Generally, rather than L_q as the basis for number of parking spaces in the system, the design should account for the maximum possible queue length. S represents the number of parking spaces and S parking spaces is equivalent to having $S + 1$ spaces in the system (queue plus wash bay). An arriving car will find a space 90% of the time if there are at most S cars in the system. This condition is equivalent to the following:

$$p_0 + p_1 + \cdots + p_S \geq 0.9$$

From Figure 2.4, cumulative p_n for $n = 5$ is 0.91221. Therefore the condition is satisfied for $S \geq 5$ spaces. The number of spaces S can also be determined from the definition of p_n,

$$(1-\rho)\left(1+\rho+\rho^2+\cdots+\rho^S\right) \geq 0.9$$

(*M/M/c*): (*GD/N/K*) Queuing Model (includes N and/or K = infinity)			
Input Data (to enter an infinite value, type i or infinity):			
λ =	4	μ =	6
c =	1		
System Limit N=	Infinity	Source limit K =	Infinity

Figure 2.3   Poisson queues transition diagram.

Output Results:			
λ_{eff}=	4.00000	ρ/c =	0.66667
Ls =	1.99999	Lq =	1.33333
Ws =	0.50000	Wq =	0.33333

n	P(n)	CumulativeP(n)
0	0.3333333351	0.3333333351
1	0.2222222234	0.5555555585
2	0.1481481489	0.7037037074
3	0.0987654326	0.8024691401
4	0.0658436217	0.8683127618
5	0.0438957478	0.9122085096
6	0.0292638319	0.9414723415
7	0.0195092213	0.9609815628
8	0.0130061475	0.9739877103
9	0.0086707650	0.9826584753
10	0.0057805100	0.9884389853
11	0.0038536733	0.9922926586
12	0.0025691156	0.9948617742

Figure 2.4   *PoissonQueuingModels.xls* output of Example 2.5.

The truncated geometric series sum equals $\dfrac{1-\rho^{S+1}}{1-\rho}$. The condition reduces to

$$1-\rho^{S+1} \geq 0.9 \quad \text{or} \quad \rho^{S+1} \leq 0.1$$

Taking logarithms of both sides (remember that $\log(x) < 0$ for $0 < x < 1$) reverses the inequality direction:

$$S \geq \frac{\ln(0.1)}{\ln(\rho)} - 1 = \frac{\ln(0.1)}{\ln\left(\dfrac{4}{6}\right)} - 1 = 4.679 \approx 5$$

2.5.4 Multiple-Server Models

This section considers models with multiple parallel servers. In the $(M/M/c):(GD/\infty/\infty)$ model there are c parallel servers, each with service rate μ. The effect of c parallel servers is a proportionate increase in service rate for the facility. Because there is no limit on the number in the system, the arrival rate λ_{eff} is λ. From the generalized model (Section 2.5), λ_n and μ_n are defined as

$$\lambda_n = \lambda, \quad n \geq 0$$

$$\mu_n = \begin{cases} n\mu, & 0 < n < c \\ c\mu, & n \geq c \end{cases}$$

and

$$p_n = \begin{cases} \dfrac{\lambda^n}{\mu(2\mu)(3\mu)\ldots(n\mu)} p_0 = \dfrac{\lambda^n}{n!\mu^n} p_0 = \dfrac{\rho^n}{n!} p_0, & n < c \\[4ex] \dfrac{\lambda^n}{\left(\prod_{i=1}^{c} i\mu\right) c\mu^{n-c}} p_0 = \dfrac{\lambda^n}{c!c^{n-c}\mu^n} p_0 = \dfrac{\rho^n}{c!c^{n-c}} p_0, & n \geq c \end{cases}$$

For $\rho = \dfrac{\lambda}{\mu}$, and $< \dfrac{\rho}{c} < 1$, p_0 is determined from the relationship $\sum_{n=0}^{\infty} p_n = 1$.

$$p_0 = \left\{ \sum_{n=0}^{c-1} \frac{\rho^n}{n!} + \frac{\rho^c}{c!} \sum_{n=c}^{\infty} \left(\frac{\rho}{c}\right)^{n-c} \right\}^{-1}$$

$$= \left\{ \sum_{n=0}^{c-1} \frac{\rho^n}{n!} + \frac{\rho^c}{c!} \left(\frac{1}{1-\dfrac{\rho}{c}}\right) \right\}^{-1}$$

From the definition of L_q:

$$L_q = \sum_{n=c}^{\infty} (n-c) p_n$$

$$= \sum_{n=0}^{\infty} n p_{n+c}$$

$$= \sum_{n=0}^{\infty} \frac{\rho^{n+c}}{n!}$$

$$= \sum_{n=0}^{\infty} n \frac{\rho^{n+c}}{c^n c!} p_0$$

$$= \frac{\rho^{c+1}}{c!\,c} p_0 \sum_{n=0}^{\infty} n \left(\frac{\rho}{c}\right)^{n-1}$$

$$= \frac{\rho^{c+1}}{c!\,c} p_0 \frac{d}{d\left(\frac{\rho}{c}\right)} \sum_{n=0}^{\infty} n \left(\frac{\rho}{c}\right)^{n-1}$$

$$= \frac{\rho^{c+1}}{(c-1)!(c-\rho)^2} p_0$$

Because $\lambda_{eff} = \lambda$, $L_s = L_q + \rho$, W_s and W_q respectively are obtained by dividing L_s and L_q by λ.

Example 2.6

Two cab companies, each owning two cabs, serve a community. Calls arriving at a rate of eight per hour to each dispatch office indicate that the market is shared equally. Average time to service a call by either company is exponentially distributed with mean of 12 minutes. Both companies were recently purchased by an investor interested in consolidating them but who is concerned about the effect on customer service.

Cabs are servers and the cab fare is the service. Each company is represented by an $(M/M/2):(GD/\infty/\infty)$ model with $\lambda = 8$ calls per hour and $\mu = 5$ fares per hour for each cab. The average time a customer waits for a cab, W_q, is the basis for comparison. Consolidation will result in the model $(M/M/4):(GD/\infty/\infty)$ with $\lambda = 2 \times 8 = 16$ calls per hour and μ remains 5 fares per hour. PoissonQueuingModel.xls comparative analysis data is in Figure 2.5.

From Figure 2.5, the waiting time for a cab is 0.356 hour ($\approx$ 21 minutes) for the two company environment and 0.149 ($\approx$ 9 minutes) after consolidation. A reduction of more than 50% is evidence that consolidation is warranted. From the preceding analysis, pooling services is more efficient. This result is true even when separate installations are "very" busy (see Problem 2.23).

(M/M/c):(GD/N/K) Queueing Model (includes N and/or K = infinity)			
Input Data (to enter an infinite value, type i or infinity):			
λ =	8	μ =	5
c =	2		
System Limit N=	infinity	Source limit K =	infinity
Output Results:			
λ eff =	8.00000	ρ/c =	0.80000
Ls =	4.44442	Lq =	2.84442
Ws =	0.55555	Wq =	0.35555

(M/M/c):(GD/N/K) Queueing Model (includes N and/or K = infinity)			
Input Data (to enter an infinite value, type i or infinity):			
λ =	16	μ =	5
c =	4		
System Limit N=	infinity	Source limit K =	infinity
Output Results:			
λ eff =	15.99999	ρ/c =	0.80000
Ls =	5.58570	Lq =	2.38570
Ws =	0.34911	Wq =	0.14911
n	P(n)	CP(n)	1 - CP(n)

Figure 2.5   Comparative summary for Example 2.6.

2.5.4.1 Special Multiple-Server Models

This section presents multiple-parallel-server models, also based on the generalized model in Section 2.5. In the following example the service rate per server, μ, is constant. However, in contrast to the $(M/M/c):(GD/\infty/\infty)$ model, c is dependent on, p_n, the number in the system. There is no limit on the number in the system, therefore the arrival rate, λ_{eff}, is λ.

Example 2.7

B&K Groceries operates with three checkout counters. As shown in the following table, the schedule for the number of service counters is dependent on the number of customers in the store:

Number of Customers in the Store	*Number of Counters in Operation*
1 to 3	1
4 to 6	2
More than 6	3

Customer arrival rate to the counters area is $\lambda = 10$ per hour. The average checkout time is exponential with mean of 12 minutes. Steady-state probabilities p_n for n customers in the system is determined as follows:

$$\lambda_n = \lambda = 10 \text{ customers per hour}, \quad n = 0,1,2,\dots$$

$$\mu_n = \begin{cases} \dfrac{60}{12} = 5 \text{ customers per hour}, & n = 0,1,2,3 \\ 2 \times 5 = 10 \text{ customers per hour}, & n = 4,5,6 \\ 3 \times 5 = 15 \text{ customers per hour}, & n = 7,8,\dots \end{cases}$$

Therefore,

$$p_1 = \left(\frac{10}{5}\right)^1 p_0 = 2p_0$$

$$p_2 = \left(\frac{10}{5}\right)^2 p_0 = 4p_0$$

$$p_3 = \left(\frac{10}{5}\right)^3 p_0 = 8p_0$$

$$p_4 = \left(\frac{10}{5}\right)^3 \left(\frac{10}{10}\right)^1 p_0 = 8p_0$$

$$p_5 = \left(\frac{10}{5}\right)^3 \left(\frac{10}{10}\right)^2 p_0 = 8p_0$$

$$p_6 = \left(\frac{10}{5}\right)^3 \left(\frac{10}{10}\right)^3 p_0 = 8p_0$$

$$p_{n\geq7} = \left(\frac{10}{5}\right)^3 \left(\frac{10}{10}\right)^3 \left(\frac{10}{15}\right)^{(n-6)} p_0 = 8\left(\frac{2}{3}\right)^{(n-6)} p_0$$

p_0 is determined from

$$p_{0+}p_0\left\{2+4+8+8+8+8+8(2/3)+8(2/3)^2+8(2/3)^3+\cdots\right\}=1$$

or

$$p_0\left\{1+2+4+8+8+8+8(1+(2/3)+(2/3)^2+(2/3)^3+\cdots\right\}=1$$

The geometric sum series

$$\sum_{i=0}^{\infty}x^i=\frac{1}{1-x},\quad |x|<1$$

is used to obtain

$$p_0\left\{31+8\left(\frac{1}{1-\frac{2}{3}}\right)\right\}=1$$

or

$$p_0=\frac{1}{55}$$

Values of p_n for $n>0$ are determined from p_0.

The probability that only one counter will be open is the probability that at most three customers are in the system, $P(n\leq 3)$, which is

$$p_1+p_2+p_3=(2+4+8)\left(\frac{1}{55}\right)=0.255$$

Also,

$$\begin{pmatrix}\textit{Expected number}\\\textit{of idle counters}\\\textit{in the system}\end{pmatrix}=3p_0+2(p_1+p_2+p_3)+1(p_4+p_5+p_6)+0(p_7+p_8+\cdots)=1$$

In Example 2.8 the service rate per server, μ, is constant. However, there is a limit on the number in the system and the upper limit on the number of servers is less than the number in the system. It is therefore possible that not all arriving customers can join the system and the effective arrival rate, λ_{eff}, is less λ. In Figure 2.6 customers arrive from the source at the rate of λ customers per hour. An arriving customer may enter the system or go elsewhere at a rate of λ_{lost}. As a result the effective arrival rate is

$$\lambda_{eff}=\lambda-\lambda_{lost}$$

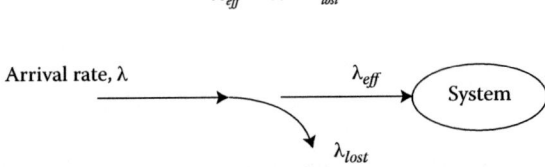

Figure 2.6 Relationship between λ, λ_{eff} and λ_{lost}.

Example 2.8

Visitor parking is limited to five spaces and the vehicle arrival rate, λ, is six vehicles per hour. Elapsed time parking is exponentially distributed with a mean of 30 minutes. Arrivals that cannot park wait in a temporary space until a parked vehicle leaves. However, the temporary space accommodates only three vehicles. Customers who cannot park or find a temporary space go elsewhere. An arriving vehicle will not be able to enter the parking lot when eight vehicles are on the lot.

Because a parking space is a server, the system has a total of $c = 5$ parallel servers. Also, the capacity of the system is $5 + 3 = 8$ vehicles. Probabilities, p_n, are determined as a special case of the generalized model in Section 2.5 using $\lambda_n = 6$ cars/hour, $n = 0,1,2,\ldots, 8$ and

$$\mu_n = \begin{cases} n\left(\dfrac{60}{30}\right) = 2n\,\text{cars/hour}, & n = 1,2,3,4,5 \\[2mm] 5\left(\dfrac{60}{30}\right) = 10\,\text{cars/hour}, & n = 6,7,8 \end{cases}$$

From Section 2.5,

$$p_n = \begin{cases} \dfrac{3^n}{n!}\, p_0, & n = 1,2,3,4,5 \\[3mm] \dfrac{3^n}{5!5^n}\, p_0, & n = 6,7,8 \end{cases}$$

p_0 is computed by substituting p_n, $n = 1,2,\ldots,8$, into

$$p_0 + p_1 + p_2 + p_3 + p_4 + p_5 + p_6 + p_7 + p_8 = 1$$

or

$$p_0 + p_0\left(\frac{3}{1!} + \frac{3^2}{2!} + \frac{3^3}{3!} + \frac{3^4}{4!} + \frac{3^5}{5!} + \frac{3^6}{5!5} + \frac{3^7}{5!5^2} + \frac{3^8}{5!5^3}\right) = 1$$

From the above equation $p_0 = 0.04812$ and values of p_1 through p_8 are

n	1	2	3	4	5	6	7	8
p_n	00.14436	0.21654	0.21654	0.16240	0.09744	0.05847	0.03508	0.02105

The average number of vehicles in the lot (those parked or waiting for a space) is L_s, the average number in the system:

$$L_s = 0\,p_0 + 1p_1 + 2p_2 + 3p_3 + 4p_4 + 5p_5 + 6p_6 + 7p_7 + 8p_8 = 3.1286\,\text{cars}$$

The proportion of vehicles unable to enter the lot is p_8. Therefore,

$$\lambda_{lost} = \lambda p_8 = 6 \times 0.02105 = 0.1263\,\text{cars/hour}$$
$$\lambda_{\mathit{eff}} = \lambda - \lambda_{lost} = 6 - 0.1263 = 5.8737\,\text{cars/hour}$$

From Little's equation,

$$W_s = \frac{L_s}{\lambda_{\mathit{eff}}} = \frac{3.1286}{5.8737} = 0.53265\,\text{hours}$$

A vehicle in a temporary space is in the queue. Therefore, waiting time is

$$W_q = W_s - \frac{1}{\mu}$$

$$W_q = 0.53265 - \frac{1}{2} = 0.03265 \, \text{hours}$$

The average number of occupied parking spaces is the average number of busy servers,

$$\bar{c} = L_s - L_q = \frac{\lambda_{\text{eff}}}{\mu} = \frac{5.8737}{2} = 2.9368$$

and

$$(\textit{Parking lot utilization}) = \frac{\bar{c}}{c} = \frac{2.9638}{5} = 0.58736$$

2.6 Jackson Network Models

Previous sections of this chapter describe *M/M/1* and *M/M/C* models. However, actual systems are usually a complex network of these basic models. This section focuses on the behavior of a complex system and network performance measures. The analysis is introduced with an algebraic description of continuous deterministic networks. Then the stochastic routing of items through a network of *M/M/1* and *M/M/C* nodes is presented.

Figure 2.7 presents a five-node system connected by links indicating flow directions. Material enters at node 1 and is routed through nodes 2, 3, or 4 until exiting at node 5. For continuous flow, such as fluid in a pipeline, simultaneous splits occur at the nodes. The following definitions characterize the network:

γ = flow rate to node 1
λ_i = flow rate through node i
R_{ij} = fraction of the flow from node i to j

and

$$\sum_{\text{all } j} R_{ij} = 1 \quad \text{for} \quad \textit{all } i$$

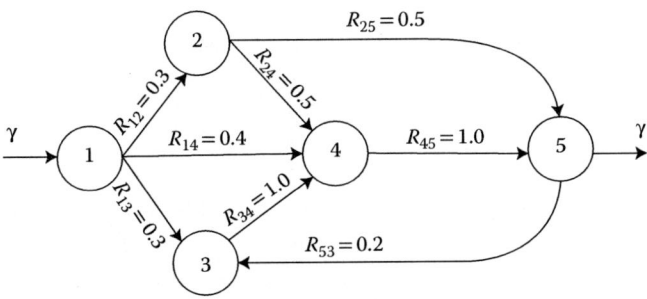

Figure 2.7 Routing network diagram.

When node i branches to nodes j and k, flow from i to j is $\lambda_i R_{ij}$ and from i to k is $\lambda_i R_{ik}$. An example of system feedback is the link from node 5 to 3 that represents the fraction R_{53} of λ_5 that returns to node 3. With feedback, flows through individual nodes may exceed γ. For a given input γ and the values for R_{ki}, flow λ_i can be determined from a simultaneous set of linear equations. At steady state the total flow, λ_i from node i, is the sum of all flows into i, therefore,

$$\lambda_1 = \gamma \quad \text{and} \quad \lambda_i = \sum R_{ki}\lambda_k \quad \text{for} \quad i \geq 2$$

Example 2.9

Steady-state flow equations for Figure 2.7 are

$$\lambda_1 = \gamma$$
$$\lambda_2 = R_{12}\lambda_1$$
$$\lambda_3 = R_{13}\lambda_1 + R_{53}\lambda_5$$
$$\lambda_4 = R_{24}\lambda_2 + R_{14}\lambda_1 + R_{34}\lambda_3$$
$$\lambda_5 = R_{25}\lambda_2 + R_{45}\lambda_4$$

In matrix notation the equations are

$$\begin{bmatrix} +1 & -0 & -0 & -0 & -0 \\ -R_{12} & +1 & -0 & -0 & -0 \\ -R_{13} & -0 & +1 & -0 & -R_{53} \\ -R_{14} & -R_{24} & -R_{34} & +1 & -0 \\ -0 & -R_{25} & -0 & -R_{45} & +1 \end{bmatrix} \begin{Bmatrix} \lambda_1 \\ \lambda_2 \\ \lambda_3 \\ \lambda_4 \\ \lambda_5 \end{Bmatrix} = \begin{Bmatrix} \gamma \\ 0 \\ 0 \\ 0 \\ 0 \end{Bmatrix}$$

These network flow equations may be solved for unknown λ_i's expressed as a constant, F_i multiplied by γ. Although flow out of the network is not explicitly considered in the above equations, it must equal γ; otherwise, there will be accumulation in the network.

Values for F_i's and λ_i's for each node i in Figure 2.7 are

Node i	F_i	$\lambda_i = F_i\gamma$
1	1.0	8/hr
2	0.3	2.4/hr
3	0.55	4.4/hr
4	1.10	8.8/hr
5	1.25	10/hr

If, in Figure 2.7, lows are discrete items, R_{ij} is the probability that an item is directed to node j, andnterarrival rate for γ is random, then the flow equations for λ_i's are the same. However, λ_i now represents an average flow through a node because the sums of random variables, λ_i, are the sum of individual expected values.

When imposing node capacities on network flow it may not be possible to solve the flow equations for a given value of γ because one or more λ_i may exceed capacity at node i. If node i has c_i servers, each server could process μ_i items per time the maximum flow of the network in Figure 2.7 is

Maximize: γ
Subject to: $\lambda_i \leq c_i \mu_i, \quad i = 1,2,\ldots,5$

and the previously defined flow equations

Because $\lambda_i = F_i \gamma$, the maximum flow becomes

Maximize: γ

Subject to: $\gamma \leq \dfrac{c_i \mu_i}{F_i}, \quad i = 1,2,\ldots,5$

The maximum γ is the smallest of the $\dfrac{c_i \mu_i}{F_i}$ values. Capacity data for Figure 2.7 are

Node i	c_i	μ_i	F_i	$\dfrac{c_i \mu_i}{F_i}$
1	1	14	1.0	14/hr
2	1	4	0.3	13.3/hr
3	1	8	0.55	14.5/hr
4	1	14	1.10	12.7/hr
5	1	14	1.25	11.2/hr

The network capacity is constrained by flow at node 5, which operates at full capacity with $\dfrac{\lambda_5}{c_5 \mu_5} = 1 = 1$. All other nodes have $\dfrac{\lambda_i}{c_i \mu_i} < 1$. The $\dfrac{\lambda_i}{c_i \mu_i}$ ratios also represent the fraction of available capacity utilized at each node. To increase network capacity, either c_5 or μ_5 must be increased. A balanced network occurs when all utilizations are approximately equal.

If interarrival and service times at each node are exponentially distributed with means of $\dfrac{1}{\lambda_i}$ and $\dfrac{1}{\mu_i}$, respectively, each network node is a queuing system and n_i is the number of items at node i waiting and being processed. Each n_i is a random variable ranging from zero to infinity. Performance characteristics of each node can be evaluated using *M/M/1* or *M/M/C* equations from previous sections in this chapter and summary node characteristics include

L_{Si} = average number at node i
W_{Si} = average waiting time at node i

$\dfrac{\lambda_i}{\mu_i}$ = average number of busy servers at node i

$\dfrac{\lambda_i}{c_i \mu_i}$ = average utilization at node i

In Jackson network analysis, performance measures for the network are based on the flow equations and the $(M/M/1):(GD/\infty/\infty)$ and $(M/M/C):(GD/\infty/\infty)$ formulas in Sections 2.5.3 and 2.5.4. In the single node queuing system, n is the number of items at the node and the probability distribution, p_n, represents the probability of n items in the system. In a complex network of queuing systems the p_n at one node is dependent on the p_n at other nodes because departures from one node are arrivals to another node, unless node departure exits the system. Therefore, a joint probability distribution for the network must be considered:

$$P(n_1, n_2, \ldots, n_k) = \text{probability of } (n_1 \text{ at node 1}, n_2 \text{ at node 2}, \ldots n_k \text{ at node } k)$$

Jackson [1] showed that this joint probability function is expressed as a product of k terms with p_n expressions similar to those presented in previous sections. As a result, the node and systemwide performance measures can be developed from the flow equations and queuing formulas for $M/M/1$ and $M/M/C$ models. The following network analysis is applied to Figure 2.7.

Step 1. Determine λ_i using γ, R_{ij} data and the network flow equations. Table 2.1 presents λ_i's calculated earlier using an input rate $\gamma = 8$/hour and equations (from Example 2.9).

Step 2. Compute $\dfrac{\lambda_i}{c_i \mu_i}$ for each node. In Table 2.1 all ratios are less than one, therefore no adjustments are necessary. If one or more ratios is ≥ 1, then system adjustments must be made. Examples of adjustments are
  - Increase c_i or μ_i at nodes with ratio ≥ 1.
  - Reduce input rate γ until all ratios are ≤ 1.
  - Reduce selected λ_i by adjusting branching probabilities.

  Of course the nature of the adjustments are application specific.

Step 3. Determine the L_{Si} and W_{Si} at each node using the values of λ_i, c_i, and μ_i and the formulas for the $M/M/1$ and $M/M/C$ queuing systems. System performance measures that describe the network include
  - PR, average production rate, the number of items that exit the network per time.

Table 2.1   Node Performance Measures

Node i	λ_i	c_i	μ_i	$\dfrac{\lambda_i}{c_i \mu_i}$	L_{Si}	W_{Si}
1	8	1	14	0.571	1.333	0.166
2	2.4	1	4	0.600	1.500	0.625
3	4.4	1	3	0.550	1.222	0.278
4	8.8	1	14	0.628	1.692	0.192
5	10	1	14	0.714	2.500	0.250

- *WIP*, average work in process, the total number of items in the network at any time (includes those being served and those waiting).
- *TT*, average throughput time, the average time an item spends in the network (includes processing and waiting time).

In Table 2.1, node 5 has the largest utilization and is the potential bottleneck, when utilization is the bottleneck indicator. However, L_{Si} and W_{Si} provide additional information on system capacity. Two nodes with the same value of $\dfrac{\lambda_i}{c_i\mu_i}$ may have very different values of L_{Si} and W_{Si}, which may be better indicators of a potential problem.

Example 2.10

System measures for Table 2.1 are

- *PR*, average production rate $= \gamma = 8/\text{hr}$.
- *WIP*, average work in process $= \sum L_{si}$ *over all nodes*. From Table 2.1, *WIP* = 8.25.
- *TT*, average lapsed time in the network for items that exit the network.

Because every item may not experience waiting at every node, *TT* is the sum of individual W_{Si}'s weighted by the fraction of flow that experiences waiting.

$$TT = \frac{1}{\gamma}\sum \lambda_i W_{si} \text{ over all nodes}$$

Using data from Table 2.1, *TT* = 1.03 hours and the formula $L = \lambda W$,

$$WIP = (\gamma)(TT) = (8)(1.03) = 8.25$$

In the Chapter 2 directory of the design environment for event-driven simulation (DEEDS) CD is an Excel/VBA program for Jackson network calculations.

2.7 Closed Form versus Discrete Event Simulation Models

The closed form queuing models presented in this chapter are characterized by the parameters

$$(a/b/c):(d/e/f)$$

where *a*, *b*, and *c* are arrival rate, service rate distribution, and the number of parallel servers, respectively. Queue discipline, maximum number in the system (queue capacity + number of servers), and number of arrivals are represented by *d*, *e*, and *f*, respectively. Taha [2] presents additional closed form models with different assumptions with respect to parameters *d*, *e*, and *f*.

Each of the model performance measures in this chapter are based on the assumption that parameters *a* and *b* represent a Poisson distribution. Also, as described in Section 2.5.1, performance measures for these closed form models reflect steady-state response and therefore ignore transient response of the system.

The development of closed form models for parameter *a* and *b* distributions other than Poisson becomes increasingly complex. To further complicate the model, in many queuing systems the system configurations are time dependent. For example the number of servers, and queue capacities are changed to reflect current time rather than number in the system. Closed form models to account for time dependent system variations are even more complex.

The pervasiveness of these issues in actual systems makes simulation an important analysis tool. Simulation also enables analysts to formulate application specific system performance metrics other than those described in this chapter. Finally, plots of system performance measures provide analysts with better insights to transient and steady-state characteristics of the system.

Program 15-9 in the Chapter 2 directory of the DEEDS CD is a Jackson network simulation program. Service rates and number of servers for each node are in the *InitialModel* worksheet. Routing percentages are in the *PDFs* worksheet. The user can easily assess the effects on network performance characteristics to changes in these parameters. Plots of performance characteristics are available by using the sampling features described in Chapter 12. Transient and steady analysis of performance characteristics is described in Chapter 13. The simulation language presented in Chapters 7, 8, and 9 is based on the generalization of Jackson network models. Chapters 15 through 25 demonstrate the flexibility of the language to model complex network systems.

Problems

2.1. Identify the customer(s) and server(s) for each of the following queuing systems:
   a. Planes arriving at an airport
   b. Taxi stand serving waiting passengers
   c. Tools checked out from a crib in a machining shop
   d. Letters processed in a post office
   e. Registration for classes in a university
   f. Legal court cases
   g. Checkout operation in a supermarket
   h. Parking lot operation
2.2. For each queuing system in Problem 2.1, describe the
   a. Nature of the calling source (finite or infinite)
   b. Nature of arriving customers (individually or in bulk)
   c. Type of interarrival time (probabilistic or deterministic)
   d. Definition and type of service time (probabilistic or deterministic)
   e. Queue capacity (finite or infinite)
   f. Queue discipline
2.3. Discuss the possibility of customers jockeying, balking, and reneging the system for each queuing system in Problem 2.1.
2.4. Describe the associated queuing system, that is, the customers, the server(s), the service time, the queue discipline, the maximum queue length, and the calling source for the following environments:
   a. Jobs are received at a workshop where the supervisor decides if it is a rush or a regular job. Some orders require only one of several identical machines for processing. Others are processed by a two-stage production line, of which two are available.
   b. Jobs arriving at one of several facilities are processed in order of arrival. Completed jobs are routed to a limited capacity shipping zone for immediate shipping.
   c. Machine specific tools are supplied from a central tool crib. When a machine breaks down, a repairperson is dispatched from the service pool to make the repair. Machines servicing rush orders receive priority in acquiring tools from the crib and repair.

2.5. a. Explain the relationship between the arrival rate λ and average interarrival time. What are the units for each?

    b. What is the average arrival rate per hour, λ, and average interarrival time in hours for each of the following?
-       i. One arrival occurs every 10 minutes.
-       ii. Two arrivals occur every 6 minutes.
-       iii. The number of arrivals in a 30-minute period is 10.
-       iv. The average interval between successive arrivals is 0.5 hours.

    c. What is the average service rate per hour, μ, and average service time in hours for each of the following?
-       i. One service is completed every 12 minutes.
-       ii. Two departures occur every 15 minutes.
-       iii. The number of customers served in a 30-minute period is five.
-       iv. The average service time is 0.3 hours.

2.6. In Example 2.2, determine
-     a. The average number of failures in one week, for service available 24 hours a day and 7 days a week.
-     b. The probability of at least one failure in a two-hour period.
-     c. The probability that the next failure will not occur within three hours.
-     d. The probability that interfailure time is at least four hours when it has been three hours since the last failure.

2.7. The time between machine breakdowns is exponential with mean six hours. If the machine has been working during the last three hours, what is the probability that it will continue without a breakdown for the next hour? That it will break down within the next 0.5 hours?

2.8. A restaurant arrival rate is 35 customers per hour and the interarrival time is exponentially distributed. The manager wants to know what percent of interarrival times will be
-     a. Less than two minutes
-     b. Between two and three minutes
-     c. More than three minutes

2.9. The time between failures of a Kencore refrigerator is exponential with mean of 9000 hours (approximately one year of operation). If the company issues a one-year warranty on refrigerators, what is the probability that a repair will be covered by warranty?

2.10. For an arrival rate to a bank operation of two customers per minute, determine
-     a. The average number of arrivals during five minutes
-     b. The probability that no arrivals will occur during the next 0.5 minutes
-     c. The probability that at least one arrival will occur during the next 0.5 minutes
-     d. The probability that the time between two successive arrivals is at least three minutes

2.11. For Example 2.4, use *PoissonQueuingSystems.xls* to compute $p_n(7)$, $n = 1, 2, \ldots, 18$, and then verify that these probabilities yield $E\{n \mid t = 7\} = 0.664$ dozen.

2.12. For Example 2.4, write the algebraic solution and then use *PoissonQueuingSystems.xls* to answer each of the following:
-     a. The probability that the stock is depleted after three days
-     b. The average number of dozen roses left at the end of the second day
-     c. The probability that at least one dozen is purchased by the end of the fourth day, given that the last dozen was bought at the end of the third day

d. The probability that the time remaining until the next purchase is at most half a day given that the last purchase occurred a day earlier

e. The probability that no purchases will occur on the first day

f. The probability that no order is placed by the end of the week

2.13. The high school band is performing a benefit concert in its new 400-seat auditorium. Local businesses buy tickets in blocks of 10 and donate them to youth organizations. Tickets go on sale for only four hours, the day before the concert. Ticket orders are Poisson with a mean of 10 calls per hour. After the box office is closed, any remaining blocks of tickets are sold at a discount as "rush tickets" one hour before the concert starts. Determine

a. The probability that it will be possible to buy rush tickets

b. The average number of rush tickets available

2.14. Each morning, the refrigerator in a small machine shop is stocked with two cases (24 cans per case) of soft drinks for the shop's 10 employees. Employees can quench their thirst any time during the eight-hour work day (8:00 a.m. to 4:00 p.m.). Each employee consumes approximately four cans a day, but the process is totally random (Poisson distribution). What is the probability that an employee will not find a drink at the start of the lunch period (noon)?

2.15. In the B&K model of Example 2.7, suppose that all three counters are always open and that the operation is set up such that the customer will go to the first empty counter. Determine

a. The probability distribution of the number of open counters

b. The average number of busy counters

c. The probability that all three counters will be in use

d. The probability that an arriving customer will not wait

2.16. First Bank operates a one-lane drive-in ATM machine. Cars arrive according to a Poisson distribution at the rate of 12 cars per hour. The time per car needed to complete the ATM transaction is exponential with mean six minutes. The lane can accommodate a total of 10 cars. When the lane is full, arriving cars seek service in another branch. Determine

a. The probability that an arriving car cannot use the ATM machine because the lane is full

b. The probability that an arriving car cannot immediately use the ATM machine

c. The average number of cars in the lane

2.17. For Example 2.5, determine

a. The percent utilization of the wash bay

b. The probability that an arriving car must wait in the parking lot prior to entering the wash bay

c. The probability that an arriving car will find an empty parking space, if there are seven parking spaces

d. How many parking spaces should be available so that an arriving car may find a parking space 99% of the time

2.18. A student performs odd jobs. On the average, job requests arrive every five days and the time between requests is exponential. Job completion time is also exponential with mean four days.

a. What is the probability that the student will be out of jobs?

b. What is the average monthly income, if the student receives about $50 a job?

c. At the end of the semester, the student decides to subcontract outstanding jobs at $40 each. What is the expected cost of subcontracted jobs?

2.19. For the $(M/M/1):(GD/\infty/\infty)$ model, explain why L_s is not equal to $L_q + 1$, in general. Under what condition will the equality hold?

2.20. Cars arrive at the Lincoln Tunnel toll gate according to a Poisson distribution, with a mean of 90 cars per hour. The time for passing through the gate is exponential with mean 38 seconds. Drivers complain of the long waiting time, and authorities are willing to reduce the average passing time to 30 seconds by installing automatic toll collecting devices, provided two conditions are satisfied: (1) the average number of waiting cars in the current system exceeds five, and (2) the percentage of gate idle time after installing the new device does not exceed 10%. Can the new device be justified?

2.21. For Example 2.6, determine
   a. The expected number of idle cabs
   b. The probability that a calling customer will be the last on the list
   c. The limit on the waiting list necessary to keep waiting time in the queue to below three minutes

2.22. For Example 2.6,
   a. Show that the 50% reduction in waiting time for the consolidated cab company will increase the percentage of time that servers are busy.
   b. Determine the number of cabs that the consolidated company must have to limit the average waiting time for a ride to five minutes or less.

2.23. In the cab company, Example 2.6, suppose that the average time per ride is 14.5 minutes. The utilization, $\dfrac{\lambda}{c\mu}$, for the two- and four-cab operations will increase to more than 96%. Is it still worthwhile to consolidate the two companies? Use average waiting time for a ride as the basis for comparison.

2.24. Determine the minimum number of parallel servers in each of the following (Poisson arrivals and departures) queuing systems necessary to ensure system stability (i.e., the queue length will not grow indefinitely):
   a. Customers arrive every five minutes and are served at the rate of 10 customers per hour.
   b. The average interarrival time is two minutes, and the average service time is six minutes.
   c. The arrival rate is 30 customers per hour, and the service rate per server is 40 customers per hour.

2.25. Customers arrive at Thrift Bank according to a Poisson distribution, with a mean of 45 customers per hour. Transactions per customer last about five minutes and are exponentially distributed. The manager wants to use a single-line multiple-teller operation, similar to the ones used in airports and post offices but knows that customers may switch to other banks if they perceive that waiting in line is "excessive." For this reason, the manager wants to limit the average waiting time in the queue to no more than 30 seconds. How many tellers are needed?

2.26. McBurger fast food restaurant has three cashiers. Customers arrive according to a Poisson distribution every three minutes and form one line that is serviced by the first available cashier. The time to fill an order is exponentially distributed with a mean of five minutes. The waiting room inside the restaurant is limited. However, the food is good, and customers are willing to line up outside the restaurant. Determine the size of the waiting room inside the restaurant (excluding those customers at the cashiers) so that the probability that an arriving customer does not wait outside the restaurant is at least 0.999.

2.27. A small post office has two open windows. Customers arrive according to a Poisson distribution at the rate of one every three minutes. However, only 80% of them seek service at the windows. The service time per customer is exponential, with a mean of five minutes. Arriving customers form one line and access the available windows on a first come, first served basis.

   a. What is the probability that an arriving customer will wait for service?
   b. What is the probability that both windows are idle?
   c. What is the average number of waiting customers?
   d. Is it possible to have only one window?

2.28. An airport services rural, suburban, and transit passengers. The arrival distribution for each group is Poisson with mean of 15, 10, and 20 passengers per hour, respectively. Passenger check-in time is exponential with mean six minutes. Determine the number of counters necessary for each of the following system requirements:

   a. The average time to check in a customer is less than 15 minutes.
   b. The percentage of counter idleness does not exceed 10%.
   c. The probability that all counters are idle does not exceed 0.01.

2.29. For Example 2.8,

   a. Compute L_q directly using the formula $L_q = \sum_{n=c+1}^{\infty} (n-c) p_n$.
   b. Compute W_s from L_q.
   c. Compute the average number of cars that cannot enter the parking lot during an eight-hour period.

2.30. The following table describes origin nodes and corresponding percent routing to destination nodes:

Origin Node	Destination Node	Percent of Items to Destination
1	2	0.3
1	3	0.3
1	4	0.4
2	5	1.0
3	5	0.4
3	6	0.6
4	6	1.0
5	7	1.0
6	7	1.0
7	2	0.15
7	6	0.05

Service characteristics for each node are

Node Number	Number of Servers	Service Rate (Items/hour)
1	2	60
2	1	60
3	1	40
4	1	50
5	1	75
6	2	50
7	2	75

The interarrival rate to the network is 1.6 minutes. Assuming exponential interarrival rates and service times, estimate the steady-state response of the system.

References

1. Jackson, J. R., Networks of waiting lines, *Operations Research*, 1957, 5, 518–521.
2. Taha, H. A., *Operations Research: An Introduction*, 9th ed. Upper Saddle River, NJ: Prentice Hall, 2010.

Chapter 3

Simulation Modeling

3.1 Introduction

Production systems are emerging as a paradigm across many industries. There are implementation variations but the essence of production systems is that a continuous one-piece flow is ideal with emphasis on optimizing and integrating systems of people, equipment, materials, and facilities, to achieve improvements in quality, cost, on-time delivery, and performance. Process improvements in these production systems are typically described as a reduction in cycle time or lead time; a reduction in the cost of space, inventory, and capital equipment; an increase in capacity utilization; and the elimination of bottlenecks. Discrete event simulation is a powerful tool for the analysis, planning, design, and operation of those facilities.

For example, aircraft arrive at an intermediate-range airport that has two runways, one for landing and the other for takeoff. Aircraft that cannot land upon arrival in the air space of the facility must circle within specified "holding stacks" until a runway becomes available. If the number of "holding" planes exceeds a certain quota, the air traffic controller must clear the takeoff runway and allow circling planes to land. The runways are connected to the passenger terminal via taxiways. A plane that cannot take off immediately upon reaching the end of the taxiway waits on a holding apron. The airport authority is currently conducting a study with the objective of improving the landing and takeoff services at the facility.

System performance measures to investigate the given situation may include

- The average time a plane waits in a holding apron
- The average number of planes in holding stacks
- The average utilization of the taxiways and runways

Considering the strict regulations under which most airports operate, these performance measures can often be determined from historical data. In conducting a study, however, the focus is on assessing the impact of proposed changes in the system (e.g., increasing the number of runways) on the desired performance measures. The issue is how to estimate performance measures on a system that does not exist. In this situation it would be naive to propose experimenting with the real system. Perhaps the only available alternative in this case is to consider representing the proposed system by a model. This is where simulation plays an important role.

Simulation modeling should be regarded as the next best thing to observing a real system in operation. The basic contribution of a simulation model is that it allows for observing performance characteristics of the system over time. These observations may then be used to estimate expected performance measures of the system. For the airport example, a simulation model will record an observation representing the number of circling planes each time a new aircraft enters or leaves the holding stacks. These observations may then be used to calculate the (time-weighted) average number of planes in the stacks.

The airport situation could be analyzed by using the results of queuing theory. However, as discussed in Chapter 2, queuing models are rather restrictive. In particular, queuing theory is only capable of describing isolated segments of the complex airport operation. This type of segmented analysis will likely fail to reflect the impact of the interaction of the various components on the overall behavior of the system. As a result, performance measure estimates may be severely biased. For example, in the airport situation, the landing and takeoff operations must be studied as two dependent queues. However, queuing theory models are not capable of representing the relationships between these two operations. This limitation is primarily due to the intractability of mathematical models for such systems.

With simulation the model typically represents the entire system rather than in a segmented fashion as in mathematical models. All cause and effect relationships among the different components of the model are accounted for in the simulation model. The purpose of a simulation model is to imitate the behavior of the real system (as closely as possible). Collecting observations is simply monitoring the behavior of the model's components as a function of simulation time.

3.2 Types of Simulation

The primary purpose of a simulation model is to gather observations about a particular system as a function of time. There are two distinct types of simulation: discrete and continuous. In discrete simulation, observations are gathered only at points in time when changes occur in the system. On the contrary, continuous simulation requires that observations be collected continuously at every point in time.

A single-server facility and an oil terminal that supplies a number of storage tanks through a pipeline network illustrate the difference between the two types of simulation. In the single-server model, changes in the status of the system occur only when a customer arrives or completes service. At these two points in time, performance measures such as queue length and waiting time in the facility will be affected. At all other points in time, these measures remain unchanged (e.g., queue length) and not yet ready for data collection (e.g., waiting time). For this reason, the model observes the system only at selected discrete points in time, hence the name discrete simulation.

In the oil pipeline example, a measure of performance could be the oil level in each tank. Because of the nature of the product, flow into and out of a tank is continuous and the system must be monitored continuously. In this case, the output must be presented as a function of the time. Because it is practically impossible to monitor a continuous system using simulation, the actual recording of observations occurs over small equal intervals of time.

Although continuous simulation has important applications, most attention has been directed toward discrete simulation. The main reason is the wide range of problems in this area. Also, continuous simulation appears straightforward, whereas discrete simulation usually requires more user creativity.

3.3 The Simulation Clock

Simulation collects observations by monitoring the "modeled" system over time. As such, the simulation model must have an internal clock. Unlike familiar time devices, this clock is designed to initiate the collecting of information at the moment before changes in the system occur. Such a clock in continuous simulation is straightforward because it implies "looking" at the system at equally spaced time intervals. However, in the discrete case, the system is observed only when changes in the state of the system occur.

To demonstrate how a discrete simulation clock works, consider a simple system with a single-server facility. Simulation, in this case, follows the movement of each customer from the instant of arrival into the system until a system departure occurs. To achieve this result, the simulator must be capable of determining the exact time each customer arrives to and departs from the service facility. Such information is available from knowledge of the times between successive arrivals and customer service time. In essence, given the time that the facility "opens for business," together with the time interval between successive arrivals and the service time per customer, the model is able to trace the flow of each customer, from system arrival to system departure.

Note that simulation does not trace individual customers. It identifies the points of arrival and departure chronologically on the timescale regardless of the identity of the customer arriving or departing the system. This approach, in essence, categorizes system changes into two specific types of simulation events, arrivals and service completions, and suppresses the specific identity of each customer.

Figure 3.1 shows a mapping of single-server events on the timescale. The simulation processes these events in chronological order and specifies actions depending on the type of the event. For example, when an arrival event takes place, the customer will either start service if the facility is available, or will wait in a queue until the facility is free. When a customer is ready to begin service, the service completion event is scheduled (and mapped onto the timescale) by adding the service time to the current simulation clock time. In a similar manner, when a service completion event occurs, a check is made for any waiting customers. Upon service completion, the waiting customer departs the queue and another engages the facility with a new service completion event mapped onto the timescale. The clock in discrete simulation stops at the time an event occurs. As a result, the simulation clock may be viewed as "jumping" to the point in time that defines the next chronological event. This explains the reason that such simulations are referred to as *discrete event simulation*.

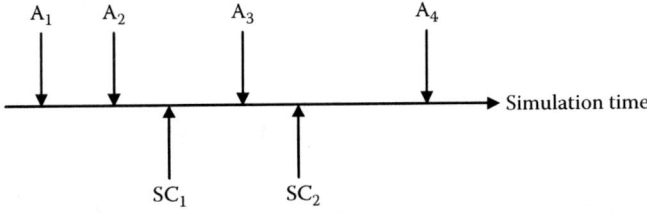

Figure 3.1   Mapping simulation events on the timescale.

3.4 Randomness in Simulation

The discrete event scheme described for the single-server facility works only if two basic data elements are known: (1) the interarrival time and (2) the service time. You may wonder how such information is obtained during the course of a simulation.

In most situations, customer interarrival times and service times occur in a random fashion. The way to represent this randomness is by using probability distributions to describe the variables being considered. For example, if the arrivals occur in a Poisson stream, then from probability theory, the interarrival distribution is exponential. In general, information on customer arrivals is based on either direct observation of the system or historical data. Otherwise, some plausible assumption must be made to describe the random process using a probability distribution.

Once the probability distribution has been selected, statistical theory provides the basis for obtaining random samples based on (0, 1) uniform random numbers. These random samples are used to map the occurrence of an event on the timescale. For example, if the interarrival time is exponential, then a random sample drawn from that distribution represents the lapsed time before the next arrival.

Randomness in simulation is not limited to time. Suppose, for example, arriving customers may choose between one of two servers according to individual preference. To represent this situation, an arriving customer's preference for one server over the other must be quantified using observational or historical data. The result is two ratios representing the allocation of customers to the two servers. This is equivalent to specifying a probability distribution from which samples can be drawn to decide on the server chosen by an arriving customer.

3.5 Discrete Simulation Languages

Discrete simulation involves a large number of repetitious computations. Consequently, the use of computers to carry out these computations is essential. However, the number of computations is not the only obstacle to simulation modeling. In the single-server simulation presented in Sections 3.3 and 3.4, simulation involves a complex logical structure that requires special expertise to develop a computer model. An example is mapping generated events chronologically on a timescale. Writing a procedural language (i.e., BASIC, FORTRAN, or Java) program for this segment of the simulation model requires above-average programming skills. In the absence of software development tools to relieve this burden, simulation as a tool would most certainly be the domain of "elite" programmers. Fortunately, the repetitive nature of simulation computations is the basis for the development of a suite of computer programs that are applicable to all simulation models. For example, routines for ordering events chronologically, as well as those for generating random samples, apply to any simulation model.

Modern computers have induced the development of full-fledged simulation languages that are applicable to any simulation situation. Some languages have been developed for either continuous or discrete simulations. Others can be used for combined continuous and discrete modeling. All languages provide certain standard programming facilities and differ in the degree of detail that must be specified to develop a model. There is a trade-off. Usually, languages that are highly flexible in representing complex situations require the user to provide a more detailed account of the model logic. This type of language is generally less user friendly. In contrast, compact, easy-to-implement languages are usually more rigid, which may make it difficult to represent complex interactions. In general, the two types of discrete simulation languages are event scheduling and process based.

Event-scheduling languages deal directly with the individual actions associated with each type of event. These languages are highly flexible in representing complex situations because the user provides most of the details. Most prominent among these languages are SIMSCRIPT and GASP.

Process-based languages are usually more compact as they are designed to represent the movement of a "customer" (commonly referred to as a transaction or entity) from the instant it enters the simulated system until it is discharged, thus relieving the user from programming most of the details. The oldest of these languages is GPSS. More recent languages include SLAM, SIMAN, and SIMULA. GPSS, SIMAN, and SLAM allow user-defined procedural language routines to be linked to the model. Additionally, SIMAN and SLAM have the capability to model continuous systems.

3.6 Simulation Projects

Simulation projects include two broad phases:

1. Construction, debugging, and running of the model
2. Interpretation of model output

The first phase is usually the most time consuming. It begins with selecting a suitable simulation language (or procedural language) to represent the logic of the model. It also includes collecting input data for the model. Debugging ensures that the model logic is correct. Finally, tracing simulation computations as a function of time verifies that results of the model "make sense." Once the model is validated, it can be run for an "appropriate" length of time to estimate the system's performance.

The second phase of modeling is the most crucial and probably the most neglected. Simulation, by the very nature of inherent randomness, is a statistical experiment. Therefore, the simulation output must be interpreted statistically. The simulation experiment has peculiar statistical properties that must be considered when interpreting the results. Otherwise, the output may be biased.

Just as in any other physical experiment, a simulation study must be subjected to all the proper statistical techniques. Failing to do so may render worthless simulation results. Unfortunately, this aspect of a simulation project is often ignored because a user is not well schooled in the use of statistical techniques. This deficiency, coupled with lack of knowledge about the peculiarities of the simulation experiment, contributes to the misuse of simulation. Fortunately, there is a trend that simulation languages include automatic features to assist with applying statistical tests on the simulation output.

3.7 Design Environment for Event-Driven Simulation

Because the majority of discrete simulation systems may be viewed in some form or another as a queuing situation, the design of available process-oriented languages has centered around the use of three basic blocks or nodes: a source for creating transactions, a facility where transactions are serviced, and a queue where transactions may wait. However, to simulate complex situations, it is necessary to develop special-purpose blocks to "direct traffic" among these three basic blocks. In serial traffic, each block contributes to satisfying the "needs" of the transaction as it flows from one basic block to the next. The disadvantage of the "serial block" design philosophy is that the language is much less user friendly because of the number of blocks required to direct "traffic flow" in the model. In reality, only the most simple of situations may be represented using the serial block approach.

The design environment for event-driven simulation (DEEDS) approach presented in the following chapters eliminates the use of special-purpose nodes and reduces the language to dealing with exactly four nodes: a source, a queue, a facility, and a delay. As shown in Section 10.6, the delay node, which may be considered as an infinite capacity facility, enhances the flexibility of the language. Each node is designed to be self-contained, in the sense that sufficient information is provided to manage the transaction within a given node. User-developed routines written in Visual Basic for Applications (VBA) manage the transaction as it enters and leaves a node. In essence, such a definition focuses on "directing traffic" without the many blocks of a process language at the expense of minimal incremental effort for "serial" traffic situations. This approach has proven to be effective and convenient in handling very complex simulation models.

Chapter 4

Probability and Statistics in Simulation

4.1 Role of Probability and Statistics in Simulation

Simulation is applied to environments in which randomness is a key element in the description of the system. As discussed in Chapter 1, randomness in simulation is represented using probability distributions. For example, in many service facilities, the random arrival or departure of customers is typically described by a Poisson process. Because randomness is synonymous with simulation, simulation output must be viewed as a *sample* in a statistical experiment. Therefore, simulation output must be subjected to proper statistical inference tests. Also, performance measures from a simulation model typically must be expressed with appropriate confidence intervals.

This chapter is not intended as a review of probability or statistics. It is offered as a quick reference for simulation users on

- The characteristics of probability distributions commonly used in simulation models
- Statistical goodness-of-fit models used to identify distributions
- Confidence intervals and the application of hypothesis testing to output data

Information sources to model random behavior may be from closed-form probability distributions or histograms derived from historical data or data collection efforts. Also, because simulation is an experimental design to estimate system performance measures, statistical inference techniques are critical to the decision-making process. Figure 4.1 summarizes the organization of this

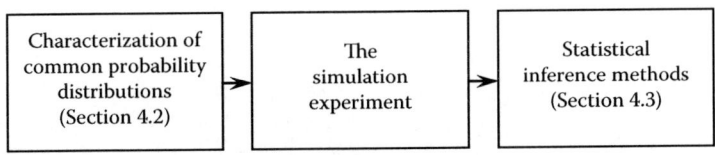

Figure 4.1   Role of probability and statistics in simulation.

47

chapter relative to the simulation experiment. Further details on these topics may be found in the references cited at the end of the chapter.

4.2 Characterization of Common Distributions in Simulation

This section introduces the definitions and properties of some continuous and discrete distributions commonly used in simulation. Section 4.2.2.2 then shows how goodness-of-fit tests are used to identify distributions from raw data.

The following notation and definitions will be used throughout the chapter. Given that x is a continuous random variable in the interval [a, b], $f(x)$ represents the probability density function (pdf) of x. The cumulative density function (CDF) is represented by the notation $F_x(X)$. The functions $f(x)$ and $F_x(X)$ have the following properties:

$$\int_a^b f(x)dx = 1 \quad F_x(X) = \int_a^x f(x)dx, \quad a < x < b$$

$$Mean = \int_a^b x f(x)dx \quad Variance = \int_a^b (x - mean)^2 f(x)dx$$

For the case where x is discrete over the values $a, a + 1,...,$ and b, $f(x)$ and $F_x(X)$ are replaced by $p(x)$ and $P_x(X)$, in which case integration is replaced by summation.

4.2.1 Properties of Common Distributions

4.2.1.1 Uniform Distribution

$$f(x) = \frac{1}{b-a}, \quad a \le x \le b$$

$$F_x(X) = \frac{X-a}{b-a}, \quad a \le x \le b$$

$$Mean = \frac{b+a}{2} \quad Variance = \frac{(b-a)^2}{12}$$

The shape of the uniform $f(x)$ is shown in Figure 4.2. The uniform [0, 1] distribution plays a key role in simulation. Specifically, uniform [0, 1] random numbers are used to generate random samples from any pdf. This technique is detailed in Section 5.3.

4.2.1.2 Negative Exponential Distribution

$$f(x) = \mu e^{-\mu x}, \quad x > 0, \mu > 0$$

$$F_x(X) = 1 - \mu e^{-\mu X}, \quad X > 0$$

$$Mean = 1/\mu \quad Variance = 1/\mu^2$$

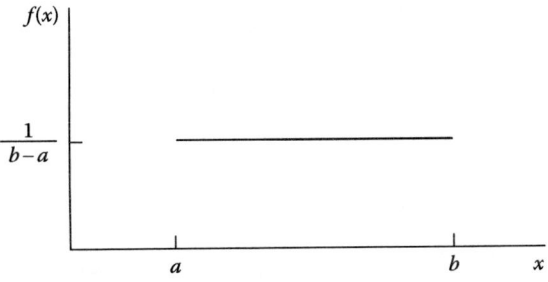

Figure 4.2   Uniform density function.

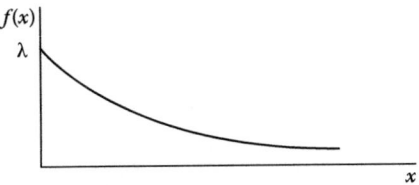

Figure 4.3   Exponential density function.

The negative exponential pdf is shown in Figure 4.3. This distribution applies frequently in the simulation of the interarrival and service times of facilities. It also applies in many reliability problems to describe the time to failure of a system's component.

Exponential random variables possess a unique *forgetfulness* or *lack of memory* property. That is, given T is the time period that elapsed since the occurrence of the last event, the time t remaining until the occurrence of the next event is independent of T. This means that

$$P\{x > T + t \mid x > T\} = P\{x > t\}$$

The forgetfulness property ensures that events whose interoccurrence times are exponential are *completely random*. The exponential distribution is related to the gamma and Poisson distributions, described in 4.2.1.3 and 4.2.1.9, respectively.

4.2.1.3 Gamma (Erlang) Distribution

$$f(x) = \frac{\mu}{\Gamma(\alpha)}\mu x^{\alpha-1}e^{-\mu x}, \quad x > 0, \mu > 0, \alpha > 0$$

$$F_x(X) = 1 - e^{-\mu x}\sum_{i=0}^{\alpha-1}(\mu X)^i, \quad X > 0, \alpha \ integer$$

$$F_x(X) \ has \ no \ closed \ form \ if \ \alpha \ is \ noninteger$$

$$Mean = \alpha/\mu \quad Variance = \alpha/\mu^2$$

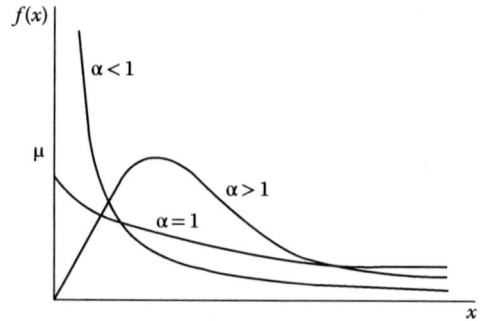

Figure 4.4 Gamma density function.

The scalars μ and α are *shape parameters*. The distribution is known as the *Erlang* when α is a positive integer. The Erlang distribution with shape parameters α and μ is the sum (convolution) of independent and identically distributed exponentials with mean 1/μ. This relationship may be useful for recognizing the use of Erlang in simulation models. For example, if jobs arriving from a source with exponential interarrival time and mean 1/μ are assigned on a rotational basis to *n* machines, the time between allocation of jobs at each machine is necessarily Erlang (μ, *n*). The shape of a gamma pdf changes with the change in shape parameter α as shown in Figure 4.4. Notice that for α = 1, the gamma distribution reduces to the exponential distribution.

4.2.1.4 Normal Distribution

$$f(x) = \frac{1}{\sqrt{2\pi}\sigma} e^{-\frac{(x-\mu)^2}{2\sigma^2}}, \quad -\infty < x < \infty$$

$$F_x(X) \text{ has no closed form}$$

$$\text{Mean} = \mu \quad \text{Variance} = \sigma^2$$

The normal pdf is shown in Figure 4.5. The customary convention $N(\mu, \sigma^2)$ refers to a normal random variable with mean μ and variance σ^2. The normal random variable describes phenomena with symmetric variations above and below the mean μ. More importantly, the normal distribution has properties that make it particularly useful for statistical analysis of simulation output. The most prominent of these is the *central limit theorem*, which states that the distribution of sample averages tends toward normality regardless of the distribution describing the population from which the samples are taken. This powerful result allows the determination of confidence intervals as well as the application of hypothesis testing to the output of the simulation experiments (see Section 4.3).

4.2.1.5 Lognormal Distribution

$$f(x) = \frac{1}{x\sqrt{2\pi}\sigma} e^{-\frac{(\ln(x)-\mu)^2}{2\sigma^2}}, \quad x > 0$$

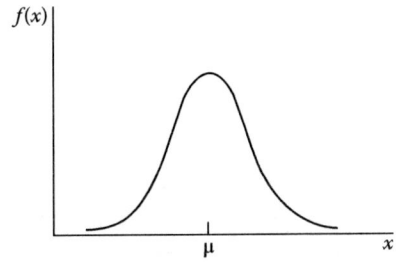

Figure 4.5   Normal density function.

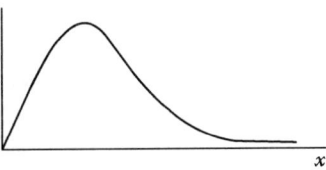

Figure 4.6   Lognormal density function.

$$F_x(X) \text{ has no closed form}$$

$$Mean = e^{\mu+\frac{\sigma^2}{2}} \quad Variance = (e^{\sigma^2}-1)e^{2\mu+\sigma^2}$$

The lognormal density function shown in Figure 4.6 is directly related to the normal distribution in the following manner: A random variable x follows a lognormal distribution with parameters μ and σ if, and only if, $\ln(x)$ is $N(\mu, \sigma^2)$. The lognormal has special applications in reliability models.

4.2.1.6 Weibull Distribution

$$f(x) = \alpha\mu(\mu x)^{\alpha-1}e^{-(\mu x)^{\alpha}}, \quad x > 0, \ \mu > 0$$

$$F_x(x) = 1 - e^{-(\mu X)^{\alpha}}, \quad X > 0$$

$$Mean \ \frac{1}{\alpha\mu}\Gamma(1/\alpha)$$

$$Variance = \frac{1}{\alpha\mu^2}\left[2\Gamma(2/\alpha) - \frac{1}{\alpha}\Gamma^2(1/\alpha)\right]$$

The Weibull density function shown in Figure 4.7 has wide applications in reliability models. For $\alpha = 1$, the Weibull distribution reduces to the negative exponential.

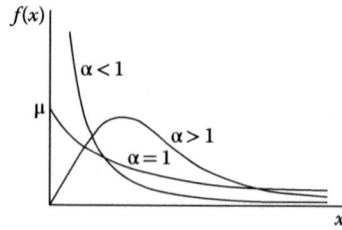

Figure 4.7   Weibull density function.

4.2.1.7 Beta Distribution

$$f(x) = \frac{\Gamma(\alpha + \beta)}{\Gamma(\alpha)\Gamma(\beta)} x^{\alpha-1}(1-x)^{\beta-1}, \quad 0 < x < 1$$

$F_x(X)$ *has no closed form*

$$Mean = \frac{\alpha}{\alpha + \beta} \quad Variance = \frac{\alpha\beta}{(\alpha+\beta)^2(\alpha+\beta+1)}$$

Figure 4.8 shows examples of the beta density function for different ratios of its shape parameters α and β, where both α and β are greater than or equal to 1. The function takes a variety of shapes for other combinations of α and β less than 1.

The beta distribution has been proposed to describe activity durations in PERT networks. It must be noted that a [0, 1] beta random variable x can be transformed to a beta random variable y on the [a, b] interval by using $y = a + (b - a)x$.

4.2.1.8 Triangular Distribution

$$f(x) = \frac{2(x-a)}{(b-a)(c-a)}, \quad a \le x \le b$$

$$f(x) = \frac{2(c-x)}{(c-b)(c-a)}, \quad b \le x \le c$$

$$F_X(X) = \frac{(X-a)^2}{(b-a)(c-a)}, \quad a \le X \le b$$

$$F_X(X) = 1 - \frac{(c-X)^2}{(c-b)(c-a)}, \quad b \le X \le c$$

$$Mean = \frac{a+b+c}{3}$$

$$Variance = \frac{a^2 + b^2 + c^2 - ab - ac - bc}{18}$$

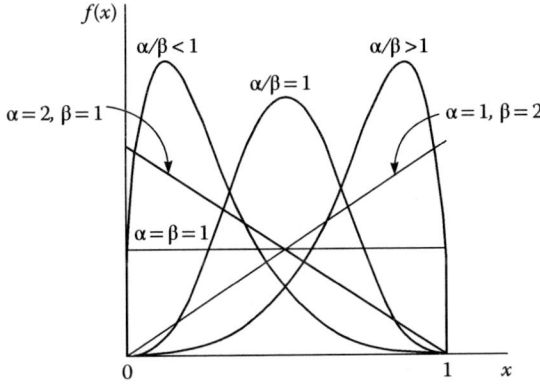

Figure 4.8   Beta density function.

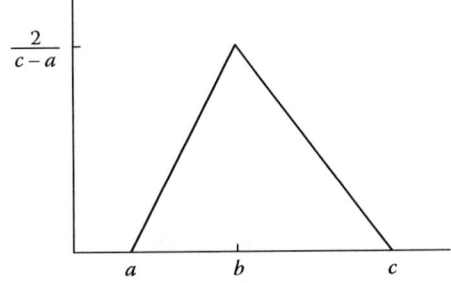

Figure 4.9   Triangular density function.

The triangular density function shown in Figure 4.9 is often used as a first approximation for situations where there is insufficient observational data. This random variable is expected to be unimodal over a finite range of variation.

4.2.1.9 Poisson Distribution

$$f(x) = \frac{\lambda^x e^{-\lambda x}}{x!}, \quad x = 0, 1, 2, \ldots$$

$F_X(x)$ *has no closed form*

Mean $= \lambda$ *Variance* $= \lambda$

The Poisson distribution describes counts of events during a specified time unit such as the number of arrivals and departures at a service facility. In this case, the parameter λ represents the number of events per unit time. The Poisson is the only distribution that has equal mean and variance. This property can sometimes be used to recognize the applicability of the distribution to a set of raw data. The Poisson distribution is related to the exponential distribution in the following manner: Given x is the random variable representing the number of Poisson occurrences with mean λ per unit time, the distribution of the time intervals between *successive* occurrences is exponential with mean $1/\lambda$ time units.

4.2.2 *Identifying Distributions on the Basis of Historical Data*

In practice, probability distributions used in simulation models are often determined from historical data. The procedure involves

1. Summarizing historic or observed data in the form of an appropriate histogram
2. Hypothesizing a theoretical distribution on the basis of the shape of the histogram and testing its goodness-of-fit using an appropriate statistical model

4.2.2.1 *Building Histograms*

Histograms are a pictorial reference for "guessing" the shape of the population from which observed data are drawn. Histogram data may be from historical records or collected observations. In either case, a histogram is constructed by dividing the range of data into nonoverlapping intervals or cells. Each data point is then assigned to one of the defined cells. The result is a tally of the number of points within each cell.

The choice of cell width is crucial to producing a histogram that is descriptive of the shape of the population from which the data are obtained. To appreciate this point, imagine the extreme case in which the cell width is defined such that at most one data point or none would fall within the cell. The other extreme would be to represent the entire range of the data by one cell. In both cases, the result would not be indicative of the shape of the population. Because there are no fixed rules governing the selection of the cell width, the user must exercise judgment in making a reasonable choice.

Excel's *Histogram* program tallies frequencies, calculates percentages, and graphs the relative frequencies and cumulative relative frequency. Examples of constructing a histogram using Excel are shown in Appendix A.

Example 4.1

Figure 4.10 provides two examples of histograms. In part (a) the data represent the service time (in minutes) in a facility, and in part (b) the data represent arrivals per hour at a service facility.

Histogram (a), which represents a continuous random variable, has cell width of three minutes (cells need not be of equal widths). Notice that cells in a histogram a do not overlap. Specifically, cell intervals are defined as [0, 3), [3, 6), [6, 9), [9, 12), and [12, ∞]. The tally for each cell is the number of data points in the cell. For example, in histogram (a) there are eight service times (out of the total of 100) that are greater than or equal to six minutes and less than nine minutes. Clearly, the best histogram is when cells are as small as possible. In histogram (b) there is no cell width because the random variable is discrete. Instead, each cell is taken to represent a value of the discrete variable.

The histograms in Figure 4.10 are converted to probability distributions by computing the relative frequency for each cell. Figure 4.11 shows the resulting probability functions for the histograms in Figure 4.10. The cell boundaries from histogram (a) lose their identity, and instead, each cell is represented by its midpoint. Probability function (a) represents a continuous random variable. The resulting function is used as an approximation of the continuous distribution. Clearly, the best histogram for continuous functions is when cells are as small as possible.

Although the piecewise linear approximation and the discrete density functions of Figure 4.11 can be used as input data to the simulation model, the modeler may prefer (perhaps it is less cumbersome) to represent these distributions by closed forms by assuming the functions in Figure 4.11 are samples from unknown populations.

The modeler may hypothesize theoretical density functions (e.g., exponential and Poisson) with shapes that "resemble" those of the samples. Section 4.2.2.2 details statistical procedures used to test the "goodness-of-fit" of proposed theoretical density functions.

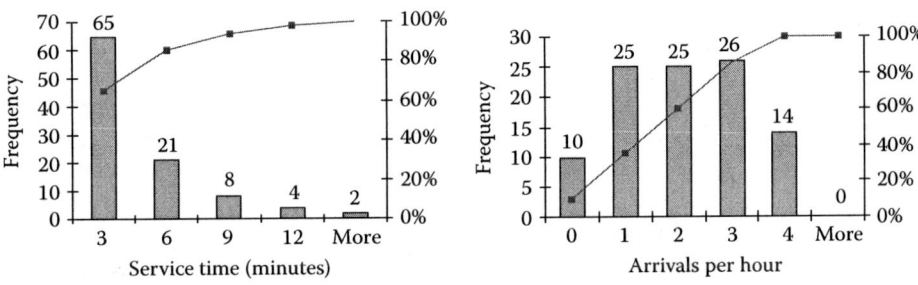

Figure 4.10 Frequency histograms for service time and arrival rates.

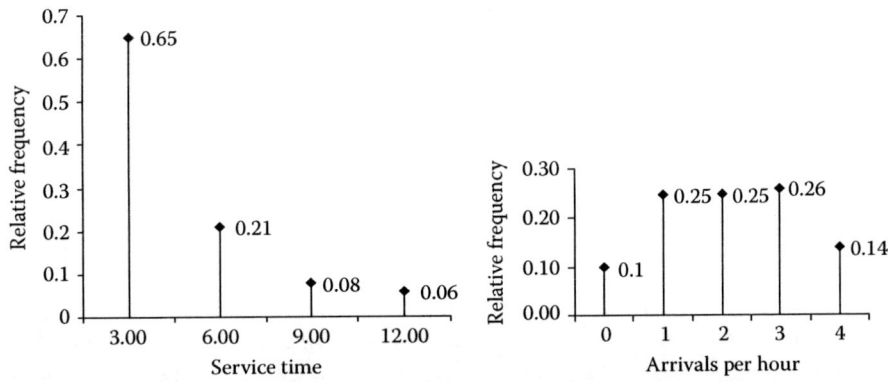

Figure 4.11 Frequency histograms for service time and for arrivals per hour.

4.2.2.2 Goodness-of-Fit Tests

As indicated in Section 4.2.2.1, histogram-based empirical density functions may be used in simulation models. Although the use of empirical distributions is straightforward, they may be cumbersome when performing sensitivity analysis for the model. Imagine, for example, an empirical distribution represents service time in a facility. To determine the impact of increasing or decreasing the service rate, it would be necessary to change the specifications of the empirical distribution for each sensitivity analysis study. Alternatively, if the empirical distribution is replaced by a closed-form function, the task of carrying out sensitivity analysis should become much simpler. Service rate changes are achieved by simply changing the distribution parameter(s).

In this section, two statistical models are presented for testing the goodness-of-fit using a theoretical distribution to represent empirical data: the chi-square (χ^2) and Kolmogrov–Smirnov (K-S) tests. The χ^2 test has more general applicability than the K-S test. Specifically, the χ^2 test applies to both discrete and continuous random variables, whereas the K-S test is applicable to continuous random variables only. Additionally, the K-S test, in its original form, assumes that the parameters of the theoretical distribution are known. The χ^2 test, on the contrary, is based on estimating distribution parameters from raw data. More recently, however, the K-S test has been modified to account for estimating the parameters of the exponential, normal, and Weibull distributions from raw data.

4.2.2.2.1 Chi-Square Test

The χ^2 test starts by representing the raw data in a histogram and probability function as demonstrated in Section 4.2.2.1. The resulting histogram provides a visual clue to the possible shape of the theoretical distribution. After selecting a theoretical distribution to be considered, distribution parameters are estimated from the sample data. The theoretical frequency associated with each cell of the empirical histogram is estimated in the following manner: Let $[a_{i-1}, a_i]$ represent the boundaries of cell i and assume that $f(x)$ represents the hypothesized theoretical pdf. If the available sample includes n data points, the theoretical frequency in cell i on the basis of $f(x)$ is given as

$$n_i = np_i = n \int_{a_{i-1}}^{a_i} f(x)dx, \quad i = 1,2,\ldots,N$$

where p_i is the probability that x falls within cell i. (The value of p_i for a discrete random variable equals the probability that the ith value of the variable is realized.)

Given O_i is the observed frequency in cell i, the test statistic

$$\chi^2 = \sum_{i=1}^{N} \frac{(O_i - n_i)^2}{n_i}$$

becomes asymptotically χ^2 with $N - k - 1$ degrees of freedom as the number of cells $N \to \infty$. The scalar k represents the number $(O_i - n_i)^2/n_i$ of distribution parameters estimated from the observed data. Given this information, the null hypothesis stating that the observed data are drawn from the pdf $f(x)$ is accepted at a level of significance α if the test statistic χ^2 does not exceed $\chi^2_{n-k-1,1-\alpha}$. Otherwise the null hypothesis is rejected.

Example 4.2

To illustrate the applications of the χ^2 test, consider histogram (a) in Figure 4.10 (Example 4.1). The histogram appears to have a shape similar to an exponential distribution. The mean service time is needed to compute the theoretical frequency ($n_i = 1, 2,\ldots, 5$). In general, the mean can be estimated directly from raw data as

$$\bar{x} = \sum_{i=1}^{n} x_i/n$$

where $n = 110$, the total number of data points. However, the mean may also be estimated from the histogram as

$$\bar{x} = \sum_{i=1}^{N} O_i\bar{x}_i/n$$

where N is the number of histogram cells, O_i the observed frequency in cell i, and $\bar{x}_i$ the midpoint of cell i.

This formula yields the average service time for histogram (a) in Figure 4.10 as

$$\bar{x} = \sum_{i=1}^{N} O_i\bar{x}_i/n$$

$$\bar{x} = (65 \times 1.5 + 21 \times 4.5 + 8 \times 7.5 + 6 \times 10.5)/100 = 3.15 \text{ minutes.}$$

Using the estimate $\bar{x} = 3.15$ minutes, the hypothesized exponential density function is written as

$$\bar{x} = \sum_{i=1}^{N} O_i \bar{x}_i / n$$

$$f(x) = (1/3.15) * e^{-x/3.15} \quad x > 0$$

The associated CDF is

$$F_x(X) = 1 - e^{-x/3.15} \quad X > 0$$

The theoretical frequency in cell $[a_{i-1}, a_i]$ is

$$n_i = np_i = n[F_x(a_i) - F_x(a_{i-1})]$$

Table 4.1 summarizes the computations of the χ^2 test given the hypothesized exponential density function with mean 3.15 minutes.

Notice that cells 4 and 5 are combined because, as a general rule, it is recommended that the expected frequency n_i in any cell be no less than five points. Also, because of the tail of the exponential density function, 2.22% of the cumulative area is above the upper interval limit of 12.0. The 2.22% is included in cell 4. Because we estimated the mean of the exponential from the observations, the χ^2 will have $4 - 1 - 1 = 2$ degrees of freedom. Assuming a significance level $\alpha = 0.05$, the critical value from the chi-square tables is $\chi^2_{2,.95} = 7.81$. Because the χ^2 value of 2.39 in Table 4.1 is less than the critical value of 7.81, the hypothesis that the observed data is from an exponential distribution with mean 3.15 minutes is accepted.

A problem with the χ^2 test for continuous random variables is the lack of definite rules to select cell width. In particular, the "grouping" of data and their representation by cell midpoints can lead to loss of information. In general, the objective is to select the cell width "sufficiently" small to minimize the adverse effect of midpoint approximation, but large enough to capture information regarding the shape of the distribution.

In Table 4.1, the user must calculate theoretical frequency n_i and the χ^2 statistic in the last column and then decide if the hypothesis about the distribution should be accepted. As shown in Appendix A, these calculations are facilitated using the Excel's CHITEST function.

Table 4.1 Chi-Square Goodness-of-Fit Test

Cell Number	Cell Boundaries	Cell Midpoint x_i	Observed Frequency O_i	Observed Frequency O_i	Sample Mean $x_i O_i/100$	$n_i \cdot F(a_i)$	Theoretical Frequency n_i	$(O_{i-ni})/n_i$
1	[0,3)	1.5	65	65	0.975	61.42	61.42	0.21
2	[3,6)	4.5	21	21	0.975	85.11	23.70	0.31
3	[6,9)	7.5	8	8	0.6	94.26	9.14	0.14
4	[9,12)	10.5	4	6	0.63	97.78	3.53	1.73
				100	3.15		97.78	2.39
6	[12,∞)			2				

4.2.2.2.2 Kolmogrov–Smirnov Test

The K-S model is another goodness-of-fit test frequently referenced in the literature. The test is applicable to continuous distributions only. It differs from the χ^2 test in that it does not require a histogram. Instead, the model utilizes the raw data $x_1, x_2, \ldots, x_n$ to define an empirical CDF of the form

$$F_n(x) = 1/n \text{ (number of } x_i\text{'s} < x)$$

The idea is to determine a corresponding CDF, $G(x)$, from an assumed fitted distribution and then define the maximum deviation D_n between $F_n(x)$ and $G(x)$ as

$$D_n = \max \left\{ |F_n(x) - G(x)| \right\}$$

Intuitively, large values of D_n adversely affect the fit. Statistical tables that provide critical values for different significance levels have been compiled for testing the null hypothesis that $G(x)$ is the CDF of the population from which the data are drawn.

The K-S test has limitations. In addition to its applicability to continuous distribution only, the test assumes that none of the fitted distribution parameters can be estimated from raw data. Exceptions to this case include the normal, exponential, and Weibull random variables. In these situations, it may be advantageous to use the K-S test because its implementation does not require a histogram that causes a possible loss of information (see [3], pp. 201–203).

4.2.2.3 *Maximum Likelihood Estimates of Distribution Parameters*

To apply the χ^2 test of goodness-of-fit, it is necessary to estimate the parameters of the theoretical distribution from sample data. For example, in the uniform distribution the limits of the interval (a, b) covered by the distribution must be estimated. For the normal distribution, estimates of the mean μ and standard deviation σ are needed. Using estimates for the parameters and a proposed theoretical distribution the expected frequencies are calculated.

Parameter estimation from sample data is based on important statistical theory that requires these parameters to satisfy certain statistical properties (i.e., unique, unbiased, invariant, and consistent). These properties are usually satisfied when the parameters are based on the *maximum likelihood estimates*. For example, in the case of the normal distribution, this procedure yields the sample mean and variance as the maximum likelihood estimates of the population mean and variance. Also, in the uniform distribution, the sample minimum and maximum values estimate the limits of the interval on which the distribution is defined. Similarly, in the exponential distribution, the sample mean provides the maximum likelihood estimate of the single parameter distribution.

Unfortunately, determining maximum likelihood estimates is not always convenient. For example, the gamma, beta, and Weibull distributions require solutions for two nonlinear equations.

A full presentation of the maximum likelihood procedure is beyond the scope of this chapter. Most specialized books on statistics treat this topic (see, for example, [1] and [2]). A concise summary of the results of applying the maximum likelihood procedure to important distributions can be found in [3] (pp. 157–175).

4.3 Statistical Output Analysis

As described in Section 4.1, simulation is essentially a statistical experiment that is subject to random error. Simulation results should therefore be presented in the form of statistical confidence intervals rather than simple point estimates. Additionally, comparison of output measures representing alternative simulation settings (i.e., different system configurations) must be performed using statistical hypothesis testing techniques. In this section, we present a summary of the procedures for constructing mean confidence intervals and for hypothesis testing of mean performance measures.

4.3.1 Confidence Intervals

Suppose that the sample $x_1, x_2,..., x_n$ represents the values of a given performance measure in a simulation experiment. The sample mean and variance of this measure are defined as

$$\bar{x} = \sum_{i=1}^{N} x_i/n \quad S^2 = \sum_{i=1}^{b} (x_i - \bar{x})^2/(n-1)$$

Suppose that μ represents the true mean of the population from which the sample $x_1, x_2,..., x_n$ is obtained. From the assumption that $x_1, x_2,...,$ and x_n are independent and drawn from a normal population, the random variable

$$t = \frac{\bar{x} - \mu}{\sqrt{S^2/n}}$$

is known to follow a t-distribution with $n - 1$ degrees of freedom. Using x and s, a confidence interval that covers the true mean, μ, $100(1 - \alpha)\%$ of the time, can be expressed by the following probability statement:

$$p\left\{-t_{n-1,1-\alpha/2} \leq \frac{\bar{x} - \mu}{\sqrt{S^2/n}} \leq -t_{n-1,1-\alpha/2}\right\} = 1 - \alpha$$

The quantity $t_{n-1,1-\alpha/2}$ represents the upper $(1 - \alpha/2)$ critical value of the t-distribution. In this case, α represents the confidence level and the desired confidence interval is defined by the following inequality:

$$-t_{n-1,1-\alpha/2} \leq \frac{\bar{x} - \mu}{\sqrt{S^2/n}} \leq -t_{n-1,1-\alpha/2}$$

which gives

$$\bar{x} - t_{n-1,1-\alpha/2}\sqrt{S^2/n} \leq \mu \leq \bar{x} + t_{n-1,1-\alpha/2}\sqrt{S^2/n}$$

The given range defines the $100(1 - \alpha)\%$ confidence interval for the true mean, μ.

Example 4.3

Consider the simulation of an environment in which TV units arrive for inspection at time intervals that are uniformly distributed between 3.5 and 7.5 minutes. The inspection station has two operators. The inspection time is exponentially distributed with a mean of nine minutes. Estimates are that 85% of the units pass the inspection. The remaining 15% are returned for adjustment and then reinspected. The adjustment time follows a uniform distribution between 15 and 30 minutes. The objective is to determine a 90% confidence interval for the average time a unit spends in the system before it is sent for packaging. The following data represent 10 observations of the time (in minutes) a unit spends in the system:

24.1, 40.2, 26.7, 42.4, 20.3, 31.6, 49.8, 39, 34.5, 31.4

The sample mean and standard deviation are

$$\bar{x} = 34 \text{ minutes}$$

$$s = 9.04 \text{ minutes}$$

Because $t_{9,.9} = 1.833$, the desired 90% confidence interval is given as

$$34 - 1.833(9.04/10) <= \mu <= 34 + 1.833(9.04/10)$$

$$28.76 <= \mu <= 39.24$$

This confidence interval is equivalent to stating that the probability that the true μ is within the range [28.76, 39.24] is 0.90.

4.3.1.1 Satisfying the Normality Assumption in Simulation

In the previously defined confidence interval, it is stated that sample data must be drawn from a normal population in order for the t-distribution to be applicable. This point was not investigated in the preceding example. Indeed, given that the inspection-adjustment times are uniform and exponential, it is doubtful that the system time is normally distributed. How can the t-distribution be used to compute the confidence interval? The answer is that in simulation, an observation is never equal to a "single occurrence" during the course of the simulation. Instead, as explained in Section 4.3, a "simulation observation" is defined as the *average* of several successive occurrences.

In other words, each x_i in the sample is an average of the system time for "several" completed TV units. The central limit theorem in statistics states that averages are asymptotically normal regardless of the identity of the parent population from which the data are drawn. This observation is key to justifying the use of regular statistical tests in simulation experiments. Keep in mind, however, that the simulation experiment has other peculiarities that may affect the validity of applying available statistical techniques to simulation output. This point is discussed further in Chapter 5.

4.3.2 Hypothesis Testing

Hypothesis testing is used in simulation experiments to test whether a mean of a given measure of performance is equal to a specified value. Additionally, one may be interested in testing whether two means obtained from different simulation environments are significantly different in the statistical sense. The basic underlying theory for hypothesis testing is the same as in the case of the confidence interval, namely, observations must be independent and drawn from a normal population.

First consider the simple case where the mean is compared to a specific value, with $\bar{x}$ and s representing the mean and standard deviation of the sample. The usual procedure is to assume a *null* hypothesis, H_0, defined as

$$H_0 : \mu = \mu_0$$

where μ_0 is a specified value. In this case, the alternative hypothesis may assume one of the following forms:

$$A_1 : \mu <> \mu_0$$

$$A_2 : \mu > \mu_0$$

$$A_3 : \mu < \mu_0$$

The basis for accepting or rejecting the null hypothesis is the following t-statistic with $n - 1$ degrees of freedom:

$$t_0 = \frac{\bar{x} - \mu_0}{S/\sqrt{n}}$$

Let α represent a desired level of significance ($0 < \alpha < 1$), and define $t_{n-1,\alpha}$ as the critical value of the t-distribution such that

$$P\{t > t_{n-1,\alpha}\} = \alpha$$

The conditions for rejecting the null hypothesis are summarized as follows:

Alternative Hypothesis	Condition for Rejecting H_0
$A_1 : \mu <> \mu_0$	$|t_0| > t_{n-1,\alpha/2}$
$A_2 : \mu > \mu_0$	$t_0 > t_{n-1,\alpha/2}$
$A_3 : \mu > \mu_0$	$t_0 > t_{n-1,\alpha/2}$

The same idea can be extended to testing the null hypothesis regarding the equality of two population means, that is,

$$H_0 : \mu_1 = \mu_2$$

In this case, the alternatives will be defined as

$$A_1 : \mu_1 <> \mu_2$$

$$A_2 : \mu_1 > \mu_2$$

$$A_3 : \mu_1 > \mu$$

The associated test statistic is then given as

$$t_0 = -\frac{\bar{x}_1 - \bar{x}_2}{\sqrt{S_1^2/n_1 + S_2^2/n_2}}$$

which represents a *t*-distribution with N degrees of freedom, where

$$N = \frac{S_1^2/n_1 + S_2^2/n_2}{\dfrac{S_1^2/n_1}{n_1 + 1} + \dfrac{S_2^2/n_2}{n_2 + 1}} - 2$$

The following table summarizes the conditions for rejecting H_0:

Alternative Hypothesis	Condition for Rejecting H_0
$A_1 : \mu_1 <> \mu_2$	$|t_0| > t_{N,\alpha/2}$
$A_2 : \mu_1 > \mu_2$	$t_0 > t_{N,\alpha}$
$A_3 : \mu_1 > \mu_2$	$t_0 > -t_{N,\alpha}$

Example 4.4

Consider the TV situation in Example 4.3 and suppose that we want to test the effect of adding a third inspector on the time a TV unit stays in the system. The data below are collected for three operators:

$$15.7, 19.5, 9.2, 15.3, 14.3, 29, 40, 23.2, 24.3, 28.6$$

These data yield $\bar{x} = 21.9$ minutes and $s = 9.05$ minutes. Using these data and those of Example 4.3,

$$\bar{x}_1 = 34, \quad S_1^2 = 9.04^2 = 81.72, \quad n_1 = 10$$

$$\bar{x}_2 = 21.9, \quad S_2^2 = 9.05^2 = 81.90, \quad n_2 = 10$$

Therefore,

$$t_0 = \frac{34 - 21.9}{\sqrt{8.172 + 8.19}} = 2.99$$

$$N = \frac{(8.172 + 81.9)^2}{8.172^2/11 + 8.19^2} - 2 \approx 20$$

For $\alpha = 10$

$$t_{N,\alpha/2} = 1.725$$

$$t_{N,\alpha} = 2.528$$

The decisions for accepting or rejecting H_0 are summarized as follows:

Alternative	Statistics	Conclusion
$A_1 : \mu_1 <> \mu_2$	$t_0 = 2.99, t_{N,\alpha/2} = 1.725$	Reject $H_0 : \mu_1 = \mu_2$
$A_2 : \mu_1 > \mu_2$	$t_0 = 2.99, t_{N,\alpha} = 2.528$	Reject $H_0 : \mu_1 \leq \mu_2$
$A_3 : \mu_1 < \mu_2$	$t_0 = 2.99, t_{N,\alpha} = 2.528$	Reject $H_0 : \mu_1 \geq \mu_2$

The results show that H_0 is accepted only for $A_3 : \mu_1 < \mu_2$. This means that $\mu_1 \geq \mu_2$ is accepted.

Although the above discussion was restricted to comparisons of one and two means, the situation can be extended to any number of means by using the analysis of variance (ANOVA) technique. The presentation of this material is beyond the scope of this chapter, however. Miller and Freund [4] provide a complete discussion on the subject.

4.4 Summary

In this chapter, the role of statistics both at the input and output ends of the statistical experiment are summarized. Also presented are statistical tests that may be used to analyze simulation experiments. However, the special peculiarities of simulation regarding sampling must be carefully observed. In particular, questions of normality and independence could mean serious limitations on the use of simulation data. These points are investigated further in Chapter 5.

Problems

4.1. Derive the mean and variance for each of the following distributions using the pdfs given in Section 4.2:
  a. Uniform
  b. Exponential
  c. Gamma
  d. Normal
  e. Lognormal
  f. Beta
  g. Triangular
  h. Poisson

4.2. Prove that the exponential distribution has no memory. Explain in your own words what "lack of memory" means.

4.3. Customers arrive at a three-server facility according to a Poisson distribution with mean A. If arriving customers are assigned to each server on a rotational basis, determine the distribution of interarrival times at each server.

4.4. Repeat Problem 4.3 assuming that arriving customers are assigned to the three servers randomly according to the respective probabilities a_1, a_2, and a_3, where $a_1 = a_2 + a_3 = 1$ and $a_i > 0$, $i = 1, 2, 3$.

4.5. For the Poisson distribution of arrivals with mean A arrivals per unit time, show that during a sufficiently small period of time h, at most one arrival can take place.

4.6. Suppose that the service times at two checkout counters in a grocery store are sampled and that the means and variances are found to be approximately equal for the two samples. Is it reasonable to conclude that service times at the two counters follow the same probability distribution? Explain.

4.7. Given the sample $x_1, x_2, \ldots, x_n$, show that its variance can be estimated as

$$S_x^2 = \frac{n \sum_{i=1}^{n} x_i^2 - \left(\sum_{i=1}^{n} x_i \right)^2}{n(n-1)}$$

4.8. Apply the formula in Problem 4.7 to the following sample representing service times in a facility:

23.0, 21.5, 19.7, 20.5, 19.9, 22.8, 23.2, 22.5, 21.9, 20.1

4.9. Consider the following two samples representing the time to failure in hours of two electronic devices:

Device 1: 32.5, 21.2, 33.7, 27.5, 29.6, 17.3, 30.2, 26.8, 29.0, 34.8

Device 2: 20.5, 22.6, 23.7, 21.1, 18.3, 19.9, 21.9, 17.8, 20.5, 22.6

Based on the observations from these two samples alone, which device exhibits a higher "degree of uncertainty" around its mean value?

4.10. The following data represent the interarrival times in minutes at a service facility:

4.3	3.4	0.9	0.7	5.8	3.4	2.7	7.8
4.4	0.8	4.4	1.9	3.4	4.1	5.1	1.4
0.1	4.1	4.9	4.8	15.9	6.7	2.1	2.3
2.5	3.3	3.8	6.1	2.8	5.9	2.1	2.8
3.4	3.1	0.4	2.7	0.9	2.9	4.5	3.8
6.1	3.4	1.1	4.2	2.9	4.6	7.2	5.1
2.6	0.9	4.9	2.4	4.1	5.1	11.5	2.6
2.1	10.3	4.3	5.1	4.3	1.1	4.1	6.7
2.2	2.9	5.2	8.2	1.1	3.3	2.1	7.3
3.5	3.1	7.9	0.9	5.1	6.2	5.8	1.4
0.5	4.5	6.4	1.2	2.1	10.7	3.2	2.3
3.3	3.3	7.1	6.9	3.1	1.6	2.1	1.9

a. Develop a histogram for interarrival time.
b. Estimate the mean and variance.
c. Hypothesize a theoretical distribution for the data and an explanation for your selection.
d. Apply the chi-square goodness-of-fit test at a 95% confidence level.

4.11. From Section 4.2, two different distributions may have similar density functions. Therefore, hypothesizing a theoretical distribution for given histogram data may not be a clear-cut choice. In this case, it is helpful to base the selection on the characteristics of the situation under study. Identify some of the characteristics associated with each of the following distributions:
a. Poisson
b. Exponential
c. Normal
d. Uniform
e. Gamma
f. Weibull
g. Beta

4.12. Consider the following sample that represents time periods (in seconds) to transmit messages:

25.8	76.3	35.2	36.4	58.7
47.9	94.8	61.3	59.3	93.4
17.8	34.7	56.4	22.1	48.1
48.2	35.8	65.3	30.1	72.5
5.8	70.9	88.9	76.4	17.3
77.4	66.1	23.9	23.8	36.8
5.6	36.4	93.5	36.4	76.7
89.3	39.2	78.7	51.9	63.6
89.5	58.6	12.8	28.6	82.7
38.7	71.3	21.1	35.9	29.2

Test the hypothesis that these data are drawn from a uniform distribution at a 95% confidence level given the following information:

a. The interval of the distribution is between 0 and 100.
b. The interval of the distribution is between a and b, where a and b are unknown parameters.
c. The interval of the distribution is between a and 100, where a is an unknown parameter.

4.13. For the sample in Problem 4.12, test the hypothesis that the data are drawn from the following theoretical distributions (use a 95% confidence level):

a. Exponential
b. Normal
c. Lognormal

References

1. Kendall, M. and S. Stuart, *The Advanced Theory of Statistics*, 3rd ed., Vol. 1. New York: Hafner Publishing Co., 1969.
2. Kendall, M. and S. Stuart, *The Advanced Theory of Statistics*, 3rd ed., Vol. 2. New York: Hafner Publishing Co., 1973.
3. Law Averill, M., *Simulation Modeling and Analysis*, 4th ed. New York: McGraw-Hill Book Company, 2007.
4. Miller, I. and J. Freund, *Probability and Statistics for Engineers*, 3rd ed. Englewood Cliffs, NJ: Prentice-Hall, 1985.

Chapter 5

Elements of Discrete Simulation

5.1 Concept of Events in Simulation

In systems terminology, the purpose of simulation modeling is to collect statistics that can be used to infer the performance characteristics of a system. As discussed in Section 3.2, changes in the system state do not occur on a continuous basis in discrete event simulation. Rather, changes occur as the result of what are called *primary events*. For example, in a single queue system, the state of the waiting line will increase by one when a customer arrives at the system (arrival event) and the facility is busy. The state will decrease by one when a customer exits the system (service completion event). Indeed, changes to the state of the queue occur at points in time when a customer arrives at the system or completes service.

The design of a discrete simulation language depends on whether the language is an *event, activity,* or *process* type. Although simulation languages differ in the manner in which the user codes the model, the use of events is fundamental to *all* discrete event simulators and all simulators use the concept of chronologically ordered primary events to advance the simulation clock. All discrete event simulators when applied to the same system (i.e., equivalent system events) must produce the same sequence of events on the timescale (barring, of course, any arithmetic discrepancies inherent to the simulator). As discussed in Section 5.2, these three approaches to simulation language have a common basis.

5.2 Common Simulation Approaches

This section introduces the *event-scheduling, activity-scanning,* and *process-simulation* approaches via examples. The section also concentrates on how the concept of events is used in all three approaches.

5.2.1 Event-Scheduling Approach

The event-scheduling approach is demonstrated by using a familiar queueing example. The system is represented graphically in Figure 5.1. Dotted lines represent the boundary on the system. Customers originating at the Source may either join the Queue or engage the Facility and begin service. After service completion, the customer leaves the system. Customers arrive according to a given distribution. The customer service time also follows a known distribution. The purpose of the simulation is to gather information on performance characteristics of the queue and facility. In this example, the specific measures are

- Length of Queue
- Waiting time in Queue
- Utilization of Facility
- Idle time of Facility

Changes in states of the system that affect performance measures are when

- A customer arrives at the system.
- A customer completes service.

A primary event is the time scheduled for these state transitions. In general, the approach to event scheduling is that when a new (arrival or service completion) event is scheduled, it is automatically placed in chronological order on the timescale with other primary events. The simulator selects the next event "scheduled" to occur and, depending on its type (arrival or departure), performs the appropriate actions. The event is then removed from the primary events list (PEL) to indicate that the event has occurred. Because the next state transition corresponds to the next primary event, the simulation clock is advanced to the time of the next event on the PEL. This process is repeated until the simulation period ends.

The chronological "list" on which primary events reside is the PEL. For reasons that will be explained later, entities on the PELs are referred to as transactions. Each transaction must have an *attribute* representing the type of the event (e.g., arrival or service completion) and another to indicate the time that the event is scheduled to occur. There may be other attributes that provide more information about the event. Transactions are in the PEL in ascending (actually, nondescending) order of their time to occur and removed according to their sequence on the PEL. The PEL also represents the automatic time-advancing mechanism (or clock) essential for any simulation language.

The following activities, dependent on the arrival and service completion events, are exhaustive in the sense that all the possibilities related to an arriving customer and a customer completing service are completely described. These actions also include the collection of statistics on the system's performance measures. These activities, *secondary events*, are dependent on their

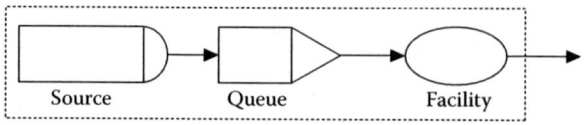

Figure 5.1   Basic simulation model.

respective primary event. In pseudo code, the primary-secondary relationships are described as follows:

Primary Event: Arrival—Related Secondary Events

```
Schedule next arrival on PEL at time =
          current simulation time + interarrival time.
If Facility is idle
Then
   Mark server busy (change state of Facility).
   Update Facility idle time statistics.
   Schedule service completion on PEL at time =
          current simulation time + service time.
Else
   Place Customer in the Queue (change state of Queue).
   Update Queue length statistics.
End If
Select the next primary event.
```

Primary Event: Service Completion—Related Secondary Events

```
If Queue is empty
Then
   (a) Declare server idle (change state of server).
Else
   (a) Remove customer from the Queue (Change state of Queue)
   (b) Update Queue waiting time statistics.
   (c) Schedule service completion on PEL at time =
          current simulation time + service time.
End If
Depart ·the system.
Discard the current event.
Select the next primary event.
```

Figure 5.2 summarizes the simulation process by depicting the general logic of advancing the clock, selecting the primary event, and implementing its dependent secondary events. Regardless of the simulation language, event scheduling is an excellent design tool. The preliminary design of the system includes pseudo code documentation of the primary events and dependent secondary events. The design documentation becomes the focus of discussions with operators, process managers, and other members of the simulation team to ensure that everyone understands the simulation effort. As the effort transitions from design to model development, the focus shifts to dividing the secondary events into those that must be managed by the model and those managed by the simulator. For example, in the pseudo code shown above, most simulation languages manage the data collection and the PEL.

Example 5.1

To appreciate the details of the event-scheduling approach, a numeric example of the single-server model is introduced here to show how the simulation is performed and how performance measures are computed. Specifically, suppose that in the single-server model, a customer is released from the Source (i.e., arrives at the Facility) every 10 minutes. For simplicity, the service time in the Facility, on a rotational basis, is either 6 or 12 minutes. The first customer arrives at time zero with

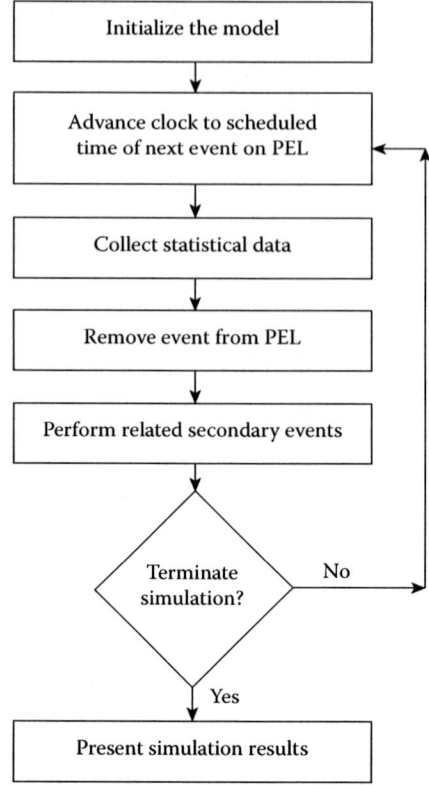

Figure 5.2 Basic simulator logic.

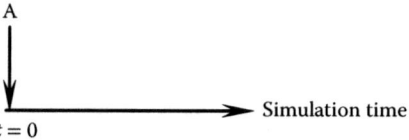

Figure 5.3 Primary event scheduled at *t* = 0.

the Facility in an idle state. To facilitate the presentation of the computations, the scheduling of the events (their order on the PEL) is summarized graphically. Figure 5.3 shows the scheduling of the first arrival, which, according to the starting conditions of the model, occurs at *t* = 0. Therefore, at *t* = 0, the arrival event A appears on the timescale and the PEL has only one event.

The simulation is started with the arrival event A at *t* = 0. Recall from the pseudo code that the first secondary event is to schedule the next arrival event on the PEL. With an interarrival time of 10 minutes, the next arrival event is scheduled at *t* = 10. Next, the simulator determines whether the customer that arrives at *t* = 0 will begin service in the Facility or join the Queue. Because the Facility is idle at *t* = 0, service is started and the Facility is declared busy. Given the service time is 6 minutes, the service completion event, SC, for the customer that has just started service will take place at *t* = 6. The updated primary events are as shown in Figure 5.4.

Two comments are in order for Figure 5.4. First, even though SC, at *t* = 6, is scheduled after A, at *t* = 10, had been scheduled, the event SC must appear first on the timescale. Second, because the primary event A, at *t* = 0, has occurred, it is removed. In any simulation language, the time-advancing mechanism of the simulator automatically accounts for these two details.

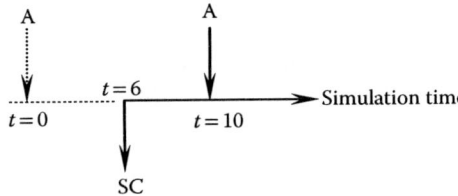

Figure 5.4   Primary events scheduled at $t = 10$.

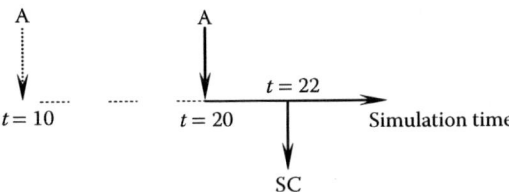

Figure 5.5   Primary events scheduled at $t = 20$.

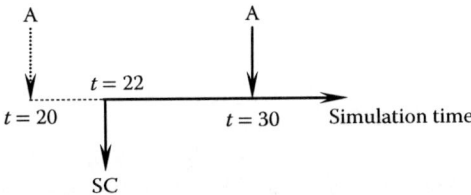

Figure 5.6   Primary events scheduled at $t = 22$.

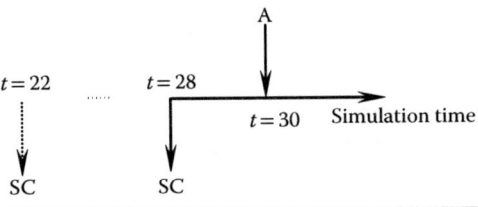

Figure 5.7   Primary events scheduled at $t = 28$.

The simulation now "advances" to the next imminent event, at $t = 6$. Because the event is a service completion, the simulation checks the Queue for waiting customers. Because the Queue is empty, the server is declared idle.

The next event on the PEL, which is an arrival, occurs at $t = 10$. The next scheduled arrival occurs at $t = 10 + 10 = 20$. Because the Facility is idle, service can start immediately, and the service completion event occurs at $t = 10 + 12 = 22$, because the service time is now equal to 12. The updated PEL *immediately after* $t = 10$ is as shown in Figure 5.5.

At $t = 20$, an arrival occurs. The next arrival occurs at $t = 20 + 10 = 30$. Because the Facility is busy, the customer joins the Queue. The PEL immediately after $t = 20$ appears as shown in Figure 5.6.

Now, at $t = 22$, a service completion takes place. Because the Queue has a waiting customer, this customer is selected to start service immediately in the Facility. Given a service time of 6 minutes, a departure event occurs at $t = 22 + 6 = 28$. The updated events list after $t = 22$ is shown in Figure 5.7.

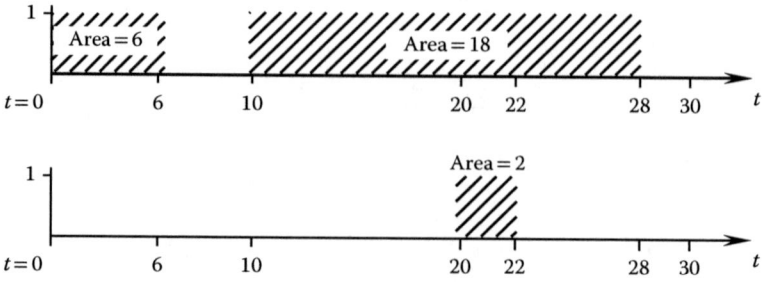

Figure 5.8   Simulation data from *t* = 0 to *t* = 30.

The mechanics of event processing in a simulation should be evident at this point. A basic feature of simulation is how to collect data on the selected performance measures for the above example. Recall that the measures include

■ Average Facility utilization
■ Average Queue length
■ Average Facility idle time
■ Average waiting time in Queue

Figure 5.8 summarizes the data for the example as a function of simulation time. The variables representing Facility utilization and Queue length differ from those of Facility idle time and customer waiting time in that only utilization and Queue length are *time-dependent observations*. The average value of a time-dependent variable is the area under the curve representing the variable divided by the simulation time. The other two variables are averaged by dividing the sum of the updates by the number of updates. (This is explained in more detail in Section 5.4.) From the data in Figure 5.8:

■ Average Facility utilization = $(6 + 18)/30 = 0.8$
■ Average Queue length = $2/30 = 0.07$ customer
■ Average Facility idle time = $[(10 - 6) + (30 - 28)]/2 = 3$ minutes
■ Average waiting time = $(22 - 20)/1 = 2$ minutes

Keep in mind that these estimates are not reliable because, as detailed in Chapter 6, the simulation must be run for a reasonable length of time to ensure reliability of data. It is important to notice that the data in Figure 5.8 are updated when a primary event occurs. This fact is also reflected in Figure 5.2.

5.2.2 Activity-Scanning Approach

The activity-scanning approach is very similar to the event-scheduling approach. The primary difference is in the manner in which secondary events are organized. In event scheduling, primary and secondary events were tightly coupled. In activity scanning, the primary and secondary events are logically separate. In fact, there may be separate primary and secondary events lists. The difference is illustrated by applying activity scanning to the single-server model. The single-server model is described using the following three events:

1. Arrival event
2. Departure event
3. Begin service event

Actions associated with each event are as follows:

1. Arrival event
   a. Place arriving customer in the Queue.
   b. Schedule the next arrival event.
2. Service completion event
   a. Declare the Facility idle.
3. Begin service event
   a. Remove a customer from the Queue and begin service in the Facility.
   b. Schedule service completion.

Contrasting these events with those of event scheduling reveals the following:

■ An arriving customer is always placed in the Queue regardless of the status of the Facility, and a service completion will always make the Facility idle regardless of the status of the Queue. Activities associated with the arrival and departure events are unconditional and therefore always executed.

■ Begin service events are dependent on two conditions being satisfied before these events occur: the Queue is not empty and the Facility is idle. If either or both of these conditions are not satisfied, the events do not occur.

Simply stated, *arrival* and *service completion* activities are primary events and *begin service* activities are secondary events. Another way of viewing the difference is that the primary events are scheduled and cause changes to the state of the system. The secondary events are not scheduled and occur as a result of other changes in the state of the system (i.e., primary or possibly other secondary events).

Activity scanning works as follows: First, consider the next primary event (arrival or service completion) on the PEL and perform its dependent events as shown above. Because the primary event will change the state of the Queue or Facility, the secondary event "begin service" is scanned for execution, provided its (two) conditions are satisfied. Following the execution of the secondary (conditional) events, the events list is updated. The process is repeated by selecting the next event on the PEL.

In a general situation, activity scanning may utilize more than one secondary event. In such cases, the list of secondary events must be scanned repeatedly (following the occurrence of a primary event) until those remaining cannot be executed. In this manner, all changes to the state of the system will be affected before the next primary event is considered. Two observations are in order regarding the way activity scanning works:

1. The primary events (arrival and departure) advance the simulation clock, exactly as in event scheduling.
2. The secondary events are viewed as dependent on the state of the system after a primary event. The main difference occurs in the manner in which secondary events are organized. A comparison of the actions between activity scanning and event scheduling will demonstrate that the two approaches accomplish identical results.

In Figure 5.2, the iterative scan is represented by "perform secondary events." Although implementing the activity-scanning approach is obviously an inefficient process because of repeated checking for secondary events, effort by the user to code an activity-scanning model may not be as extensive as in event scheduling. However, event scheduling is a powerful communications tool in the design process. Also the pseudo code as a design tool is easily translated to event-scheduling routines.

5.2.3 Process-Simulation Approach

Section 5.2.1 presented details of event-scheduling simulation. A simulation language based on this approach normally provides procedures that allow the user to (1) store and retrieve events (with their pertinent attributes) on the PEL, (2) keep track of customers waiting in queues, and (3) update statistics of performance measures. However, it is usually the user's responsibility to invoke these procedures at the appropriate time during the simulation.

Process simulation differs from event-scheduling simulation in that the user can represent the model in a rather "compact" fashion. As an illustration, the single-server system shown in Figure 5.1 can be modeled in a process-oriented simulation language in the following manner:

Source; intercreation time; time offset creation
Queue; queue capacity; queue discipline
Facility; number of parallel servers; service time/server

In process-simulation terminology, *Source*, *Queue*, and *Facility* represent nodes. Now *transactions (entities)* flow sequentially from one node to another. Each transaction has its own *attributes* as it moves between successive nodes. Transaction attributes include the node and the transaction's departure time from the node. Notice the absence of any reference to event processing.

A process-simulation language utilizes node specifications to determine the disposition of transactions from "arrival" at *Source* to "service completion" at *Facility*. At any time between primary events, transactions must reside in one of the model's nodes. Although transparent to the user, a transaction residing in a node has an associated primary event on the PEL and when the primary event occurs, the transaction automatically attempts to enter the succeeding node. The exception to this rule is with *Queue*. A transaction resides in a queue until its movement is caused by a transaction leaving another node (i.e., some primary event). As such, *Queue* is an *inactive* node because, unlike other nodes, there is no associated primary event for *Queue* and it is therefore unable to initiate movement of its resident transactions.

The single-server model above is used to demonstrate that process languages are based on the same concept of events in event scheduling. Starting with the (active) *Source* node, its specification includes the creation time of the first transaction as well as the time between successive creations. Given these data, the processor of the simulation language, with no interference from the user, generates and stores the first transaction in the PEL. However, rather than designating the created transaction as an arrival (as in event scheduling), the transaction is simply associated with the *Source* node. Now, during the course of the simulation, every time a *Source* transaction is encountered, the processor will automatically create the next *Source* transaction and store it chronologically in the PEL. These actions are exactly the same as the arrival event in the event-scheduling approach.

The destination node of a transaction leaving *Source* is determined by the simulation language logic in the following manner:

1. If *Facility* is idle, the transaction skips *Queue* and immediately begins service and experiences a time delay in *Facility* equal to its service time. The transaction is stored chronologically in the PEL (compare with scheduling a service completion event in the event-scheduling approach).
2. If *Facility* is busy, the transaction enters *Queue* and remains there until *Facility* is available.

When service is completed, a *Facility* transaction is released. This is equivalent to encountering a service completion event in the event-scheduling approach. At this point, *Facility* examines *Queue* for possible waiting transactions. If any exist, the first eligible transaction is moved into *Facility*, and its service completion time is computed. A new *Facility* transaction (service completion event) is then stored in the PEL. If *Queue* is empty, *Facility* is declared idle.

The signature of a process-simulation language is the sequential flow of transactions through the model that is enabled by secondary events embedded in the node specification. Essentially, speaking in terms of the single-server model, the arrival and service completion events in event scheduling are now embedded in the *Source* and *Facility* specifications. If both models are executed with the same input data (and the same sequence of random numbers), the associated PEL entries will be identical in a chronological sense with a one-to-one correspondence between arrival/service completion events and *Source/Facility* transactions.

As evident from this discussion, modeling in process simulation is more compact than in event scheduling. The reason for this convenience is that most of the logic specified by the user in event scheduling is embedded as a process-simulation feature. However, the inevitable result of embedded features is a degree of language rigidity that makes it difficult to model complex decisions that affect transaction flow. To address that limitation, process languages have a set of specialized nodes used to encode complex decisions. For process languages, this effort is not as straightforward as event scheduling. In the extreme case, process-simulation languages allow for procedural language "hooks" such as BASIC, FORTRAN, or Java, to model complex decisions.

5.3 Computations of Random Deviates

In the single-server model of Example 5.1, the simulation was performed using deterministic interarrival and service times. Continuing this simulation for a sufficiently long time will show that all the simulation observations and statistics repeat in a cyclic fashion. As such, simulation under deterministic conditions becomes a trivial exercise with predictable results.

For simulation modeling to be of interest, at least one of the events describing the model must occur in a random fashion. In the single-server model, the interarrival time (at *Source*) or the service time (at *Facility*), or both, must be random. For example, the interarrival or service times may follow uniform or exponential probability density functions (pdfs).

To accommodate randomness in simulation, a simulation language must have procedures for generating random samples from any probability distribution. The most common methods for sampling distributions include

- Inverse method
- Convolution method
- Acceptance–rejection method
- Methods based on special statistical properties of the sampled random variable

5.3.1 Inverse Method

The inverse method is based on the fact that for any pdf $f(x)$ of the random variable x, the cumulative density function (CDF) $F(X)$ defined as

$$F_X(X) = \int_{-\infty}^{x} f(y)dy, \quad 0 \le F(X) \le 1$$

is always uniformly distributed on the (0, 1) interval. This result can be proven as follows. Let $y = F(X)$ be a new random variable; then, for $0 < Y < 1$,

$$P\{y \le Y\} = P\{F(X) \le Y\} = P\{X \le F^{-1}(Y)\} = F[F^{-1}(Y)] = Y$$

shows that $y = F(X)$ is uniform on the (0, 1) interval.

The result can be used to sample from a distribution $f(x)$ as follows:

Step 1. Obtain a random number R from the (0, 1) uniform distribution; that is, $0 \le R \le 1$.
Step 2. Determine the random sample X by solving

$$R = F(X)$$

which yields

$$X = F^{-1}(R)$$

This process is shown graphically in Figure 5.9.

Example 5.2

The general uniform pdf is defined as

$$f(x) = \frac{1}{b-a}, \quad a \le x \le b$$

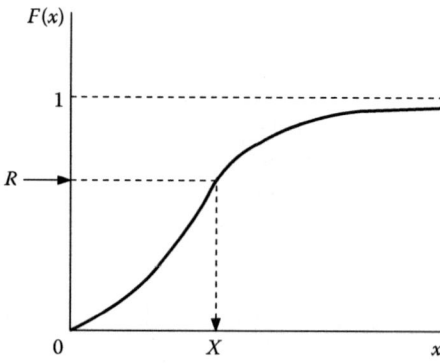

Figure 5.9 Illustration of the inverse method.

The CDF of x is given by

$$F_x(X) = \frac{X-a}{b-a}, \qquad a \le x \le b$$

Given R is a $(0, 1)$ random number, a sample X from the general uniform distribution is determined from

$$R = \frac{(X-a)}{(b-a)}$$

which yields $X = a + R(b - a)$.

Example 5.3

The negative exponential pdf is defined as

$$f(x) = \lambda e^{-\lambda x}, \qquad x > 0$$

Its CDF $F(X)$ is given by

$$F(X) = 1 - e^{-\lambda X}, \qquad X > 0$$

For a $(0, 1)$ random number R, the exponential sample is

$$X = -(1/\lambda)\ln(1 - R)$$

Because $(1 - R)$ is the complement of R, the above formula may be written as

$$X = -(1/\lambda)\ln(R)$$

Example 5.4

Consider the following discrete distribution of the random variable x whose pdf is $p(x)$ and CDF is $P(X)$:

x, X	0	1	2	3	4
$p(x)$	0.1	0.2	0.4	0.1	0.2
$P(X)$	0.1	0.3	0.7	0.8	1.0

A graphical summary of the CDF $P(X)$ is shown in Figure 5.10.

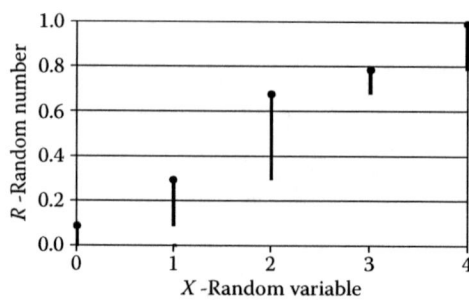

Figure 5.10   Inverse method applied to a discrete pdf.

A sample from the discrete distribution can be determined according to the following table:

If R Is in the Range	Then X Is
$0 \leq R \leq 0.1$	0
$0.1 < R \leq 0.3$	1
$0.3 < R \leq 0.7$	2
$0.7 < R \leq 0.8$	3
$0.8 < R \leq 1.0$	4

The inverse method works nicely when the inverse function $F^{-1}(R)$ is available in closed form. There are distributions, however, for which the closed form of $F^{-1}(R)$ cannot be determined. For these distributions, it is necessary to use one of the three methods described in Sections 5.3.2, 5.3.3, and 5.3.4, respectively. However, the inverse method is usually the most efficient computationally.

5.3.2 Convolution Method

In the convolution method, the random variable is expressed as the statistical sum of other easy-to-sample (independent and identically distributed) random variables. A typical example is the m-Erlang (gamma) random variable, which is the sum of m independent and identically distributed negative exponentials. Sampling from the Poisson distribution can also be used because the interevent time in a Poisson stream is the negative exponential.

Example 5.5

The m-Erlang random variable with parameter μ is the convolution of m independent and identically distributed negative exponential random variables with parameter μ. Given the negative exponential pdf,

$$f(x_i) = \mu e^{-\mu x_i}$$

and the results from Example 5.2,

$$X_i = -(1/\mu)\ln(R_i), \quad i = 1, 2, \dots, m$$

the Erlang random variable

$$Y = X_1 + X_2 + \cdots + X_m$$

is the sum of m independent and identically distributed negative exponentials with parameter μ. Then,

$$\begin{aligned} Y &= -(1/\mu)[\ln(R_1) + \ln(R_2) + \cdots + \ln(R_m)] \\ &= -(1/\mu)\ln(R_1 R_2 \ldots R_m) \end{aligned}$$

where $R_1, R_2, \ldots, R_m$ are (0, 1) random numbers. As a result, a single sample from an Erlang distribution requires m (0, 1) random numbers.

Example 5.6

Another distribution that can be sampled on the basis of exponential is the Poisson. If the Poisson has a mean λ per unit time, the time interval between occurrences of successive events becomes negative exponential with mean $1/\lambda$. Thus, a Poisson sample n is obtained by sampling successively from the exponential until the *sum* of the exponential times generated exceeds t. Then, n is the number of generated exponentials, less one.

The computations can be simplified considerably as follows: Mathematically, Poisson sampling is equivalent to determining an n that satisfies

$$t_1 + t_2 + \cdots + t_n \leq t \leq t_1 + t_2 + \cdots + t_{n+1}$$

where $t_1 = -(1/\lambda)\ln(R_1)$ is an exponential sample. Therefore,

$$-(1/\lambda)\ln(R_1 R_2 \ldots R_n) < t < -(1/\lambda)\ln(R_1 R_2 \ldots R_{n+1})$$

or

$$R_1 R_2 \ldots R_n < e^{-\lambda t} < -R_1 R_2 \ldots R_{n+1}$$

This indicates that (0, 1) random numbers are generated until their product exceeds $e^{-\lambda t}$. At that point, n equals the number of random numbers generated, less one. For example, if $\lambda =$ four events per hour and $t = 0.5$ hour, then $e^{-\lambda t} = 0.1353$. For the (0, 1) random number sequence 0.67, 0.48, 0.95, 0.61, 0.59, 0.92, notice that

$$(0.67)(0.48)(0.95)(0.61) < 0.1353 < (0.67)(0.48)(0.95)(0.61)(0.59)$$

yields the sample $n = 4$. Of course, the computational effort increases for larger values of λt.

The convolution method for the last two examples becomes quite inefficient computationally as m (or n) becomes large. The acceptance–rejection method in Section 5.3.3 is designed to alleviate this problem.

5.3.3 Acceptance–Rejection Method

In the acceptance–rejection method, the pdf $f(x)$ that we want to sample is replaced by a "proxy" pdf $h(x)$ that is easily sampled. We now show how $h(x)$ is defined such that samples from $h(x)$ can be used to represent random samples from $f(x)$.

Define the so-called *majorizing* function $g(x)$ such that

$$g(x) > f(x), \quad -\infty < x < \infty$$

and then normalize $g(x)$ to obtain the pdf $h(x)$ as

$$h(x) = \frac{g(x)}{\int_{-\infty}^{\infty} g(y)dy}, \quad -\infty < x < \infty$$

The acceptance–rejection procedure starts by obtaining a random sample Y from $h(x)$ (recall that $h(x)$ is chosen with the stipulation that it can be sampled easily). The sample Y can be determined by the inverse method. Let R be a $(0, 1)$ uniform random number. The steps of the procedure then are given as follows:

1. Obtain a sample $Y = y$ from $h(x)$.
2. Determine R as a $(0, 1)$ random number independent of Y.
3. If $R \leq f(y)/g(y)$, accept $X = y$ as the sample. Otherwise, return to step 1.

The validity of the procedure is based on a proof of the following statement:

$$P\{Y \leq x \mid Y = y \text{ is accepted}, -\infty < y < \infty\} = \int_{-\infty}^{x} f(z)dz, \quad -\infty < x < \infty$$

This statement states that samples $X = y$ and $h(x)$ that satisfy the condition $R \leq f(y)/g(y)$ are in reality samples from the original distribution $f(x)$. The efficiency of the acceptance–rejection method is enhanced as the probability of rejection in step 3 decreases. This probability depends on the choice of the majorizing function $g(x)$.

Example 5.7

Consider the gamma pdf

$$f(x) = \beta^{\alpha} x^{\alpha-1} \frac{e^{-x/\beta}}{\Gamma(\alpha)}, \quad x > 0$$

where $\alpha > 0$ is the shape parameter and $\beta > 0$ is the scale parameter. In the case where α is a positive integer, the distribution reduces to the Erlang case (Example 5.4). In particular, $\alpha = 1$ produces the negative exponential (Example 5.2).

Available acceptance–rejection methods apply to either $0 < \alpha < 1$ or $\alpha > 1$. We demonstrate the case for $\alpha > 1$ due to Cheng [3]. For $0 < \alpha < 1$, the reader may consult [1]. A summary of the method is given in [5] (pp. 255–258). For $\alpha > 1$, the "proxy" pdf $h(x)$, defined for the acceptance–rejection method, can be shown to be

$$h(x) = ab \frac{x^{a-1}}{(b + x^a)^2}, \quad x > 0$$

where $a = \sqrt{2\alpha - 1}$ and $b = \alpha^a$. The function $h(x)$ is actually associated with the majorizing function:

$$g(x) = \left[\frac{4\alpha^{\alpha} e^{-\alpha}}{a\Gamma(\alpha)} \right] h(x)$$

The inverse method can be used to obtain samples from $h(x)$ (recall that the main reason for using the acceptance–rejection method is that the original pdf cannot be sampled directly by the inverse method). For the (0, 1) random number R,

$$y = H^{-1}(R) = [bR/(1-R)]^{1/\alpha}$$

The steps of the procedure are as follows:

1. Set $i = 1$.
2. Generate $y = H^{-1}(R_i)$.
3. Generate R_{i+1}.
4. If $R_{i+1} < f(y)/g(y)$, accept $Y = y$. Otherwise, *set $i = i + 1$* and go to step 1.

The actual algorithm by Chen [3], though based on the outline above, is designed for computational efficiency by eliminating redundant computations associated with the application of the acceptance–rejection method from the algorithm.

5.3.4 Other Sampling Methods

Some distributions can be sampled by expressing the random variable as a function of other easy-to-sample random variables. The usefulness of the approach depends, of course, on the existence of such functions. Example 5.8 illustrates sampling from a normal distribution.

Example 5.8

The procedure is due to Box and Muller [2]. Let R_1 and R_2 be two uniform (0, 1) random variables and define

$$x_1 = \sqrt{-2\ln R_1}\,\cos(2\pi R_1)$$
$$x_2 = \sqrt{-2\ln R_2}\,\cos(2\pi R_2)$$

Then, it can be shown that

$$f(x_1, x_2) = \left[\frac{1}{\sqrt{2\pi}}e^{-x_1^2/2}\right]\left[\frac{1}{\sqrt{2\pi}}e^{-x_2^2/2}\right]$$

which indicates that X_1 and X_2 are independent standard normal distributions with mean 0 and standard deviation 1; that is, each is $N(0, 1)$.

To obtain a sample (actually two samples) from a normal distribution with mean μ and standard deviation σ, first generate the random numbers R_1 and R_2 from which x_1 and x_2 are computed using the above formulas. Next, obtain the associated samples Y_1 and Y_2 of $N(\mu, \sigma^2)$ as

$$Y_1 = \mu + \sigma x_1$$

$$Y_2 = \mu + \sigma x_2$$

A "last-resort" sampling method approximates the continuous pdf by a discrete function. Simply use the inverse method of Example 5.3, and sample from the approximated distribution.

The method is straightforward and was used in older versions of the GPSS simulation language. However, it is generally inconvenient and should be used only when everything else fails.

5.3.5 Generation of (0, 1) Random Numbers

Because the sampling of any random variable is based on the use of (0, 1) random values, simulation models must have a (0, 1) uniform random number generator. Also, to validate simulation computations or make comparison runs, the random number generator must be able to reproduce the same sequence of random numbers when needed. The most common method is the so-called *multiplicative congruential* formula given by

$$R_{i+1} = R_i b (mod\ m), \quad i = 0, 1, 2, \dots$$

where R_0, b, and m are constants. The result of successive applications of the formula is the sequence $R_1, R_2, R_3, \dots$. The generated sequence has the inherent property of being *cyclic* in the sense that it may contain repeatable subsequences. However, a quality generator, in addition to producing *independent uniform* random numbers, must produce very long cycles. Obviously, the cycle length must be long enough to cover any simulation run.

The quality of the generator depends on the choice of constants R_0, b, and m. However, the specific values that produce good-quality random numbers are necessarily computer dependent [4]. Most programming languages have a (pretested) procedure that produces random numbers. As a result, it is rarely necessary for a user to develop a routine.

5.4 Collecting Data in Simulation

As indicated in the opening statement of this chapter, the purpose of simulation is to collect data that describe the system's behavior. In this section, we define the types of variables that are typically used in collecting simulation data and then show how statistics are computed for these variables during the simulation. Methods for computing queue and facility statistics in the context of simulation are also presented.

5.4.1 Types of Statistical Variables

In simulation, there are two types of statistical variables:

1. Observation based
2. Time based

It is convenient to start with the definition of a time-based variable because it is the more subtle of the two. A time-based variable is by definition associated with the time. For example, a queue length is a time-based variable because a queue length must be qualified with respect to the length of time this specific length has been maintained. There is a difference between having four customers waiting in a queue for 20 minutes and the same number waiting for 30 minutes. The time-element property is therefore fundamental to the definition of time-based variables.

An observation-based variable is a variable that is *not time based*. Examples of such variables are the waiting time in a queue and the time between order placements in an inventory situation.

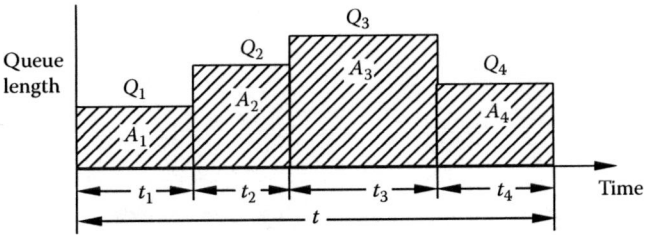

Figure 5.11 Example of a time-based variable.

The main difference between observation- and time-based variables is the way their means and variances are computed.

In the case of an observation-based variable, suppose $W_1, W_2, \ldots, W_n$ are the waiting times of customers 1 through n. Define $\overline{W}$ and σ^2 as the mean and variance of W_i, then,

$$\overline{W} = \frac{\sum_{i=1}^{n} W_i}{n} \quad \sigma_W^2 = \frac{\sum_{i=1}^{n}(W_i - \overline{W})^2}{n-1} = \frac{\sum_{i=1}^{n} W_i^2 - n\overline{W}^2}{n-1}$$

To calculate the mean and variance of a time-based variable, the variation of the variable is recorded as a function of time. Suppose that Figure 5.11 represents the change in queue length with time over a simulated period of T time units.

The figure shows that the queue length Q_i is maintained for t_i time units, $i = 1, 2, 3, 4$. Let $\overline{Q}$ and σ_Q^2 be the mean and variance of Q_i. Then

$$\overline{Q} = \frac{\sum_{i=1}^{4} Q_i t_i}{T} = \frac{A_1 + A_2 + A_3 + A_4}{T}$$

$$\sigma_Q^2 = \frac{\sum_{i=1}^{4}(Q_i - \overline{Q})^2 t_i}{T} = \left[\frac{\sum_{i=1}^{4} Q_i^2 t_i}{T}\right] - Q^{-2}$$

Notice that $Q_i t_i$ represents the area segment A_i under the curve resulting from the length Q_i for a time period t_i. This interpretation shows that, in general, the average value of a time-based variable is the total area under the curve representing the variable divided by the time base T, which in essence is the mean value of the curve.

The formulas suggest that the data for computing the means and variances can be accumulated in segments during the course of the simulation. To achieve this result, the simulator maintains two accumulators SX and SSX, for each statistical variable, to maintain the sum of the observations and the sum of the squares of the observations. These two accumulators are initialized to zero. For observation-based variables, whenever a new variable X_i is observed, the accumulators are updated using

$$SX_i = SX_{i-1} + X_i, \quad SX_0 = 0$$

$$SSX_i = SSX_{i-1} + X_i^2, \quad SSX_0 = 0$$

For time-based variables, t_i is automatically defined for the new value X_i by the lapsed time since X_{i-1}. For example, in Figure 5.11, the point in time at which Q_1 changes to Q_i automatically defines the length of time period t_i associated with Q_i. Therefore, at the ith change in value, accumulators are updated using

$$SX_i = SX_{i-1} + X_i * t_i$$

$$SX_i^2 = SX_{i-1}^2 + X_i^2 * t_i$$

At the *end* of the simulation, the accumulator values are used to compute the mean and variance. Fortunately, all simulation languages maintain the automatic accumulators SX and SSX for all the variables.

5.4.2 Histograms

The mean and variance provide information about the central tendency and spread of a statistical variable. However, additional information is available from a visual profile of the variable distribution. A convenient way to achieve this result is to plot the variable's raw data using a histogram. A histogram is constructed by dividing the entire range of the raw data into equal and nonoverlapping (disjoint) intervals. For example, in the range $[a, b]$, $[b_{i-1}, b_i)$ represents the ith interval. A counter n_i for the ith interval is initialized to zero. The interval counter is incremented by one each time a data value x satisfies the condition

$$b_{i-1} \leq x < b_i$$

After all the data have been recorded, the histogram of x is the discrete function

$$h(x) = \begin{cases} n_i, & \text{for } b_{i-1} \leq x \leq b_i, \text{ all } i \\ 0, & \text{otherwise} \end{cases}$$

The above histogram definition applies only to observation-based variables. For time-based variables, the one-unit increase in the counter n_i must be weighted by the length of time the value x is maintained.

Example 5.9

Suppose a histogram is needed for the following data in an Excel spreadsheet that represent the waiting time in a queue:

5.9	8.3	4.8	2.6	7.7	6.4	3.5
7.3	1.8	3.6	7.1	6.6	3.7	7.7
7.9	7.5	6.5	8.9	2.4	0.3	8.8
9.8	5.6	7.1	7.6	6.7	6.3	3.9
6.1	4.2	7.3	1.7	5.5	9.3	7.9

Bin	Frequency	Cumulative %	Relative f_i
1	1	2.86	0.0286
2	2	8.57	0.0571
3	2	14.29	0.0571
4	4	25.71	0.1143
5	2	31.43	0.0571
6	3	40.00	0.0857
7	6	57.14	0.1714
8	10	85.71	0.2857
9	3	94.29	0.0857
10	2	100.00	0.0571
More	0	100.00	

Figure 5.12 Histogram.

Note that the units are in minutes. The minimum and maximum values of 0.3 and 9.8, respectively, have a range of 9.5 minutes. Therefore, choose an interval size of 1. Boundaries of the successive intervals are given by [0, 1], [1, 2], [2, 3], ... , [9, 10].

Relative and cumulative frequency entries are shown in Figure 5.12.

$$f_i = n_i/n = n_i/35, \quad i = 1, 2, \ldots, 10$$

$$C_i = \sum_{j=1}^{i} f_j, \quad i = 1, 2, \ldots, 10$$

The relative frequency column is easily obtained from the cumulative frequency percentages using the equations

$$f_i = C_i = C_{i-1} \text{ and } f_1 = C_1$$

Information in Figure 5.12 provides an estimate of the waiting time pdf and CDF. An estimate of the probability that the waiting time is less than seven minutes is given as

$$P\{\text{waiting time} < 7\} \leq f_1 + f_2 + \cdots + f_7 = C_7 = 0.571$$

Linear interpolation may be used for estimating probabilities of waiting time ranges that do not coincide with interval boundaries. For example,

$$P\{2.5 < \text{waiting time} < 6\} = C_6 - (C_2 + C_3)/2 = .3998 - .11415 = .28565$$

The accuracy of the pdf and CDF estimates increases with an increase in the number of observations. Additionally, the choice of the interval width is crucial in the development of the histogram (the estimate of the pdf and CDF). If the interval is too small or too large, the histogram may not capture the information needed to reflect the shape of the original (or parent) distribution.

It must be noted that because observations are generated one at a time during the simulation run, it may not be possible to determine the range of the variable in advance of the simulation. To accommodate this occurrence, histograms are constructed with underflow and overflow cells. The underflow cell is at the lower end of the histogram, and it captures all the observations whose values are less than the lower limit of the first interval. Similarly, the overflow cell is above the highest

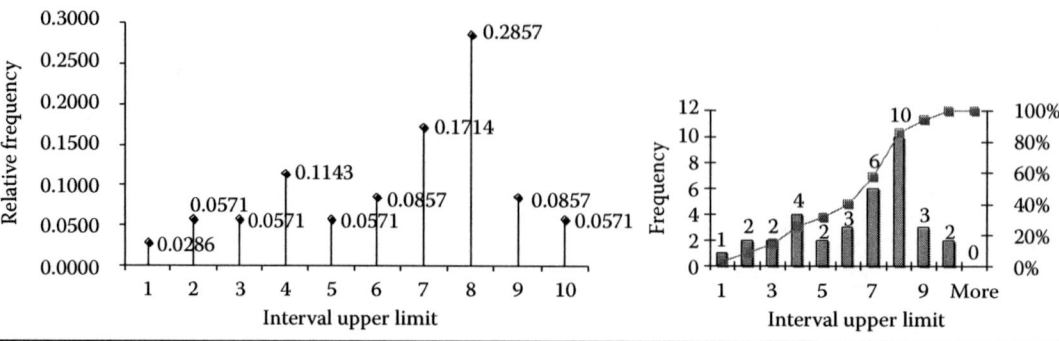

Figure 5.13 Histogram and probability function.

interval to capture all other observations (see Figure 5.13). A disproportionately large number of observations in the underflow or overflow cells indicates the need to revise the histogram limits. However, with actual values of the statistical variables, Excel's *Histogram* program can be used to tally frequencies, calculate percentages, and graph the relative frequencies and cumulative relative frequency. Examples of constructing a histogram from Excel are given in Appendix A.

The approach to developing the above histogram applies only to observation-based variables. For time-based variables, the observed frequency n must be adjusted to reflect the effect of the time element. Example 5.10 demonstrates the idea of associating time weights with frequency.

Example 5.10

Suppose the simulated queue length changes with time in the following manner:

Queue Length	Time Interval
0	$0 < t \leq 5.5$
1	$5.5 < t \leq 12$
2	$12 < t \leq 13$
1	$13 < t \leq 16$
0	$16 < t \leq 24$
2	$24 < t \leq 25$

A histogram of queue length that has not been weighted by time is shown in the following table:

Queue Length	Observed Frequency n_i	Relative Frequency f_i
0	2	0.3333
1	2	0.3333
2	2	0.3333

This is incorrect because, as in Section 8.4.1 the mean and variance of time-based variables must account for the length of time associated with each value. Instead of counting each value as a single entry, it must be weighted by the length of time the observed queue length is maintained. This definition produces the following table:

Queue Length	Weighted Frequency n_i	Relative Frequency f_i
0	13.5	0.54
1	9.5	0.38
2	2.0	0.08

These results are dramatically different from those without time weights.

5.4.3 Queue and Facility Statistics in Simulation

Using terminology from Section 5.2, the majority of simulation models include two basic nodes: (1) queues where transactions (customers) can wait if necessary and (2) facilities where customers are served. As a result, most simulation languages, particularly those based on the process-oriented approach, automatically produce performance statistics for these two nodes as standard reports. Although a facility differs from a queue in that its delay time is usually scheduled by sampling from a service time distribution (the length of stay in a queue is dependent on the state of the system), the nodes share similarities that help with interpreting performance statistics. How to compute and interpret these statistics is described in this section.

5.4.3.1 Queue Statistics

Common performance measures for queues are

■ Average queue length
■ Average waiting time in queue

These measures are determined from the variation in queue length over the duration of the simulation. A numeric example is used to demonstrate the explanation.

Example 5.11

Figure 5.14 shows a variation in queue length over a simulated period of 25 time units. Because queue length is a time-based variable, the average queue length is

$$L_q = \frac{Area}{T} = \frac{36}{20} = 1.8 \text{ transactions}$$

To obtain the average waiting time in queue, W_q, note that each unit increment in queue length in Figure 5.14 is equivalent to a transaction entering the queue. Also, each unit decrement in queue length signifies queue departure. Assuming a first-in, first-out (FIFO) queue discipline, the individual waiting times of the six transactions that enter the queue can be computed as shown in Figure 5.14; that is, W_1 is the waiting time of the first transaction and W_6 is the waiting time of the sixth transaction.

Because state changes in queue length occur in discrete units that correspond to the transaction entering or leaving the queue, observe that

$$W_1 + W_2 + \cdots + W_6 = \text{area under queue length curve}$$

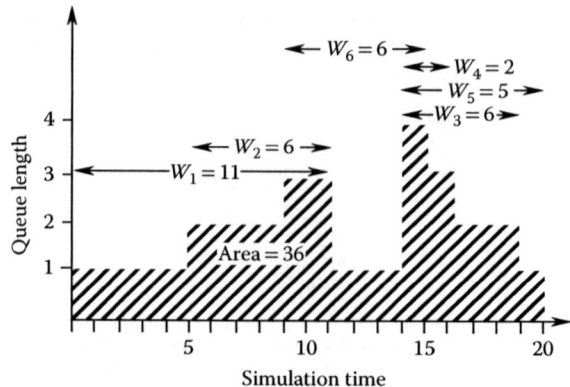

Figure 5.14   Plot of queue length versus time.

As a result, the average waiting time is

$$W_q = \frac{\text{Sum of waiting times}}{\text{Number of transactions entering queue}}$$

$$= \frac{Area}{6} = \frac{36}{6} = 6 \text{ time units}$$

The above-average waiting time calculations apply only to those transactions that *must wait* (i.e., transactions with positive waiting time). To calculate the average waiting time for all transactions, add those with *zero* wait time to the number of transactions entering the queue.

The formula to compute W_q, the average waiting time in queue, actually confirms the following well-known result from queueing theory (see [6]):

$$\overline{L}_q = \lambda \overline{W}_q$$

where λ is the arrival rate of the transactions. To verify this relationship in terms of the computations given above, observe that for n as the number of transactions entering the queue,

$$L_q = \frac{Area}{T} = \left(\frac{Area}{n} \right)\left(\frac{n}{T} \right) = \overline{W}_q \lambda$$

When n represents the number of transactions with positive wait time, λ represents the rate at which transactions enter the queue. On the contrary, when n includes transactions with zero wait time, λ represents the rate at which transactions are generated at the source.

5.4.3.2 Facility Statistics

A facility is designed with a finite number of parallel servers. The basic performance measures for a facility include

■ Average utilization
■ Average busy time per server
■ Average idle time per server

Average utilization is a time-based variable that measures the average number of busy servers. Busy time per server is measured from the instant the server becomes busy until it again becomes idle. A server's busy time may span the completion of more than one service depending on the degree of facility utilization. Idle time per server represents uninterrupted periods of idleness for the server. The calculation of these measures is demonstrated in Example 5.12.

Example 5.12

In a four-server facility, suppose that the number of busy servers changes with time as shown in Figure 5.15. Performance measures are calculated over a period of 20 time units. As in the computations of queue statistics, first compute the areas A_1 and A_2 from which

$$Average\ utilization = \frac{A_1 + A_2}{20} = 1.55\ servers$$

To compute the average busy time, notice that $A_1 + A_2$ is equal to the sum of the busy times of busy servers. Also notice that the number of times the status of servers is changed from idle to busy is five. As a result,

$$Average\ busy\ time\ per\ server = \frac{Sum\ of\ busy\ times}{Number\ of\ observations}$$

$$= \frac{A_1 + A_2}{5} = \frac{31}{5} = 6.2\ time\ units$$

Because idleness is the opposite of busy, the sum of idle times for all servers must equal area A_3 in Figure 5.15. Additionally, the number of occurrences of idleness should equal the number of times the servers become busy during a specified simulation period. As a result,

$$Average\ idle\ time\ per\ server = \frac{A_3}{5} = \frac{49}{5} = 9.8\ time\ units$$

Example 5.13

In a simulation model, facilities may be arranged in sequence as shown in Figure 5.16. In this case, if all the servers in facility 2 are busy, a transaction completing service in facility 1 will not be able to depart and is *blocked*. Under blocking conditions, a server in facility 1 is idle and unable to serve another transaction because its "server" in the facility is occupied by the blocked transaction. Facility blocking affects the server's average utilization in the sense that the average facility

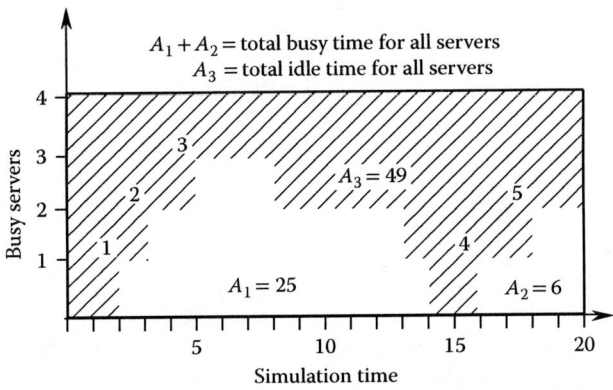

Figure 5.15   Plot of busy servers versus time.

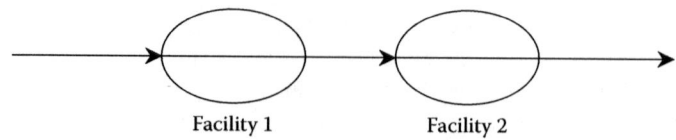

Facility 1   Facility 2

Figure 5.16 Two facilities in sequence.

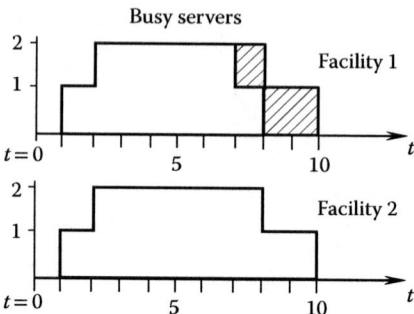

Figure 5.17 Example of gross facility utilization with blocking.

utilization also includes the average number of blocked servers. As a result, it is more appropriate to use the term *average gross utilization* to reflect the effect of blocking. To illustrate the effect of blocking, suppose that there are two servers in each of facility 1 and facility 2. In facility 1, server 1 starts at $t = 1$ and ends at $t = 8$, whereas server 2 starts service at $t = 4$ and terminates at $t = 7$. Server 1 in facility 2 starts service at $t = 0$ and terminates it at $t = 10$, whereas server 2 commences service at $t = 2$ and completes it at $t = 8$. Figure 5.17 illustrates changes in the number of busy servers for the two facilities.

Notice in Figure 5.17 that although server 2 in facility 1 finished at $t = 7$, it is blocked until $t = 8$, when server 2 in facility 2 completes the service. Similarly, server 1 in facility 1 finishes at $t = 8$ but is blocked until $t = 10$, when server 1 of facility 2 becomes idle. At $t = 10$, the following statistics for facility 1 are

$$Average\ blockage = \frac{1+2}{10} = .3\ \text{servers}$$

$$Average\ gross\ utilization = \frac{9+6}{10} = 1.5\ \text{servers}$$

$$Average\ net\ utilization = 1.5 - 0.3 = 1.2\ \text{servers}$$

The sum of blockage times in facility 1 is equal to the shaded area in Figure 5.17. Because blockage occurs twice,

$$Average\ blockage\ time = \frac{1+2}{2} = 1.5\ \text{time units}$$

It may be tempting to think that the average busy time (gross) is equal to the average service time plus the average blockage time. In facility 1 above, the average gross busy time is $15/2 = 7.5$, the average service time is $(7 + 5)/2 = 6.0$, and the average blockage time is 1.5. This observation is true only when every transaction that completes service in facility 1 experiences a blockage delay before entering facility 2. In general, the number of blockages is less than the number of service completions. In such a case, the sum of the average blockage time and average service time should not be equal to the average gross busy time (per server).

5.5 Summary

Presented in this chapter is the framework for using events as the basis for the development of the three known types of simulation: event scheduling, activity scanning, and process simulation. Also presented are techniques for sampling from pdfs within the context of simulation. Time-based and observation-based variables are defined, and methods for collecting their statistics (including histograms) are provided.

Finally, the chapter closes with the presentation of methods for computing the basic statistics of queues, facilities, and resources. The details presented in this chapter are the basis for the design environment for event-driven simulation (DEEDS) environment.

Problems

5.1. Identify the discrete events needed to simulate the following situation: Two types of jobs arrive from two different sources. Both types of jobs are processed on a single machine with priority given to jobs arriving from the first source.

5.2. Jobs arrive at a constant rate to a carousel conveyor system. Three service stations are spaced equally around the carousel conveyor. If the server is idle when a job passes in front of its station, the job is removed from the conveyor for processing. Otherwise, the job continues to rotate about the carrousel until a server becomes available. A job is removed from the system after processing has been completed. Define the discrete events needed to simulate the situation.

5.3. Cars arrive at a two-lane drive-in bank for service. Each lane can house a maximum of four cars. If the two lanes are full, arriving cars must seek service elsewhere. If at any time the number of cars in a lane is at least two cars longer than the other, the last car in the longer lane will jockey to the last position in the shorter lane. The bank operates the drive-in facility from 8:00 a.m. to 3:00 p.m. each workday. Identify the discrete events that describe this situation.

5.4. Carry out the simulation of Example 5.1 given that the interarrival time is six minutes. The service time at the Facility is either four or eight minutes assigned on a rotational basis to arriving customers. Conduct the simulation for 10 arrivals, then compute the following statistics:
   a. Average Facility utilization
   b. Average Queue length
   c. Average Facility idle time
   d. Average waiting time in Queue

5.5. Consider the following sequence of (0, 1) random numbers:
   0.93, 0.23, 0.05, 0.37, 0.24, 0.66, 0.35, 0.15, 0.94,
   0.90, 0.61, 0.08, 0.14, 0.69, 0.79, 0.18, 0.44, 0.32
   Suppose that customers arrive at a facility in a Poisson stream with mean 10 per hour. Determine the arrival time of the first five customers.

5.6. Repeat Problem 5.5 assuming that only every other arriving customer is admitted into the facility.

5.7. Demand for a certain item per day can be described by the following discrete pdf:

Demand	0	1	2
Probability	0.1	0.6	0.3

Generate the demand for the first three days using the sequence of (0, 1) random numbers in Problem 5.5.

5.8. Customers arrive at a service installation according to a Poisson distribution with mean 10 per hour. The installation has two servers. Experience shows that 60% of the arriving customers prefer the first server. By using the (0, 1) random sequence given in Problem 5.5, determine the arrival times of the first three customers at each server.

5.9. In Problem 5.6, the interarrival time at the facility is gamma distributed with parameters $\alpha = 2$ and $\mu = 10$. Apply the acceptance–rejection method to the problem and compare the results of the two procedures.

5.10. Consider the normal distribution with mean 10 and standard deviation 2. Obtain a sample from the distribution by each of the following methods, using the (0, 1) random numbers given in Problem 5.5:
   a. Discrete approximation
   b. Box–Muller procedure
   c. Central limit theorem

5.11. Classify the following variables as either observation based or time based:
   a. Time to failure of an electronic component
   b. Inventory level of an item
   c. Order quantity of an inventory item
   d. Number of defective items in an inspected lot
   e. Time needed to grade examination papers
   f. Number of cars in the parking lot of a car rental agency

5.12. The utilization of a service facility with three parallel servers can be summarized graphically as shown in Figure 5.18. Compute the following:
   a. The average utilization
   b. The percentage of idleness
   c. The average idle time
   d. The average busy time

5.13. Two facilities operate in series. The first facility has two servers, the second has only one. The service time in facility 1 is 6 minutes per customer. In facility 2 the service time is 10 minutes. Currently the two facilities are fully occupied. The two servers in facility 1 started on their jobs at $t = 5$ and $t = 7$, respectively. The second facility started at $t = 8$. Transactions arrive at facility 1 every 8 minutes. Compute the following at $t = 50$:
   a. Average gross utilization of each facility
   b. Average blockage of facility 1
   c. Net utilization of facility 1

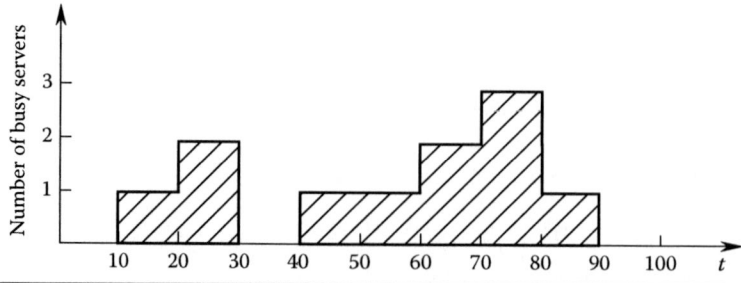

Figure 5.18   Number of busy servers versus time.

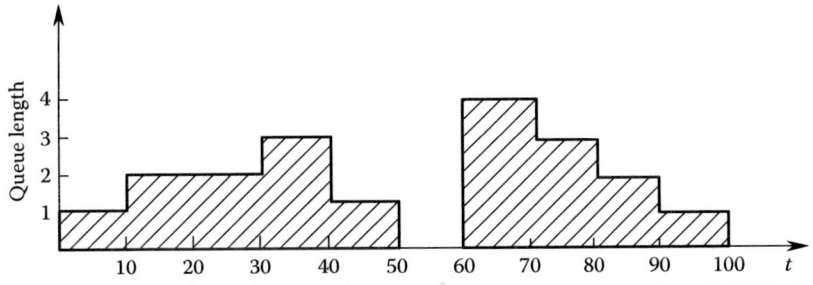

Figure 5.19 Queue length versus time.

5.14. Figure 5.19 summarizes the changes in queue length with time. Compute the following:
 a. Average queue length
 b. Average waiting time in queue for those who must wait

References

1. Ahrens, J. and V. Dieter, Computer methods for sampling from the exponential and normal distributions, *Communications of the Association for Computing Machinery*, 1972, 15, 873–882.
2. Box, G. E. P. and M. F. Muller, Note on the generation of random normal deviates, *Annals of Mathematical Statistics*, 1958, 29, 610–611.
3. Cheng, R. C., The generation of gamma variables with nonintegral shape parameters, *Applied Statistics*, 1977, 26(1), 71–75.
4. Fishman, G. S., *Principles of Discrete Event Simulation*. New York: Wiley, 1979.
5. Law Averill, M., *Simulation Modeling and Analysis*, 4th ed. New York: McGraw-Hill Book Company, 2007.
6. Stidham, S., A last word on $L = \lambda W$, *Operations Research*, 1974, 22, 417–421.

Chapter 6

Gathering Statistical Observations in Simulation

6.1 Introduction

In this chapter, peculiarities that distinguish a simulation experiment and a traditional laboratory experiment are presented. These peculiarities require special procedures for gathering experimental observations. This chapter also stresses the difficulties encountered while developing these procedures and recommends approaches to overcome them.

6.2 Peculiarities of the Simulation Experiment

Simulation experiments exhibit certain peculiarities that must be taken into account before applying statistical inference methods. Specifically, implementation of statistical methods is based on the following three requirements:

1. Observations are independent.
2. Observations are sampled from identical distributions.
3. Observations are drawn from a normal population.

In a sense, the output of the simulation experiment may not satisfy any of these requirements. However, certain restrictions can be imposed on gathering the observations to ensure that these statistical assumptions are not grossly violated. This section explains how simulation output violates these three requirements. Methods for overcoming these violations are then presented in Section 6.3.

6.2.1 Issue of Independence

The output data in a simulation experiment are represented by the time series $X_1, X_2,..., X_n$ that corresponds to n successive points in time. For example, X_i may represent the ith recorded queue length or the waiting time in queue. By the nature of the time series data, the value of X_{i+1} may be influenced by the value of its immediate predecessor X_i. In this case the time series data are *autocorrelated*. In other words, the data points $X_1, X_2,\ldots, X_n$ are not independent.

To study the effect of autocorrelation, suppose that μ and σ^2 are the mean and variance of the population from which the sample $X_1, X_2,\ldots, X_n$ is drawn. The sample mean and variance are given by

$$\bar{X} = \sum_{i=1}^{n} X_i / n \qquad S^2 = \sum_{j=1}^{n} \left(X_j - \bar{X}\right)^2 / (n-1)$$

The sample mean $\bar{X}$ is an unbiased estimator of the true mean μ; that is, $E\{X\} = \mu$ even when data are correlated. However, because of the presence of correlation, S^2 is a biased estimator of the population variance, in the sense that it underestimates (overestimates) σ^2 if the sample data $X_1, X_2,\ldots, X_n$ are positively (negatively) correlated. Because simulation data usually are positively correlated, underestimation of variance could lead to underestimation of the confidence interval for the population mean, which in turn could yield the wrong conclusion about the accuracy of $\bar{X}$ as an estimator of μ.

6.2.2 Issue of Stationarity (Transient and Steady-State Conditions)

The peculiarities of the simulation experiment are further aggravated by the possible absence of stationarity, particularly during the initial stages of the simulation. Nonstationarity generally implies that $\bar{X}$ and S^2 are not constant (stationary) throughout the simulation. In addition, because $X_1, X_2,\ldots, X_n$ are autocorrelated, the covariances are time dependent. As a result, estimates based on the "early" output (i.e., the first few observations) of the simulation may be heavily biased and therefore unreliable.

Nonstationarity means that the data $X_1, X_2,\ldots, X_n$ are not from identical distributions. It is usually the result of starting the simulation experiment with conditions that are not representative of those that prevail after an extended operation of the system. In a more familiar terminology, the simulation system is said to be in a *transient state* when its output is time dependent or nonstationary. When transient conditions subside (i.e., the output becomes stationary), the system operates in *steady-state* conditions.

To illustrate the serious nature of stationarity effects on the accuracy of simulation output, consider the example of a single-server queueing model with Poisson input and output. The interarrival and service times are exponential with means 1 and 0.9 minutes, respectively. Simulation is used to estimate W_s, the total time a transaction spends in the system (waiting time + service time). The accuracy of W_s from the simulation model is compared with the theoretical value from queueing theory.

The simulation model is executed for three independent runs with each utilizing a different sequence of random numbers. The run length in all three cases is 5200 simulated minutes. Figure 6.1 summarizes the variation of cumulative average $\overline{W}_{s(t)}$ for three different runs as a function of the simulation time. Cumulative averages, $\overline{W}_{s(t)}$, are defined as

$$\overline{W}_{s(t)} = \sum_{i=1}^{n} W_i / n_i$$

where W_i is the waiting time for customer i and n the number of customers completing service during the simulation interval $(0, t)$.

Although the three runs differ only in the streams of random numbers, the values of $\overline{W}_{s(t)}$ vary dramatically during the early stages of the simulation (approximately $0 < t < 2200$). Moreover, during the same interval, all runs yield values $\overline{W}_{s(t)}$ that are significantly different from the theoretical values of W_s (obtained from queueing theory). This initial interval is considered to be part of the transient (warm-up period) of the simulation.

To stress the impact of the transient conditions, a plot of the (cumulative) standard deviation associated with the points in Figure 6.1 is shown in Figure 6.2 for the same three runs. The figure shows that the standard deviation also exhibits similar dramatic variations during the transient period of the simulation.

From Figures 6.1 and 6.2, it can be concluded that the steady state has not been reached in any of the three runs during the entire 5200 minutes of the simulation experiment. In general,

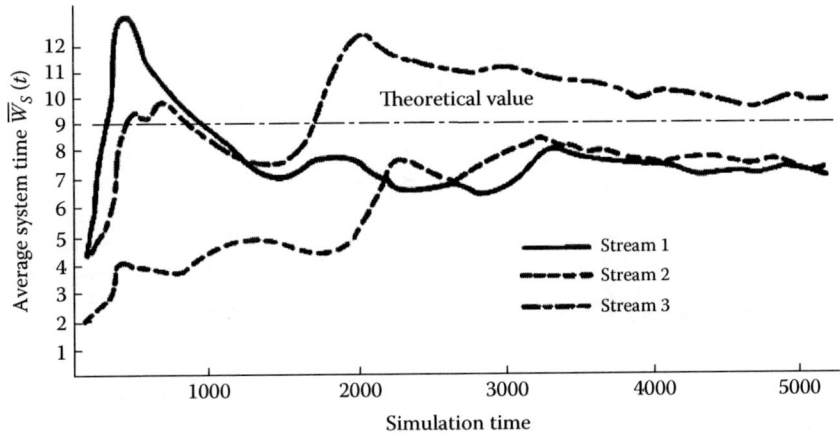

Figure 6.1 Plot of $\overline{W}_{s(t)}$ versus simulation time.

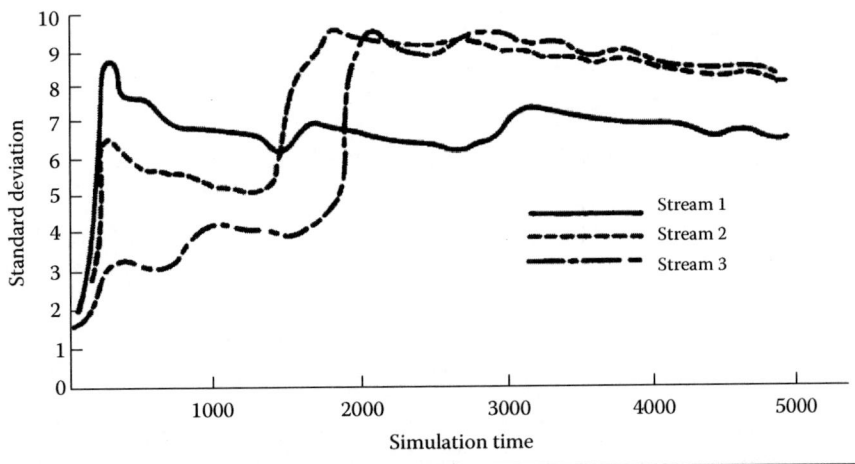

Figure 6.2 Plot of standard deviation versus simulation time.

the steady state is achieved when, regardless of the random number sequence, the distribution mean and variance of a performance measure becomes stationary. Notice that steady state does not mean that variations in the system around the true mean W_s are eliminated. Rather, all that steady state tells us is that the distribution of a performance measure has become time independent.

6.2.3 Issue of Normality

The remaining issue is the requirement that all sample data must be drawn from normal populations. Although, generally, simulation data do not satisfy normality, a relaxation of this requirement can be achieved with a form of the well-known central limit theorem. In other words, correlated time series data can be modified to (reasonably) satisfy the normality condition. This issue is discussed further in Section 6.3.

6.3 Accounting for the Peculiarities of the Simulation Experiment

This section demonstrates how simulation data can be managed to account for the peculiarities of the simulation experiment. In particular, the effects of lack of independence, absence of normality, and lack of stationarity are considered.

6.3.1 Normality and Independence

As indicated in Section 6.2, although $\overline{X}$ is an unbiased estimate of μ, correlation of the time series data may yield a biased estimate of the variance, S^2. In addition, the individual data points are not usually drawn from a normal population. As a result, the raw data points $X_1, X_2,..., X_n$ should not be used as sample observations in a statistical test. To alleviate this problem, a single simulation observation is defined in the following manner, based on whether the measurement variable is observation based or time based (see Section 5.4.1). Let X_j be the jth value of the observation-based variable and suppose that $X(t)$ is the value of the time-based variable at time t. Suppose further that $X(t)$ is observed for the time interval $(0, T)$ and that n values of X_j have been collected. Then, for statistical analysis in simulation, a single observation is defined as

$$Y = \begin{cases} \sum_{j=1}^{n} \dfrac{X_j}{n}, & X_j \text{ is observation based} \\ \displaystyle\int_0^T \dfrac{X(t)\mathrm{d}t}{T}, & X(t) \text{ is time based} \end{cases}$$

The definitions above are in effect batch averages. These averages possess two important properties, provided that n (or T) is *sufficiently large*:

1. Successive observations $Y_1, Y_2,..., Y_N$ from the same run, though not completely independent, are reasonably uncorrelated.
2. The sample averages of $Y_1, Y_2,..., Y_N$ are asymptotically normal.

Keep in mind that the two results depend on having n (or T) "sufficiently" large. Although procedures exist for the determination of large n (or T), these procedures cannot be implemented easily in a simulation experiment (see [2] for details).

The previous discussion may give the impression that the suggested definition of Y is a guaranteed remedy to the problems of independence and normality. However, the definition of the simulation observation as a batch average raises a problem of its own. Batch averages as defined above are unbiased estimates so long as the batch size (n in observation-based variables and T in time-based variables) is fixed for all Y_j, $i = 1, 2,..., N$. A typical simulation experiment usually includes both observation- and time-based variables. If the batch size T is fixed to accommodate time-based variables, then batch size n (corresponding to T) of the associated observation-based variable will necessarily be a random variable. In a similar manner, T becomes a random variable if n is fixed. Therefore, either the observation- or time-based batch average is biased.

One suggestion for solving this problem is to fix the value of T and count n within T. When n reaches a prespecified fixed size, data collection for the observation-based variable is terminated. The difficulty with this approach is that there is no guarantee that every batch size T will contain (at least) n observation-based data points.

Another suggestion is to define the batch size for each variable (observation- or time-based) independently of all other variables. To implement this procedure, the time intervals defining the batch sizes for the different variables will likely overlap. Additionally, it will not be possible to fix the simulation run length in advance because the length will be determined by the largest sum of all batch sizes needed to generate N batch averages among all variables. This requirement in itself is a major disadvantage because one would not be able to decide in advance whether the simulation should be executing for one minute or one day before termination. Additionally, in some situations it may be necessary to maintain a very long run simply to generate a desired number of batches for just one variable, even though all the other variables may have reached their allotted target very early during the run.

In practice, the problem of random variability of n or T is all but ignored, because it has no satisfactory solution. As shown in Section 6.4, batch size is usually decided by fixing the time-based interval T. Then n will assume whatever value the observation-based variable accumulates during T. By selecting T sufficiently large, it is hoped that the variation in n will not be pronounced, particularly when the data are collected during the steady-state response.

6.3.2 Transient Conditions

The above definition of a simulation observation Y in Section 6.3.1, "reasonably" settles the questions of normality and independence in simulation experiments. The remaining issue is the transient system response conditions. In practice, the effect of the transient state is accounted for simply by collecting data after an initial warm-up period. The warm-up period length is of course model dependent. For example, in the single-server model presented in Section 6.2, the transient period is expected to increase as the value of the arrival rate approaches that of the service rate. Indeed, the steady state is never achieved if the two rates are equal. The length of the transient period also depends on the initial conditions for which the simulation run is started. For example, in the single-server model, experimentation has shown that the best way to reach the steady state "quickly" is to start the run with the queue empty and the server idle. This rule, however, does not produce similar favorable results for more complex models. Some modelers call for initializing complex systems with conditions that are similar to steady state. The problem here is that such conditions are usually not known *a priori*. Otherwise, why simulate?

Unfortunately, methods for detecting the end of the transient state are heuristics and as such do not yield consistent results. Besides, the practical implementation of these procedures within the context of a simulation experiment is usually quite difficult. It is not surprising that simulation languages do not offer these procedures as a routine feature. Indeed, it is commonly advised in practice to "observe" the output of the simulation at equally spaced time intervals as a means for judiciously identifying the approximate starting point of the steady state. Steady state is supposed to begin when the output ceases to exhibit excessive variations. However, this approach, not being well defined, is not likely to produce dependable results in practice.

In this section, an approach is suggested for estimating the end of the transient state. However, before doing so, the general ideas governing the development of currently available heuristics are discussed. These ideals are the basis for the recommended procedure.

A heuristic by Conway calls for collecting observations (in the form of Y_i as defined above) at the first batch that is neither a maximum nor a minimum among all future batches. Gafarian et al. [4] propose a procedure that appears to take the opposite view of Conway's heuristic. It calls for deleting the first n raw data points if X_n is neither the maximum nor the minimum of $X_1, X_2, \ldots, X_n$. Schriber [6] suggests that the transient period ends approximately at the point in time when the batch means of the K most recent Y_is fall within a preset interval of length L. Another heuristic, advanced by Fishman [2], calls for collecting observations after the raw data time series $X_1, X_2, \ldots, X_n$ has oscillated above and below the cumulative average $(X_1 + X_2 + \ldots + X_n)/n$ a specified number of times.

These heuristics are all based, more or less, on the following general strategy. The transient state ends when oscillations (variations) in batch means Y_i remain within "controllable" limits. Another way of looking at this strategy is to consider the following two measures for each simulation output variable:

1. The cumulative mean from the start of the simulation up to times $t_1, t_2, \ldots, t_N$, where $t_1, t_2, \ldots, t_N$ are equally spaced points over the simulation run whose length is $T = t_N$.
2. The cumulative standard deviations associated with these means.

These definitions are essentially the same as those developed in Section 6.2.2 for the single-server queueing model. These two quantities are plotted for points $t_1, t_2, \ldots, t_N$. A steady state has prevailed when both measures approximately have reached constant (stable) values. At that point the output of the simulation process has likely become stationary (i.e., time independent).

The proposed procedure can be implemented as follows: Perform an "exploratory" run and plot both the mean and standard deviation of the variable as a function of simulation time. With a visual examination of these plots, estimate the length of the transient period. A fresh run is made and data collection begins after the estimated transient period.

The proposed "visual" procedure, though not sophisticated, is based on sound principles that are consistent with available heuristics. It has the advantage of being easy to implement. The utility of this method is further enhanced with plots for all performance variables in a single graph. The user can determine the start of the steady state, by viewing plots of all of the statistical variables of the model. This is particularly important because, in general, variables of the same model may reach steady state at different points in time. In this case, the length of the transient period must take sufficiently long to account for the last variable to reach the steady state. A typical illustration of the use of the proposed procedure is depicted in Figure 6.3. The figure shows the changes in system time and facility utilization together with their standard deviations for a single-server queue. Steady state begins when all four quantities exhibit "reasonable" stability.

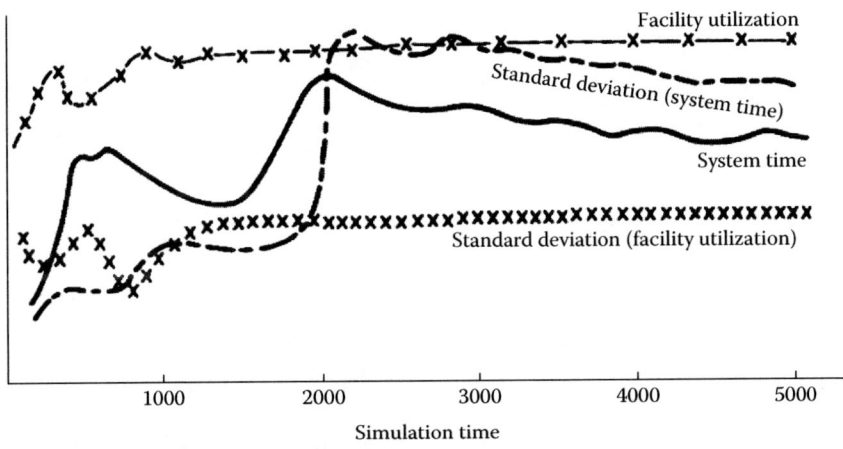

Figure 6.3   **Steady state for multivariable simulation output.**

6.4 Methods of Gathering Simulation Observations

As defined in Section 6.3, a single simulation observation is a batch average. In order to carry out desired statistical tests, it is necessary to compute N batch averages, where N is the sample size. In this section, three practical methods for defining batches are

1. The subinterval method
2. The replication method
3. The regenerative (or cycles) method

From the statistical standpoint, each method has advantages and limitations. However, the first two methods are widely used in practice.

6.4.1 Subinterval Method

In the subinterval method, only one simulation run is executed. After deleting the initial transient state (warm-up) period, the remainder of the run is divided into N equal batches, with each batch average representing a single observation. The idea of identifying batches for a hypothetical performance measure (e.g., waiting time in a queue) is illustrated in Figure 6.4.

The advantage of the subinterval method is the long simulation run that dampens the initial effect of the transient state, particularly toward the end of the simulation run. The disadvantage is that successive batches may exhibit a degree of correlation during early stages of the run. There is another disadvantage that relates to the boundary effect of the batches that will be discussed following Example 6.1.

Example 6.1

Assume an interest in estimating the mean and variance of queue length Q and waiting time W in a single-server model. Figure 6.5 shows the changes in Q over a run length of 35 time units. The first five time units represent the transient period. The remaining 30 times are divided equally among five batches, that is, sample size $N = 5$.

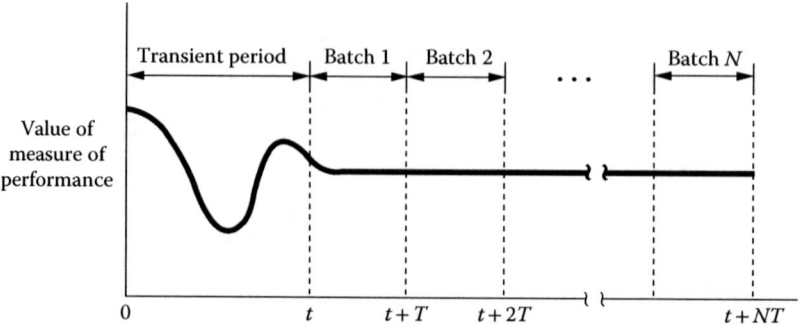

Figure 6.4   Demonstration of the subinterval method.

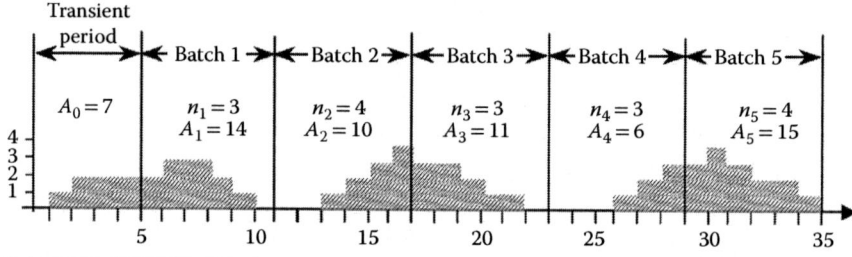

Figure 6.5   Example of the subinterval method.

If Q_{ik} and W_{ik} represent the ith data point of the kth batch, then kth batch averages are given by

$$\overline{Q}_k = A_k/T$$

and

$$\overline{W}_k = \sum_{i=1}^{n_k} W_{ik}/n_k$$

where $T (= 6)$ is the batch size for the queue length and n_k is the number of waiting customers during the interval of the kth batch. As indicated in Section 6.3.1, n_k will not be constant even though T is fixed.

Given $N (= 5)$ batches, the measures of means and variances of Q_k and W_k, $k = 1, 2,..., N$, are defined as

$$\overline{\overline{Q}} = \sum_{i=1}^{n_k} \overline{Q}_k/N \quad \text{and} \quad \overline{\overline{W}} = \sum_{i=1}^{n_k} \overline{W}_k$$

$$S_Q^2 = \sum_{k=1}^{N} \left(\overline{Q}_k - \overline{\overline{Q}}\right)^2 /(N-1)$$

and

$$S_W^2 = \sum_{k=1}^{N} \left(\overline{W}_k - \overline{\overline{W}}\right)^2 /(N-1)$$

In fact, the batch averages $\overline{Q}$ (or $\overline{W}_k$) are serially correlated. As a result, the formulas given above should usually underestimate the variances. A practical remedy, which appears to be the approach for the majority of simulation users, is to choose the batch size (T or n_k) sufficiently large to minimize the impact of autocorrelation. This approach is assumed throughout the discussion.

Table 6.1   Subinterval Data

Observation k	$\overline{W}_k$	$\overline{Q}_k$
1	4.67	2.33
2	2.50	1.67
3	3.67	1.83
4	2.00	1.00
5	3.75	2.50
Mean	3.32	1.87
Variance	1.14	0.35

These formulas are applied to the data in Figure 6.5 to obtain the results in Table 6.1. Computations in Table 6.1 indicate a possibly serious flaw in the definition of batches for the subinterval method. An examination of batches 2 and 3 shows that during batch 2, four customers joined the queue and only one departed. The remaining three customers departed during batch 3. However, computations in Table 6.1 are based on the assumption that four customers ($n_2 = 4$) departed in batch 2 and three ($n_3 = 3$) departed in batch 3.

This procedure introduces a bias because the waiting time of a customer is split into two smaller values that are now counted as two points. This type of error occurs on the boundaries of batches and will, in essence, result in an underestimation of the mean. A way to avoid this problem is to define the starting and ending batch boundary where the queue is empty. For example, in Figure 6.5, a batch can start at time 11 and end at time 26. This idea is the basis for the *regenerative method* introduced below in Section 6.4.3. However, that implementation poses serious difficulties.

Batch boundaries undoubtedly introduce a bias in the computation of batch averages. One way of partially overcoming this bias is to simply ignore all data points common to adjacent batches. This procedure, however, in addition to the disadvantage of discarding pertinent data, may actually distort the information. For example, in batches 2 and 3 of Figure 6.5, discarding common data leaves $n_2 = 1$ and $n_3 = 0$. Notice, however, that if the batch size T is large enough, then n_k of batch k will be sufficiently large that the deletion of common data is inconsequential. The only reservation in this approach is that there is no guarantee that n_k increases proportionately to the increase in T.

The procedure proposed above, under the best conditions, may be very difficult to implement. Specifically, keeping track of which data points to delete because of boundary effects is not a simple task. A straightforward solution is to have T sufficiently large and double count boundary data points (as in Example 6.1). If the number of double-counted points is small relative to n_k, then the batch boundary effect may be negligible. Obviously, there is no assurance that implementation of any of the proposed ideas will produce the desired results. However, increasing the batch size T improves the chances of achieving the desired result.

6.4.2 Replication Method

Figure 6.6 illustrates the replication method. Several runs are made, each with a different stream of random numbers. After deleting the initial warm-up period, each run represents a single batch.

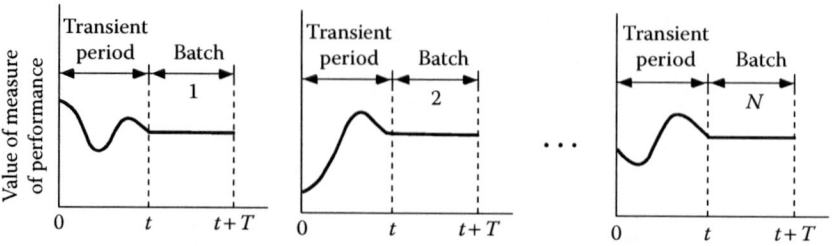

Figure 6.6 Demonstration of the replication method.

The advantage of the replication method is that it produces truly independent observations. Also, the effect of batch boundaries is not as crucial as in the case of the subinterval method. The disadvantage is that a transient warm-up period must be removed from each run. Therefore, the influence of the transient state will likely be more pronounced when compared with the subinterval method.

There are situations where there is no choice but to use the replication method for collecting the sample data. For example, a bank operation opens at 8:00 a.m. and closes at 5:00 p.m. daily. It is nonsensical to execute simulation runs beyond the 8:00 a.m. to 5:00 p.m. operating hours. This type of model is sometimes referred to as a *terminating simulation* as compared to a steady-state simulation in which the system is assumed to operate uninterruptedly for an indefinite period of time.

6.4.3 Regenerative Method

The regenerative method starts each batch with similar initial states. For example, if the queue length is the performance measure of interest, each batch may start with a zero queue length. In Figure 6.5, the three batches are defined by the intervals (0, 12), (12, 25), and (25, 35), respectively. The fact that all three batches start with zero queue length guarantees independence.

The regenerative method is designed to alleviate the problems associated with both the subinterval and the replication methods. Specifically, the method produces independent batches and eliminates the batch boundary effect. Also, by the nature of the design method, transient state has no effect on the data. However, the user will most likely prefer to bypass a warm-up period before selecting a repeatable state of the system.

The major drawback of regenerative batches is that batch sizes will usually be different. As a result, the batch average is the ratio of two random variables (compare with the discussion introducing the subinterval method). This means that batch averages are biased estimates of the population means. To overcome this difficulty, an estimator, called the *jackknife estimator* (see [5], pp. 311–312), can be used to reduce the effect of the bias. The estimator is determined using

$$Y = \begin{cases} \sum_{j=1}^{n} \dfrac{X_j}{n}, & X_j \text{ is observation based} \\ \int_{0}^{T} \dfrac{X(t)\mathrm{d}t}{T}, & X(t) \text{ is time based} \end{cases}$$

where the batch size b_k is equal to n_k when the variable is observation based and t_k when it is time based. The less biased (jackknife) estimator of the average is then given by

$$\overline{Y} = \sum_{k=1}^{N} W_k / N$$

where

$$W_k = \frac{N\overline{z}}{\overline{b}} - \frac{(N-1)(N\overline{z} - b_k Y_k)}{N\overline{b} - b_k} \quad k = 1, 2, \ldots, N$$

$$\overline{z} = \sum_{k=1}^{N} b_k Y_k / N$$

$$\overline{b} = \sum_{k=1}^{N} b_k / N$$

In this case, the variance of Y_k is given by

$$S_Y^2 = \sum_{k=1}^{N} (Y_k - \overline{Y}) / (N-1)$$

Example 6.2

In the data of Figure 6.5 (Example 6.1), regeneration points are where the queue length is zero. Therefore, the points 0, 12, and 25 represent the starting points of the batches and the simulated period has three batches. The estimate of average queue length and the average waiting time are on the basis of these three batches. Notice that b_k is given as shown in Table 6.2. Also, $\overline{z}$ is computed for both the waiting time and the queue length as

$$\overline{z} = (21 + 21 + 21)/3 = 21$$

Therefore,

$$W_k = \begin{cases} \dfrac{3*21}{3.67} - \dfrac{2(3*21 - z_k)}{(3*3.67 - b_k)} & \text{for waiting time} \\[3mm] \dfrac{3*21}{11.76} - \dfrac{2(3*21 - z_k)}{(3*11.67 - b_k)} & \text{for queue} \end{cases}$$

Table 6.2 Regenerative Method Parameters

K	$b_k = n_k$	$b_k = t_k$
1	3	12
2	4	13
3	4	10
b	3.67	11.6

Table 6.3   Regenerative Method Waiting Time and Queue Length

	Waiting Time			Queue Length		
K	$b_k Y_k$	b_k	W_k	$b_k Y_k$	b_k	W_k
1	21	3	6.67	21	12	1.74
2	21	4	5.17	21	13	1.58
3	21	4	5.15	21	10	2.038
$\bar{Y}$			5.67			1.786
$S_{\bar{Y}}^2$			0.67	6		0.232

A summary of the values of W_k along with the mean and variance is given in Table 6.3. It is interesting to compare the application of the subinterval and regeneration methods to the same data of Figure 6.5. The average waiting time from the regeneration method (= 5.67) is almost twice that for the subinterval method (= 3.32). This pronounced difference is due to the batch boundary effect inherent to the subinterval method. However, the boundary effect is reduced as we increase the batch size.

The difficulty with applying the regeneration method is in the ability to identify the regeneration points during the simulation run. It is conceivable that no points exist, or if they do, the run will not produce a reasonable sample size for proper statistical analysis.

6.5 Variance Reduction Technique

The accuracy of the mean value estimator for a variable can be improved by increasing the sample size because an increase in sample size reduces the variance for the average of the sample. However, an increase in sample size will typically increase the cost of simulation. The special characteristics of the simulation experiment can be used to effect a variance reduction in certain situations with little additional computation. This result is achieved by using the so-called *antithetic method*.

The idea of the antithetic method is to maintain two sequences of random numbers: $(R_1, R_2,...)$ and its complement $(1 - R_1, 1 - R_2,...)$. For a given performance measure, let $(x_1, x_2,..., x_n)$ and $(y_1, y_2,..., y_n)$ be time series data corresponding to the given two random number sequences. Assume that x_j is a transformation of a single random number R_j and that y_j is based on the same transformation using $1 - R_j$. In this case, x_j and y_j will be negatively correlated. Also, $E\{x_j\}$ and $E\{y_j\}$ are the same μ because they both represent the same measure of performance.

Define

$$z_j = (x_j + y_j)/2$$

Then $E\{z_j\} = \mu$. However, z_j has the superior property in that its variance is less than one-fourth the sum of the variances of x_j and y_j. To establish this result, consider

$$\text{Var }\{z\} = \tfrac{1}{4}\big(\text{Var }\{x\} + \text{Var }\{y\}\big) + \tfrac{1}{2}\text{cov }\{x, y\} \leq \tfrac{1}{4}\big(\text{Var }\{x\} + \text{Var }\{y\}\big)$$

The latter inequality follows because cov $\{x, y\} < 0$. The proposed antithetic method works well when there is a one-to-one correspondence between the output (x_j and y_j) and the random numbers (R_j and $1 - R_j$). In most simulation models, a typical performance measure (e.g., waiting time in a queue) is usually a function of more than one random number. Additionally, once the two sequences R and $1 - R$ are implemented, there is no guarantee that x_j and y_j will be generated by complementary sequences of random numbers. As such, the negative correlation between x_j and y_j may not be maintained.

As a practical matter, the implementation of the antithetic method as a variance reduction technique will be computationally complex, except for extremely simple cases such as the single-server queueing model. Some simulation languages allow the user to execute independent runs using the random sequence R and its complement $1 - R$. However, for complex models, this should not be expected to produce appreciable improvements in the results.

6.6 Summary

This chapter addressed the peculiarities of the simulation experiment and the fact that simulation output violates all the requirements of statistical theory. Redefining the simulation observation as a batch average "dampens" the adverse effects resulting from the peculiarities of the simulation experiment. The discussion shows that the choice of batch size is a crucial factor in the design of the simulation experiment and that it should be selected "sufficiently large" to produce the desired statistical properties.

Problems

6.1. Solve Example 6.1 assuming a sample size $N = 4$.
6.2. Solve Example 6.1 assuming that each batch is defined to include exactly five occurrences, that is, $n = 5$. (*Hint*: The time intervals defining the different batches will not be the same.)
6.3. Consider a single-server queueing situation. The utilization of the facility with time over an interval of 100 time units is shown in Figure 6.7. Apply the regenerative method to estimate the mean and variance of facility utilization.
6.4. Solve Problem 6.3 using the subinterval method with a sample size $N = 5$ and compare the results with those from the regenerative method.

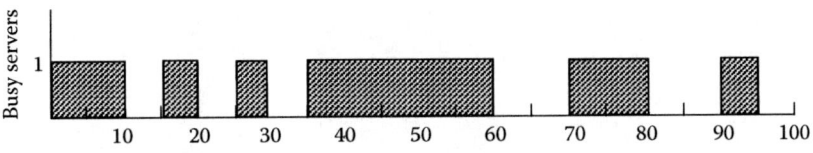

Figure 6.7   Server status versus time.

References

1. Conway, R., Some tactical problems in digital simulation, *Management Science*, 1963, 10, 47–61.
2. Fishman, G. S., Grouping observations in digital simulation, *Management Science*, 1978, 24, 510–521.
3. Fishman, G. S., *Principles of Discrete Event Simulation*. New York: Wiley, 1978.
4. Gafarian, A., C. Ancher, and T. Morisaku, Evaluation of commonly used rules for detecting steady state in computer simulation, *Naval Research Logistics Quarterly*, 1978, 25, 511–529.
5. Law Averill, M., *Simulation Modeling and Analysis*, 4th ed., New York: McGraw-Hill Book Company, 2007.
6. Schriber, T., *Simulation Using GPSS*. New York: Wiley, 1974.

EXCEL/VBA AND DESIGN ENVIRONMENT FOR DISCRETE EVENT SIMULATION (DEEDS)

II

Chapter 7

Overview of DEEDS

7.1 Introduction

Chapter 5 details a discrete simulation model, including the concept of events, time management by using the primary events list (PEL), sampling from different probability distributions, and computations of the performance measure statistics. Chapter 6 stressed the peculiarities of the simulation experiment that bias performance measures, including autocorrelated observations, non-normality of output data, and the effect of the transient state. Sampling methods were proposed to reduce adverse effects resulting from the peculiarities of a simulation experiment.

Material in Chapters 5 and 6 also describes the essential simulation program functionality needed for discrete event simulation modeling. This chapter introduces the design environment for event-driven simulation (DEEDS) environment that is based on those chapters.

7.2 Modeling Philosophy

Most discrete event systems may be viewed as a network of basis queueing systems described in Chapter 6. As such, simulation models of these systems may be represented by a network of three basic *nodes*:

1. A *source* that puts new transactions into the system
2. A *queue* where transactions wait, if necessary
3. A *facility* where transactions are serviced

Ideally, the components of a discrete event simulation model should be limited to the above three nodes. Unfortunately, only very simple systems can be modeled in this manner. In effect, the inadequacy of a "simple" three-node network in modeling general simulation systems is the lack of mechanisms for controlling the flow of transactions between nodes.

The flow of transactions must be controlled by operational restrictions of the modeled system. As a result, the simulation language must be capable of selectively directing a transaction to any other node in the network. It must also allow a transaction to alter the state of the system with changes in the data, location, or destination of other transactions in the network.

"Transaction flow" simulation languages (e.g., GPSS, SIMAN, and SLAM) overcome this difficulty by creating *special-purpose* nodes (or blocks) placed strategically to affect changes in the state of the system. These languages also complement the special-purpose nodes with the ability to incorporate a high-level programming language, such as FORTRAN, Visual Basic (VB), or Java segments, into the model to facilitate modeling complex decisions.

Special-purpose nodes in transaction flow languages enable the user to represent the system being modeled and offer the flexibility of incorporating new special nodes as needed. However, the approach has the drawback that the simulation language is more complex because the user must deal with a "large" number of special-purpose nodes (or blocks). Also, special nodes, to account for special modeling requirements, reflect the need for continuous change to the basic structure of the language. In contrast, modeling complex decisions is relatively easy in an event-scheduling simulation language. However, the obvious limitation of event scheduling is the absence of the intuitive nature of a "transaction flow" language.

The DEEDS framework limits the network nodes to (1) a source, (2) a queue, (3) a facility, and (4) a delay. The delay node may be viewed as an infinite-capacity facility that enhances the modeling capability of the language. Each node type has sufficient information to specify the conditions under which a transaction enters, resides in, and leaves the node. The simulation program is a collection of VB procedures, one for each source, facility, and delay. These procedures control the flow of transactions between nodes. These nodes and their corresponding procedures are the basis for our primary and related secondary events described in Chapter 5. In essence, DEEDS is a hybrid transaction flow/event-scheduling modeling environment because it has a "transaction flow" framework embedded in an event-scheduling environment. In DEEDS, the flow of transactions between these nodes is defined in VB procedures that identify primary events and detail their related secondary events. The collection of nodes and VB subs describes the *network* that models the system. An illustration of this concept is in Figure 7.1, where *Q1* and *Q2* represent the input queues to facility *F1*. Upon service completion, the transaction leaving *F1* must choose one of the succeeding queues, *Q3* or *Q4*, as its next destination.

The DEEDS model segment in Figure 7.1 must ensure that the procedure for facility *F1* has sufficient information so that upon service completion

1. The shorter of the two queues *Q3* and *Q4* is selected to receive the transaction.
2. If transactions are waiting in *Q1* or *Q2*, the transaction from the queue with the largest number of waiting transactions is chosen.

These requirements are accounted for in the DEEDS service completion subroutine for node *F1* in Figure 7.2. Sub *F1FacilityEvent* shows that a transaction representing the finished job is disengaged from the facility and joins *Q3* or *Q4*, depending on which queue is shorter. If *Q1* and *Q2*

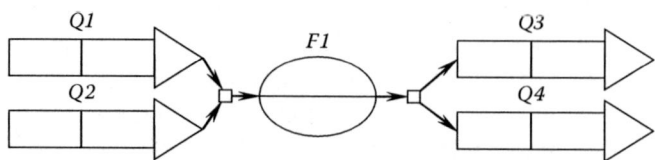

Figure 7.1   Node representation of a queueing environment.

```
Sub F1FacilityEvent(FinishedJob As Transaction)
  Dim _
    NumberInQueue as Integer, _
    NextJob As Transaction
  F1.Disengage FinishedJob
  If Q3.Length <= Q4.Length _
  Then
    Q3.Join FinishedJob
  Else
    Q4.Join FinishedJob
  End If
  NumberInQueue = _
   Worksheet.Application.Max(Q1.Length, Q2.Length)
  If NumberInQueue > 0 _
  Then
    If Q1.Length = NumberInQueue _
    Then
      F1.EngageAndService Q1Queue.Depart, 1
    Else
      F1.EngageAndService Q2Queue.Depart, 1
    End If
  End If
End Sub
```

Figure 7.2 Sample DEEDS program segment.

are empty, *F1* remains idle. If there are one or more waiting transactions in *Q1* and *Q2*, a transaction departs the longest queue and is scheduled for service completion.

The simulator has Excel spreadsheet information that indicates the number of servers and service time distribution for *F1*. Notice that the *Max* function from Excel is used to determine the comparative states of *Q1* and *Q2*.

As described in Section 7.4, transactions, queues, and facilities are instances of DEEDS-defined VB classes. *EngageAndService* and *Disengage* are VB subroutines for class *Facility*. *Join* and *Depart* are a subroutine and function, respectively, for class *Queue*. Therefore, the expression *Q1.Depart* indicates a transaction departure from *Q1*, and *F1.EngageAndService* indicates that facility *F1* is beginning service on a transaction. These details are explained in Chapters 8 and 9.

7.3 Basic Elements

As indicated in Figure 7.3, DEEDS implementation is in Excel and Visual Basic for Applications (VBA). The major advantage of this environment is the ubiquitous nature of Excel and VBA. Many of the embedded features of Excel/VBA were used to develop DEEDS, and as shown in the Figure 7.2 sample program, these features are also available for programming the model.

The simulator reads, among other things, data describing sources, queues, facilities, and delays from the spreadsheet and writes summary performance measures on sources, queues, facilities, and delays to their respective spreadsheet. There are also simulator-accessible spreadsheets that enable the modeler to specify additional model input or output.

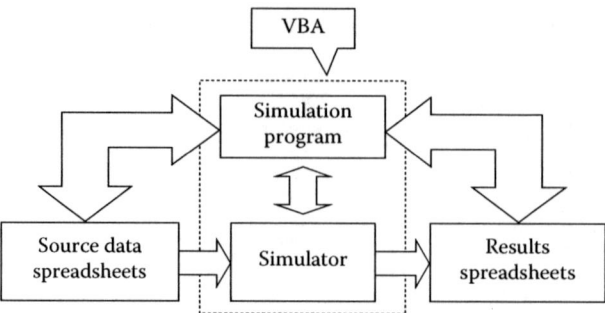

Figure 7.3   DEEDS modeling environment.

The DEEDS simulator includes the following three basic components described in Chapter 5:

1. Time and list management routines
2. Housekeeping routines
3. Model logic routines

Time and list management routines manage the PEL as well as the lists representing queues, facilities, and delays. Housekeeping routines retrieve data from the spreadsheets, initialize the simulation model, perform computations, generate random samples, and produce the final results of the simulation. The simulation program in Figure 7.3 is a collection of routines developed in VBA by the user to control model logic. Each source, facility, and delay has a primary event subroutine for its related secondary events. The VB sub in Figure 7.2 describes the processing of transactions in queues is the result of a primary event.

7.4 Basic Features

This section provides an overview of the basic features of DEEDS. Additional details of these features are presented on an "as needed" basis.

7.4.1 Network Representation

A DEEDS model is characterized by a network that consists of a combination of four nodes (source, queue, facility, and delay) connected by VB procedures. Flow between nodes is controlled by decisions based on logical algebraic expressions to assess the state of selected model components (e.g., are there transactions in the queue?). A VB procedure can perform arithmetic operations, collect statistics, and perform a complete range of node manipulations on transactions that reside anywhere in the network.

7.4.2 Time Management (Simulation Clock)

Time management routines administer the PEL to effect proper storage and retrieval of the model's primary events. It automatically advances the simulation clock from one event to the next in the proper chronological order. The PEL is completely system controlled in the sense that the user is not allowed to change the order in which events are stored or retrieved from the list. However, there are features described in Section 12.8 that enable the contents of the PEL to be examined.

7.4.3 DEEDS Class Definitions

In object-oriented design terminology, the DEEDS implementation has a total of eight classes. Four of the classes are node definitions. A transaction class is defined for entities that move between the nodes. Transactions may be anything from customers in line for tickets or waiting for the bus to microelectronic circuits being fabricated. The other three classes are for managing information used by the model. A brief description of each class is shown in Table 7.1.

In VBA, instances are specific occurrences of a class. As an example, consider a simple queuing model where transactions enter the model from a source and, as necessary, wait in a queue for service, receive service, and then exit the system. In a program for this model, there will be one instance each of source, queue, and facility. Each entity in the model is also represented by an instance of transaction. Of course, the number of instances of transactions depends on the number of active transactions in the model. VBA procedures, subs, and functions manage these class instances.

7.4.4 User's List Management

Transactions associated with source and delay nodes reside on the PEL. Each facility and delay also has an internal list of its transactions. Depending on the transaction's status in the facility, the transaction may or may not be on the PEL. Transactions in a facility are on the PEL unless a facility blocking condition exists. Each queue has an internal list of transactions that reside in its queue. Source nodes introduce transactions into the model as an arrival on the PEL. Figure 7.4 shows the relationships among the various simulator lists.

Table 7.1 DEEDS Classes

VBA Simulation Class	Represent
Source	Transactions arriving into the model
Queue	Storage area for transactions waiting for processing
Facility	Server(s) for processing transactions
Delay	Infinite capacity resource for delaying transactions
Transaction	Entities that move through the model
User-defined statistic	Observed statistical performance characteristics of a system component
User-defined probability distribution function	Empirical random variable describing a component of the system being modeled
User-defined lookup tables	Relationships between dependent and independent variables in the model

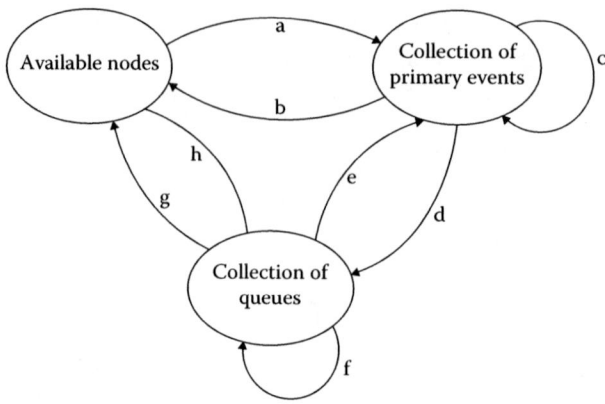

Figure 7.4 Transaction processing in DEEDS.

The primary event is determined by a transaction associated with a source, facility, or delay node. A source transaction may exit the system, return to the PEL in a facility or delay node, or become a secondary event transaction in a facility or queue. In each case, the transaction resides on that list until its movement is initiated by program logic that dictates transaction flow in the model. All transaction movement is characterized as one of eight different paths. The following are examples of a transaction traversing each path:

- An arriving transaction traverses.
- An arrival that is destroyed attempting to join a full queue is an example of traversing.
- A transaction entering the model and going directly into service or a delay is traversing.
- If that transaction had joined a queue or facility, it would have traversed.
- When a transaction is removed from a queue and placed into service it traverses.
- A transaction jockeying queues traverses.
- Combining two transactions from a queue will result in one exiting the model.
- Placing a duplicate transaction into a queue is an example of a transaction traversing.

The simulator maintains the internal lists for queues, facilities, and delays. However, unlike the PEL, the user is able to control which transactions are stored into and released from these lists. Typically, a queue discipline is specified to control the release of transactions from queues and from facilities after a service completion. However, DEEDS allows the user to override these defaults. Examples of list manipulations controlled by the user include

- Locating transactions that satisfy specific conditions
- Releasing selected transactions into the model
- Swapping transactions between two queues
- Reordering the contents of a queue according to a desired discipline
- Inserting a new transaction in a desired order in a queue
- Deleting specific transactions from a queue
- Matching or assembling transactions prior to leaving queues
- Examining the attributes of a transaction
- Modifying the attributes of a transaction

7.4.5 Generation of Random Samples

DEEDS provides for easy access to sampling from commonly used probability distributions, including uniform-continuous, uniform-discrete, exponential, normal, gamma, Weibull, beta, lognormal, triangular, and Poisson density functions. The language also provides for sampling from empirically determined discrete distributions and the approximation of continuous random variables by piecewise-linear density functions.

7.4.6 Gathering Statistical Observations

Section 6.2 discussed how simulation must deal with a peculiar statistical experiment because of (1) the existence of transient conditions, (2) the lack of independence, and (3) the lack of normality. An important feature of a simulation language is to provide means for identifying and bypassing transient conditions before collecting statistics. Additionally, the user must be able to implement the special subinterval and regenerative sampling methods (see Section 6.4) designed to alleviate the problems of non-normality and lack of independence. DEEDS provides coding facilities for identifying the start of the steady state in simulation output. It also allows the user to gather statistical observations based on either the subinterval or the replication method.

7.4.7 Interactive Debugging and Trace

Simulator tools available for testing and verifying programs include simulator messages, the trace report, and the VBA debugger. DEEDS also detects fatal logic errors and gracefully terminates the program before the user receives the dreaded "real-time execution" message. Instead, the user receives a simulator message describing the error and the time when the error was detected. Otherwise unusual circumstances are enumerated with a time stamp so that the programmer knows when the circumstances occurred. For example in Figure 7.5, these simulator messages indicate that there was no activity in the rework queue during several sampling intervals.

For a user-specified time interval, the trace report enumerates the sequence of primary events as they occur during program execution. Figure 7.6 shows a report for the time between 0 and 10. The programmer can also incorporate programmer-defined event descriptions to be included in the report.

In *interactive* mode, the VBA debugger provides instantaneous information on the execution state of the model. The user can display variable and resource statistics, file contents, and switch status. In addition to viewing statements of the input model, the debugger also allows advancing the simulation clock and changing the model parameters. The *TraceReport* is also accessible from the spreadsheet while in interactive mode.

Simulator Messages	
Warning: No entries in queue: Rework For Observation: 1	1000.00
Warning: No entries in queue: Rework For Observation: 4	4000.00
Simulation terminated successfully	4000.00

Figure 7.5   Sample DEEDS simulator messages.

Trace Report	
Time	**Event**
0.000	S1 arrival; First arrival at 0.000
0.000	S1 arrival; next arrival at 5.068
0.000	Engage F1 with 1 Server, Busy servers = 1
	SourceID = S1 SN = 2 A(1) = A(2) =
0.000	Begin Service in: F1; Completion at: 10.555
	SourceID = S1 SN = 2 A(1) = A(2) =
5.068	S1 arrival; next arrival at 7.713
5.068	Join: Q1 Length = 1
	SourceID = S1 SN = 3 A(1) = A(2) =
7.713	S1 arrival; next arrival at 33.839
7.713	Join: Q1 Length = 2
	SourceID = S1 SN = 4 A(1) = A(2) =

Figure 7.6   Sample DEEDS trace report.

7.4.8 Mathematical Expressions

Because DEEDS is implemented in the Excel–VBA environment, all the mathematical and logical VB functions are available to the user. In addition, many of the Excel functions are accessible in VBA. DEEDS also has many standard features to assist with complicated decisions.

7.4.9 Initialization Capabilities

DEEDS allows the user to initialize the contents of user-defined lists. Also, all of the VBA data types (i.e., Integer, Single, Boolean, Arrays, etc.) are available as control parameters in the model. The simulator performs its initialization procedure and then invokes a user-defined sub that performs any user-requested initializations.

7.4.10 Output Capabilities

The output produced in Figure 7.7 is in table format. However, because the output is in spreadsheets, the user has all of the graphing, charting, and analysis tools from Excel readily available to prepare professional presentations.

The standard output includes a global statistical summary with the mean of each of the model's variables. The user may also request a separate list of both observation-based and time-based variables. Additionally, a summary count of transaction movements during the simulation is helpful with interpreting and validating the model. When the subinterval method (Section 6.4.1) is used for sampling statistical variables, the global statistical summary automatically provides confidence limits at a user-chosen confidence level for all the variables of the model.

7.4.11 Model Documentation

Simulation frequently involves modeling complex logic. To facilitate the design and development of the model, the actual system must be well documented. An important part of the documentation process is for the simulation language to allow "comments" in the model that describe the program logic. More importantly, a graphical flowchart that shows interaction among the model's components and transaction flow in the model must be the basis for the language. All of these features are included in DEEDS.

Queue	Q1		
Number In	1588		
Number Out	1579		
Residing	9		
Destroyed	0		
Removed	0		
Detached	0		
Minimum	0		
Maximum	13		
	Lq		Wq
Average	1.04		16.56

Figure 7.7   Sample DEEDS queue output.

7.5 Develop and Execute a DEEDS Model

The modeled system may be represented as a network of nodes with identifier information and pseudo code to document each primary event and related secondary event. Remember that secondary events related to queued transactions are affected by the primary event of a source, facility, or a delay node because transactions in queues do not have a primary events sub. Pseudo code translates to VB instructions within subroutines for each primary event. The following briefly describes the steps to develop and execute a simulation program in DEEDS:

1. Enter the node description information (source, queue, etc.) into the spreadsheet *InitialModel*.
2. Produce *ProgramManager*-generated VBA code.

Now in the VBA environment:

1. Copy generated code to the *EventsHandler* module.
2. For each primary event sub, convert related secondary events to VB code.
3. Run the simulation program.
4. View simulation results.

7.6 Summary

In this chapter, the DEEDS modeling philosophy and an overview of its important features have been presented. DEEDS represents a "transaction flow" within an "event-scheduling environment." The network representation using four nodes closely corresponds to the transaction flow in the real system. Primary event subs administer transaction routing and allow a transaction to alter the state of the system with changes in the data, location, or destination of other transactions in the network. As demonstrated in Chapter 11, the DEEDS design philosophy facilitates the modeling of complex situations using four (self-contained) nodes: source, queue, facility, and delay.

Chapter 8

DEEDS Network Representation

8.1 Introduction

This chapter begins by identifying design environment for event-driven simulation (DEEDS) nodes and associated lists. Then, using simple examples, an introduction to network design and model development is presented. Chapters 9, 10, and 11 present development tools and advanced features of the language that facilitate model design and development.

8.2 Nodes

As indicated in Chapter 7, DEEDS utilizes four types of nodes:

1. Source
2. Queue
3. Facility
4. Delay

The first three nodes (source, queue, and facility) are familiar components of any queueing system. The delay node is introduced to enhance DEEDS's modeling flexibility. It allows a countable infinite number of transactions to reside in the node simultaneously. A facility has a maximum number of simultaneous transactions equal to the number of servers, and a transaction in a facility occupies one or more servers.

Figure 8.1 presents formal graphic representations of the four nodes. The source, facility, and delay nodes have an associated event procedure. In further discussions of these nodes, the event procedure label is omitted because event procedures are an integral part of the node. The partitions in each node are used to record node parameters. For example, as shown in Figure 8.2, a queue is uniquely defined by queue discipline and capacity.

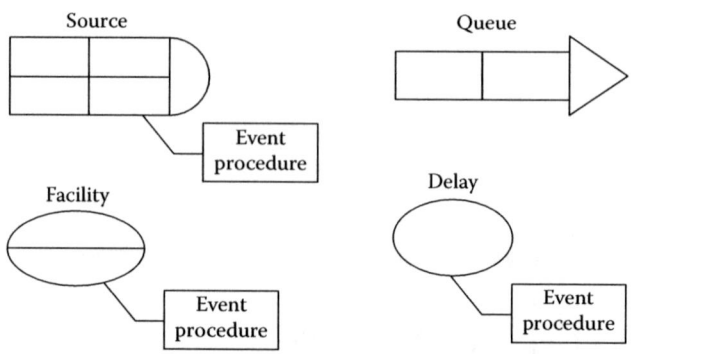

Figure 8.1   DEEDS network nodes.

Name	Priority	Limit	Offset	Distribution	Parm#1	Parm#2	Parm#3	RnID
S1	1	Infinite	0	Exponential	10	0	0	1

Name	Discipline	Capacity	Comment
Q1	LIFO	Infinite	

Name	Nservers	Distribution	Parm#1	Parm#2	Parm#3	RnID
F1	1	Exponential	7	0	0	1

Figure 8.2   DEEDS node representation of basic model.

8.3 Transactions

Transactions "flow" through the DEEDS network. The most common use of transactions is to represent customers in the system. However, transactions may also represent system resources. For example an operator moving between a workstation and a storage area is easily represented by a transaction.

In Table 8.1 each transaction has identifier information as well as a number of *attributes* that enable the user to describe characteristics of the transaction. Generally, the transactions on a list may be ordered according to an attribute value or simply based on the order of insertion into the list. Because transaction attributes are *Variant* data type, permissible attribute values, for example, are integer, Boolean, string, and double. In fact, permissible values are any Visual Basic for Applications (VBA) data type. Additional details on data types are presented in Sections 9.3 and 9.4.

8.4 Lists

DEEDS has two broad classifications of lists:

1. Lists of nodes (by type)
2. Lists of transactions

Recall from Chapter 7 that DEEDS classes include a class for each type of node (i.e., source, queue, etc.). *Transaction* is also a simulator class. VBA *Collections* are used to maintain separate lists for

Table 8.1 Transaction Attributes

Identifier	Represents
SourceID	Source node for the transaction
TransID	Transaction sequence number for the source
SerialNumber	Unique transaction identifier
Priority	Priority level (0 to 127)
CreationTime	Time the transaction entered the model
OnPEL	Status flag for transactions on the primary events list (PEL)
EventType	Identifier for the most recent event the transaction represented
TimeStamp	Time that the most recent event occurred (or will occur when transaction is on the PEL)
Attributes	Descriptive transaction information

source, queue, facility, and delay nodes. Within each queue, facility, and delay instance, there is also a *Collection* of transactions that reside in the node. The simulator maintains these transaction lists. However, as indicated in Chapter 7, changes to these lists are controlled by the simulation program.

In addition to the list of transactions in queues, facilities, and delays, the simulator administers a primary events list, or PEL, that is a time-ordered list of transactions that represent primary events in the model. As shown in Figure 5.2, the PEL is the basis for advancing the time during the simulation execution and ensuring that primary event transactions are scanned in chronological order (events that occur at the same time are processed on a first-in, first-out, or FIFO, basis).

8.5 Classes and Procedures

Subroutines and functions for the source, queue, facility, delay, and transaction classes are the basis for acquiring information about the node and specifying conditions under which a transaction enters, resides in, and leaves the node. To avoid searching the node *Collection* each time a node is referenced, a reference variable is declared and assigned to the appropriate instance in the *Collection* for each node. For consistency and simplicity, reference variable names are the node name concatenated with node type. For example, a source S1 has reference variable *S1Source*, queue Q1 has reference *Q1Queue*, and facility F1 has reference name *F1Facility*. The modeler is relieved of these details because VBA code to define and initialize nodes is automatically produced by a program managed by the user interface *ProgramManager*.

The node reference variable enables the user to apply class subs and functions to specific instances of a node. For example, the VBA statement

```
Q1Queue.JoinBy ArrivingCustomer
```

calls subroutine *JoinBy* to insert transaction *ArrivingCustomer* into instance *Q1Queue* of class *Queue* using *Q1Queue's* current queue discipline.

For those not familiar with the VBA syntax shown, subroutines may be invoked without the call statement by identifying the subroutine and a list of parameters, separated by commas. For example

```
F1Facility.EngageAndService Arrival, 1
```

is equivalent to

```
Call F1Facility.EngageAndService (Arrival, 1)
```

A third alternative that focuses on the parameter list is

```
F1Facility.EngageAndService _
          Trans :=Arrival, NumberOfServers :=1
```

A brief list, with descriptions, of class subroutines and functions is presented in Table 8.2. A complete list, with documentation of parameters and error-checking features, is in Appendix A. Details on the distinction between a Visual Basic (VB) subroutine and function are presented in Chapter 9. The primary difference is that a function has a return value and therefore can be used on the right side of an assignment statement.

Table 8.2   DEEDS Classes and Methods

Class	Type	Name	Description
Source	S	Suspend	Removes arrival from PEL
		Restart	Places arrival on PEL
Queue	F	Length	Returns the current queue length
	F	Full	Returns *True* when queue is full
	S	Join	Puts transaction into the queue based on the queue discipline
	F	Depart	Returns the transaction departing the queue
Facility	F	NumberOfBusyServers	Returns the number of busy servers in facility
	S	EngageAndService	Transaction engages facility server(s) and schedules a service completion
	S	Disengage	Transaction returns its servers and disengages the facility
Delay	S	Start	Transaction starts its delay
	S	Finish	Transaction finishes its delay
Transaction	F	Copy	Produces a duplicate transaction
	S	Assign	Assigns an attribute value
	F	A	Returns attribute value

S = sub, F = function

8.6 Simulation Program

In the DEEDS modeling environment represented in Figure 7.3, the simulator reads data describing sources, queues, facilities, and delays from the *InitialModel* spreadsheet and writes summary simulation results on sources, queues, facilities, and delays to their respective spreadsheet. The simulation program has the following components:

- Excel spreadsheets for initializing the model
- Excel spreadsheets for program output
- Program variable definitions
- Sub *Initial*
- Sub *Parser*
- Subroutines for each primary event
- Other user-defined subroutines
- Sub *ShutDown*

The simulation program is developed in VBA. In the program are variable definitions that enable communication between the simulation program, the simulator, and the debugger. Other program variables are used exclusively by the simulation program, primarily for monitoring states of the system during the simulation. There are two initialization subs in DEEDS, one in the simulator and one in the simulation program. Sub *Initial* is available to the user for initializing user-defined variables.

Sub *Parser* coordinates administration of the PEL with the simulator. Primary event subs are the heart of the simulation program. These subroutines describe related secondary events for each primary event. Other user-defined subs and functions are used to facilitate the design of primary event subs. In a manner similar to the sub *Initial*, sub *ShutDown* enables the user to present user-defined simulation results on spreadsheet *UserOutput*, before the simulator terminates the model.

The remainder of this chapter uses examples to demonstrate the fundamentals of the DEEDS programming environment.

Example 8.1

A simple single-source and single-server system model is used to present the features of the simulation environment. The interarrival rate for customers has the exponential distribution with a mean of 10 time units. Service time is also exponentially distributed with a mean of 7 time units per customer. Arriving customers wait in a queue when the server is busy. One source, one queue, and one single-server facility are used to model this system. The two primary events in Figure 8.3 are an arrival and a service completion.

In Figure 8.3 is the source's area of the *InitialModel* spreadsheet. Source description includes name, arriving transaction priority, how many arrivals will occur, when the first arrival occurs, and a description of the interarrival rate. A distribution is defined using three parameters and a random number generator. The actual number of parameters is dependent on the type of distribution. For example, the exponential has only one parameter.

Queue node parameters in the *InitialModel* spreadsheet are name, discipline, and capacity. Parameters for the facility node include name, number of servers, and service rate per server. Facility node parameters include the number of servers and probability distribution for the service time.

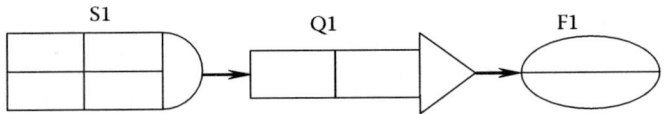

Figure 8.3 *Initial* **spreadsheet for basic model.**

Spreadsheet node descriptions are the basis for node references in the simulation program. The fact that network nodes are uniquely defined by appending the node type to the node name allows duplicate names between sources, queues, facilities, and delays. For example, the model could have *WorkStationSource*, *WorkStationQueue*, and *WorkStationFacility* in the same model.

Figure 8.4 shows subs for an arrival and service completion based on the following three assumptions:

1. When the server is busy, the arrival waits in a queue for the server.
2. Customers are served in the order of arrival.
3. Upon a service completion, the server checks to ensure that no other customers are waiting for service.

Recall that *S1Source*, *F1Facility*, and *Q1Queue* are instances of VBA classes. *EngageAndService* and *Disengage* are subroutines for the *Facility* class. In a similar manner, *Join* and *Depart* are a sub and function, respectively, for the *Queue* class. In sub *S1SourceEvent*, the primary event coincides with an arriving transaction. For *F1FacilityEvent*, the primary event represents a transaction-completing service.

Figure 8.5 shows the *Parser* subroutine that coordinates activity between the simulation program and the simulator to ensure that primary events occur in the proper sequence. The simulator

```
Sub S1SourceEvent(ByRef ArrivingCustomer As Transaction)
  F1Facility.NextFacility ArrivingCustomer, 1, Q1Queue
End Sub
Sub F1FacilityEvent(ByRef ServicedCustomer As Transaction)
  F1Facility.Disengage ServicedCustomer
  Set ServicedCustomer = Nothing
  If Q1Queue.Length > 0 _
  Then
    F1Facility.EngageAndService Q1Queue.Depart, 1
  End If
End Sub
```

Figure 8.4 Basic model—VBA program.

```
Public Sub Parser(ByRef CurrentEvent As Transaction)
  Dim EventType _
    As String
  EventType = CurrentEvent.EventType
  Time = CurrentTime
  Select Case EventType
    Case "S1Source"
      S1SourceEvent CurrentEvent
    Case "F1Facility"
      F1FacilityEvent CurrentEvent
   Case "SteadyState"

   Case "NewObservation"

   Case Else
     SimulatorMessage "Undefined Event Type:" & EventType
     End
  End Select
End Sub
```

Figure 8.5 Basic model—Parser sub.

passes to sub *Parser* the next transaction scheduled on the PEL. Subroutine *Parser* determines which subroutine has responsibility for the primary event and notifies the appropriate subroutine to route the transaction to its next destination and process-related secondary events.

Notice that the *Case* for each primary event is the same as the variable name for the node reference and the sub is the variable name concatenated with "Event." The "Else" clause is for an unrecognized primary event. In that case, a *SimulatorMessage* advises the user what happened and when it occurred. The program is then terminated. In *Parser*, an unidentified *EventType* causes the program to terminate. As presented in Chapter 9, there are straightforward VB language constructs to assess the states of the system and to model decisions.

8.7 Program Initial Conditions

Reference variables are declared and initialized at the beginning of the program so that the simulation program avoids searching a *Collection* of nodes each time a node reference is needed during execution. To initialize each reference variable, a simulator sub *Instance* searches the appropriate *Collection* and matches the variable name with its class instance, on the basis of the node name. There are comments in the simulation program that designate where user-defined variable declaration and initialization statements reside. Figure 8.6 shows reference variable declarations and respective initialization for Example 8.1.

The set of declarations below *Option Explicit* in the Figure 8.6 VBA code establishes communication links between the debugger, the simulator, and the simulation program. It is the same

```
Option Explicit
Private _
  InSources As New Collection, _
  InQueues As New Collection, _
  InFacilities As New Collection, _
  InDelays As New Collection, _
  InStatistics As New Collection, _
  InPDFs As New Collection, _
  InTables As New Collection, _
  PrimaryEventList As New Collection, _
  Time As Single, _
  S1Source As Source, _
  Q1Queue As Queue, _
  F1Facility As Facility

      ' User defined input

Public Sub Initial()
  GetResources InSources, InQueues, InFacilities, _
            InDelays, InStatistics, InPDFs, InTables
  Set PrimaryEventList = ViewPEL()
  Set S1Source = Instance("S1", InSources)
  Set Q1Queue = Instance("Q1", InQueues)
  Set F1Facility = Instance("F1", InFacilities)

      ' User initialized input
End Sub
```

Figure 8.6 Basic model—data definitions and initialization.

for every simulation program. Below those declarations are *Instance* definitions. For the simple network, *S1Source*, *F1Facility*, and *Q1Queue* are declared as instances of the *Source, Facility,* and *Queue* classes, respectively. Fortunately, all of the code described is produced by *ProgramManager*. The user simply copies it from spreadsheet *Parser* to the *EventsManager* module.

The simulator initialization activity includes creating separate *Collections* for source, queue, facility, and delay nodes (based on node information in the spreadsheet) and initializing the PEL. By calling sub *Initial*, the simulator offers the user an opportunity to perform any application-specific initializations. It is also in sub *Initial* that reference variables are initialized. Sub *Instance* searches a *Collection* and links the reference variable with its corresponding class instance in the *Collection*.

ProgramManager also produces VBA code for sub *Initial*. This code is also copied from spreadsheet *Parser* to the *EventsManager* module. In sub *Initial*, there is an area reserved for the user to initialize other application-specific variables. These features are demonstrated in problems presented in Chapter 9.

8.8 Model Development

As an overview, a DEEDS simulation model is developed in the following steps:

1. Design the network representing the system being modeled.
2. Develop node identifier information.
3. Develop pseudo code for each primary event to document related secondary events.
4. Translate the network into data and the language syntax recognizable by VB and DEEDS.

Details of these activities are presented in Chapters 9 and 10.

8.9 Summary

A complete simulation model in DEEDS has been designed and developed. In the process of developing the model, fundamental DEEDS features that have been demonstrated include how the simulator manages network nodes, how to initialize network nodes, and how to use class functions and subs to effect system changes. In the next chapter, fundamental concepts of VBA programming needed to develop a DEEDS simulation model are presented.

Chapter 9

VBA Programming

9.1 Introduction

In this chapter selected features of programming in Visual Basic for Applications (VBA) are presented. Also introduced are design environment for event-driven simulation (DEEDS) procedures for classes of network nodes and transactions as well as for statistical variables and tables. Sample code demonstrates how these features are applied to simulation modeling.

9.2 Names

The following four rules apply for names of procedures, constants, and variables:

1. The first character must be a letter.
2. The period (.), exclamation mark (!), and the characters @, &, $, and # are not allowed in a name.
3. The name cannot exceed a length of 255 characters.
4. The name should not conflict with names of Visual Basic (VB) functions, statements, and methods.

VB names are not case sensitive, but VB does preserve capitalization from the statement where the name is declared.

9.3 Data Types

The data types to define variables and arrays of variables in the simulation program are shown in Table 9.1. Each of the data types, with the exception of *Collection*, is commonly used in a DEEDS model. However, collections are ubiquitous in the DEEDS simulator. Similar to the idea of user-chains in General Purpose Simulation System (GPSS), an advanced user may define a *Collection* to manage a list of transactions or instances of other DEEDS defined classes that include *Source*, *Queue*, *Delay*, *Facility*, *Statistic*, *Table*, *PDF*, and *Transaction*.

Table 9.1   VBA Data Types

Data Type	Description
String	Anything within double quotes
Integer	−32768 to 32767
Long	Large integers
Boolean	True or false
Single	Single-precision floating point
Variant	Any of the above data types
Objects	Sheets, range, and user-defined classes
Collection	List of objects

9.4 Variable Definitions

During program execution, sub *StartSimulation* in the *Controller* module switches control between DEEDS procedures in the simulator and the simulation program. Within the simulation program, sub *Parser* determines which primary event procedure in the simulation program processes the next primary event. Within all of the simulation program modules, functions, and subs are variable definitions. Variables defined at the top of *EventsManager* are global (i.e., accessible by all procedures in the module). Other variables defined within a procedure are only accessible within that procedure. Global variables ensure that when control returns to the module, these variables have retained their values. A format to define these variables is

```
Private varname As type
```

Examples of defining a variable and an array of variables as *Private* are

```
Private _
    S1Source As Source, _
    Q1Queue As Queue, _
```

These are object variables for instances of the *Source* and *Queue* classes, respectively. Another example is

```
Private _
    ShuttingDown As Boolean, _
    PerformingMaintenance As Boolean
```

These are Boolean variables that indicate the current state of a workstation. Permissible values are *True* or *False*. The "_" is the VBA continuation character to improve readability of the program. In the next example are definitions of a String variable and two Boolean variables.

```
Private _
    TugDestination As String, _
    TugBusy As Boolean, _
```

The following examples are variable definitions for an inventory system model:

```
Private _
    InventoryOnHand As Integer, _
    InventoryOnOrder As Integer, _
    InventoryBackLog As Integer, _
    OrderAmount As Integer, _
```

Variables in each of the above examples may be viewed as global state variables accessible by all subs and functions in the *EventsManager* module. To define variables in the *EventsManager* module where retaining the value of a variable is not important when exiting a sub, the *Dim* statement format is as follows:

```
Dim varname As type
```

The following are examples of variables defined in a sub:

```
Dim _
    Index As Integer, _
    NextCustomer As Transaction, _
```

Each time the sub is entered for the above example, *NextCustomer* and *TempTrans* are transaction reference variables with value *Nothing* and *Index* has the value 0.

9.5 Constants

Constants can be used to make programs self-documenting and easy to maintain. Unlike variables, constants cannot be inadvertently changed during program execution. The syntax is

```
Private Const constname As type = expression
```

Constant names follow standard variable name conventions. Common constant types include *Byte, Boolean, Integer, Long, Single, Double, String,* and *Variant*. Expression is a literal, other constant, or any combination that includes all arithmetic or logical operators except *Is*. Constants are private by default.

If the constant type is not explicitly declared using *As type*, it has the data type most appropriate for the *expression*. Constants declared in a sub or function are local to that procedure. A constant declared outside a procedure is defined throughout the module in which it is declared. Constants may be used anywhere an expression is used. Recall from Table 8.1 that transactions have attributes that enable the user to characterize a transaction. VB constants are recommended for naming attributes. For example, an automatic guided vehicle (AGV) is represented by a transaction with the attributes shown in Figure 9.1.

```
        ' AGV attributes
     Private Const _
        Origin As Integer = 1, _
        Destination As Integer = 2, _
        SegmentIndex As Integer = 3, _
        SegmentID As Integer = 4, _
        VehicleID As Integer = 5, _
        Status As Integer = 6
```

Figure 9.1 VBA constants.

The advantage of using VB constants in this manner is that attributes have a symbolic reference that also provides program documentation. These references cannot be changed during program execution. There are several examples of attribute references in Section 9.8.

9.6 Expressions

DEEDS provides for the use of general mathematical expressions in assignment statements and logical conditions. The expression format is

$$A_1 \oplus A_2 \oplus \ldots \oplus A_m$$

where m is a finite positive integer and A_i is a variable or constant representing the ith term in the expression, $i = 1, 2, \ldots, m$. Data types allowed in an expression include *Integer, Single, Long*, and *Variant*. However, *Variant* data must be numeric.

The symbol $\oplus$ represents addition (+), subtraction (−), multiplication (*), division (/), or exponentiation (^). VB rules for evaluation of the expression are from left to right with priority given to exponentiation, followed by multiplication and division, and then addition and subtraction. Parentheses can be used to alter the left-to-right evaluation of an expression. For example, the expression A/B*C is not the same as A/(B*C).

Examples of common types of expressions are presented in the context of assignment statements in Section 9.7.

9.7 Assignment Statements

In VB there are two assignment formats. The following format is valid for all data types except *Objects* and *Collections*:

$$\text{VariableName} = \text{Expression}$$

As presented in Chapter 10, variables global to the *EventsManager* module are initialized in sub *Initial*. Examples of assignment statements that initialize global variables are

```
NumberOfBusyStations = 0
PerformingMaintenance = False

ShuttingDown = False
TugDestination = "Port"
MaximumInventoryLevel = 100
```

Examples of more complex assignment statements are

```
ServiceTime = _
    Job.A(CurrentProcessingTime) + _
        Normal(0, 0.3 * Job.A(CurrentProcessingTime))
```

Although this one may be somewhat confusing now, it will become clear after completing this chapter. The assignment format for *Objects* is

SetObjectReference = Object

For example, if *NextVehicle* is a transaction reference that will reference a transaction departing *LineQueue*, the following assignment statement will affect that result:

```
Set NextVehicle = LineQueue.Depart
```

The transaction is no longer in the queue and is now referenced by object variable *NextVehicle*. In the following statement, the transaction referenced by *CompletedJob* exits the system:

```
Set CompletedJob = Nothing
```

A typical example of an assignment in sub *Initial* is

```
Set VehicleSource = Instance("Vehicle", InSources)
```

In the above statement, the object reference is for an instance of class *Source*. The assignment statement links *VehicleSource* with an instance of *Source* named *Vehicle*. The instance resides in Collection *InSources*.

Recall that the *UserInitialed* spreadsheet offers the user the opportunity to generalize a simulation program for modeling different values of decision variables. Also, sub *ShutDown* enables the user to define output beyond the standard DEEDS output and send the results to the *UserOutput* spreadsheet.

The remainder of this section introduces the use of assignment statements to retrieve initialization data from the *UserInitialed* spreadsheet and send summary modeling results to the *UserOutput* spreadsheet. The user has the choice of absolute or relative cell reference. Relative cell references are more general in the sense that if the data's location on the spreadsheet is changed, only the reference must be changed. For absolute cell reference, each cell reference in an assignment statement must be changed.

For example, the user may prefer to have the simulation program initialize a parameter from data in the *UserInitialed* spreadsheet. *MinimumOperations* is defined as a VB variable and read from cell *B1* of the *UserInitialed* spreadsheet.

```
MinimumOperations = sheets("UserInitialed").Cells(1,2).Value
```

The above assignment is in sub *Initial*. The *Offset(0, 0)* is 0 rows and 0 columns offset from reference cell *B1*. In a similar manner, sub *ShutDown* is available for the user to send data to the *UserOutput* spreadsheet during program execution. The syntax is the opposite of initializing data from the spreadsheet. For example, with the following statement in sub *ShutDown*, *TotalNumberOfFailures* that occurred during the simulation is in cell *B10* of the *UserOutput* spreadsheet at the end of the simulation.

```
Worksheets("UserOutput").Range("B10").Offset(0, 0).Value _
                 = TotalNumberOfFailures
```

The following statements achieve the same results as above using an absolute cell reference:

```
MinOperations =
        Sheets("UserInitialed").Cells(1, 2).Value
```

```
Worksheets("UserOutput").Cells(1, 2).Value = _
                              TotalNumberOfFailures
```

In these examples, *Cells(1, 2)* refers to cell *B1*. There are many variations to these two basic formats. The most versatile of the two formats is the relative reference. Variations of the relative reference are presented in Section 9.10 in the context of reading arrays. It is particularly convenient that that spreadsheet input/output is fairly standard so the code may be copied and edited for a specific application.

9.8 Control Structures

VB control structures use expressions to evaluate complex relationships between model parameters that affect decisions on routing transactions and altering other network states. These structures determine which and when assignment statements should be executed. The assignment statements actually route transactions and change states of the model.

In Sections 9.8.1 through 9.8.4, commonly used VB control structures with examples are presented. Also introduced are class methods that affect transaction routing and changes to network nodes. Detailed descriptions of these methods are in Chapter 11.

9.8.1 If

The general format for this structure is shown in Figure 9.2. As shown, *Else Statements2* is optional.

A condition is a logical conclusion evaluated as true or false. The basis for a condition one of the comparison operators in Table 9.2.

Compound conditions have the following format:

$$\text{condition1} \oplus \text{condition2} \oplus \ldots \oplus \text{condition } m$$

```
If condition _
Then
    Statements1
[Else
    Statements2]
End If
```

Figure 9.2   VBA If structure.

Table 9.2   VBA Logical Operators

Symbol	Definition
=	Equal
<	Less than
<=	Less than or equal
>	Greater than
>=	Greater than or equal
<>	Not equal

The operator ⊕ is either *And* or *Or*.

The following are examples of transaction routing and changing states of the system. In this example, transaction *NextCustomer* departs *Q1Queue* and engages an *F1Facility* server, when *Q1Queue* is not empty.

```
If Q1Queue.Length > 0 _
Then
   Set NextCustomer = Q1Queue.Depart
   F1Facility.EngageAndService NextCustomer, 1
End If
If Q1Queue.Length > 0 _
Then
   Set NextCustomer = Q1Queue.Depart
   F1Facility.EngageAndService NextCustomer, 1
End If
```

In the following example an *Arrival* engages an idle server or joins a queue to wait for service:

```
If F1Facility.NumberOfIdleServers > 0 _
Then
   F1Facility.EngageAndService Arrival, 1
Else
   Q1Queue.Join Arrival
End If
```

Compound conditions are demonstrated in the next example. For a *RawMaterial* transaction to engage a server in *WorkStationFacility* and begin service, there must be an idle server and the workstation must not be blocked because of maintenance or waiting for maintenance. Otherwise, the *RawMaterial* must wait in the *WorkStationQueue*.

```
If WorkStationFacility.NumberOfIdleServers > 0 And _
   PerformingMaintenance = False And ShuttingDown = False _
Then
   WorkStationFacility.EngageAndService RawMaterial, 1
Else
   WorkStationQueue.Join RawMaterial
End If
```

A format for nested *If* conditions is shown in Figure 9.3. Examples of how the nested *If* is used in a simulation model are deferred until Chapter 11. There nested *Ifs* are presented in the context of class methods.

An example of a nested *If* structure is as follows:

```
If Not InventoryOrdered _
Then
  If InventoryOnHand < ReorderLevel _
  Then 'order inventory
    InventoryOnOrder = _
         MaximumInventoryLevel - InventoryOnHand
    AvgOrderSizeStatistic.Collect InventoryOnOrder
    ShippingDelay.Start deeds.NewTransaction
    InventoryOrdered = True
Else
    'inventory adequate
End If
```

The above example is a segment of code from Chapter 15 and will be explained in detail in the context of the problem. Because *InventoryOrdered* is a *Boolean* variable, the effect of the program segment is to perform certain activities when there is no inventory order outstanding. The activities are further divided by the *InventoryOnHard* status when compared to *ReorderLevel*.

```
If ConditionA _
Then
  If ConditionB _
  Then
    Statements1
  Else
    Statements2
  End If
End If
```

Figure 9.3 VBA nested If structure.

9.8.2 Case

The *Case* structure is a preferred alternative to complex *Ifs*. The format is in Figure 9.4.

Variable is any function or system parameter and Value*i* is a specific value of the variable. Value*i* formats include

```
expression,
expression TO expression and
Is comparison operator expression
```

In the following example, tanker attribute *ServiceTime* is assigned a value on the basis of the value of tanker attribute *Class*. As presented in Section 9.5, *Class* is a defined constant for referencing an attribute. On the basis of the *Case* structure, permissible values of attribute *Class* are 1, 2, and any other value.

```
Select Case Tanker.A(Class)
   Case 1
     Tanker.Assign ServiceTime, UniformContinuous(16, 20)
   Case 2
     Tanker.Assign ServiceTime, UniformContinuous(21, 27)
   Case Else
     Tanker.Assign ServiceTime, UniformContinuous(32, 40)
  End Select

Select Case Tanker.A(Class)
   Case 1
     Tanker.Assign ServiceTime, UniformContinuous(16, 20)
   Case 2
     Tanker.Assign ServiceTime, UniformContinuous(21, 27)
   Case Else
     Tanker.Assign ServiceTime, UniformContinuous(32, 40)
  End Select
```

In the following example sub *Parser* routes several service completions to the same *PrimaryEventHandler*. Of course, this version of *Parser* replaces the one produced by *ProgramManager*.

```
Select Case Variable
   Case Value1
      Statements1
   Case Value2
      Statements2
      . . .
   Case Else
      Else statements
End Select
```

Figure 9.4   VBA Case structure.

```
Select Case EventType
  Case "S1Source"
    S1SourceEvent CurrentEvent
  Case "F1Facility", "F2Facility", _
       "F3Facility", "F4Facility", _
       "F5Facility", "F6Facility"
    ServiceCompletionEvent CurrentEvent
  Case "SteadyState"
  Case "NewObservation"
  Case Else
    SimulatorMessage "Undefined Event Type: " &
EventType
    End
End Select
```

The following example uses a range of values returned by *RandomNumber(1)* to assign attribute 1 of *Customer*. Recall that because attributes are *Variant* data types, strings are permissible values.

```
Select Case Random(1)
  Case < 0.25
    Customer.Assign 1, "Red"
  Case 0.25 to 0.49
    Customer.Assign 1, "Yellow"
  Case 0.5 to 0.74
    Customer.Assign 1, "Blue"
  Case>= 0.75
    Customer.Assign 1, "Violet"
  Else
End Select
```

9.8.3 For

The *For* format is shown in Figure 9.5.

The *Index* counter indicates the maximum number of times *Statements* are executed. Without the optional *StepIncrement*, the default increment is 1. *Exit For* is optional and typically is dependent on a condition inside the *For*.

In the following *For*, a new transaction, *Arrival*, is created and engages a *SystemFacility* server for a *ServiceTime* that has an exponential distribution with mean of 10. A total of *NumComponents* transactions will be created. Of course, the number of servers in System must be greater than or equal to *NumComponents*.

```
For Index = 1 To NumComponents
  Set Arrival = NewTransaction
  SystemFacility.EngageAndService Arrival, 1, Exponential(10)
Next Index
```

```
For Index = First To Last [StepIncrement]
  Statements
  [Exit For]
Next Index
```

Figure 9.5 VBA For loop structures.

```
Do While Condition          Do
    Statements                  Statements
Loop                        Loop While Condition
```

Figure 9.6   VBA Do loop structures.

In the next example, *Q1Queue* is searched for a transaction that has attribute *Color* equal to "*Red.*" When the *For* is finished, reference variable *Original* is either *Nothing* or it references the found transaction.

```
Set Original = Nothing
For Index = 1 To Q1Queue.Length
   If Q1Queue.View(Index).A(Color) = "Red" _
   Then
      Set Original = Q1Queue.View(Index)
      Exit For   'found it
   End If
Next Index
```

9.8.4 Do

The *Do* structure has characteristics similar to the *If*. A condition must be true for assignment statements to be executed. *Do* condition(s) are tested either before or after each pass through the loop. The formats are shown in Figure 9.6. The format on the left tests condition(s) before the loop. Because there is no counter to limit the iteration count, the user must be careful to avoid infinite loops.

The following example simply empties the queue of persons waiting when the bus departs. When the queue is empty, the loop is exited. If the queue was empty when the loop is encountered, then the statement inside the *Do* is not executed.

```
Do While (BusStopQueue.Length > 0)
    Set WalkAway = BusStopQueue.Depart
Loop
```

9.9 Procedures

In general, good procedures organize a program's functional requirements into manageable units of production. Admittedly, "manageable" is a vague concept, but in general a good procedure performs related tasks and, as necessary to ensure maintainability, delegates related tasks to other procedures. Using this description, a procedure may be a called procedure, a calling procedure, or both. Procedures in VB are subs or sub functions.

From Section 8.6, the subroutines common to all simulation models are *Initial, Parser,* and *ShutDown*. Each has a specific responsibility in the model. There is also a procedure in the program for each source, facility, and delay node in the network. These procedures link the node's primary event and its related secondary events. As shown in previous examples of programs, these procedures also in effect route transactions between nodes and alter states of the system.

An experienced VB programmer will quickly conclude that the following discussion is an introduction because there are many more options than those presented. However, you will find that for most simulation models these are the only features necessary to develop *EventManager* procedures.

9.9.1 Subs

The *Sub* format is

```
[Private] Static Sub name (arglist)
  Statements
End Sub
```

Of course, statements are any combination of VB comments, data definitions, assignment statements, and control structures.

Although not an essential option, *Private* indicates that the procedure is accessible only to other procedures in the module. For a procedure to be accessible outside the module, the option is *Public*. Notice that subroutines *Parser, Initial,* and *ShutDown* are *Public* because they must be accessible in the *Controller* module.

The *Static* option preserves the procedure's locally defined variables between calls. In general, this should not cause a problem, but the trade-off is additional memory to preserve the values of these variables.

An interesting module design question is which system parameters should be accessible to all procedures in the module. A parameter is accessible if you can view and change it. Of course it is much easier to make everything global. However, when parameters are global, the possibility of inadvertent changes to a parameter increases. To reduce the possibility of unwanted changes, parameters are passed and received as an *arglist* (a list of parameters separated by commas) to and from procedures. When everything is global, *arglist* is not needed. For the called procedure that receives *arglist*, the format for each parameter is

```
[Optional] [ByVal | ByRef] varname As type
```

ByVal preserves the parameter in the calling procedure and *ByRef* allows a calling procedure parameter to be changed by the called procedure. *Optional* indicates that the parameter may be omitted in the call statement.

As an example, assume in Figure 9.7 that subroutine *TestSub* receives variables X and Y from the calling program that corresponds to parameters a and b. If X and Y are equal to 0 before *TestSub* is called, when control returns to the calling program, X is 12.5 and Y is 0 because X was passed by reference and Y was passed by value. Even though the parameter names are different, the data types must match (i.e., pairs (X, a) and (Y, b) must be the same data type).

As shown in Figure 9.8, there are three alternatives to call *TestSub*, assuming X and Y are the parameters passed to *TestSub*. The first example is the most commonly used. To distinguish

```
          Private Sub TestSub(ByRef a as single, _
                              ByVal b as integer)
              a =12.5
              b = 3
          End Sub
```

Figure 9.7   Sample VBA sub.

```
Call TestSub(X, Y)
TestSub X, Y
TestSub a:=X, b:=Y
```

Figure 9.8   Formats for VBA sub calls.

```
Public Sub Parser(ByRef CurrentEvent As Transaction)
  Dim EventType _
    As String
  EventType = CurrentEvent.EventType
  Time = CurrentTime
  Select Case EventType
    Case "S1Source"
      S1SourceEvent CurrentEvent
    Case "F1Facility"
      F1FacilityEvent CurrentEvent
   Case "SteadyState"
   Case "NewObservation"
   Case Else
     SimulatorMessage "Undefined Event Type:" & EventType
     End
  End Select
End Sub
```

Figure 9.9   Sample VBA sub calls.

types of procedures, the convention is that user-defined subs are called using the first syntax and DEEDS procedures are called using the second. Although not required, the last example is particularly useful when there are optional parameters. The format ensures that the called sub parameters correctly match the calling sub parameters. As discussed in Chapter 11, several DEEDS procedures are defined with optional parameters.

Parser and the primary event subroutines demonstrate the relationship between calling and called subs. *Parser* is a *Public* subroutine called by *StartSimulation* in the *Controller* module. The parameter passed to *Parser* by *Controller* is the transaction representing the next primary event to be processed. Using the transaction's event type, *Parser*, as shown in Figure 9.9, determines which primary event subroutine has responsibility for the transaction and then calls that subroutine with transaction as the only parameter.

As shown below, depending on event type, sub *S1SourceEvent* or *F1FacilityEvent* receives the transaction and processes the related secondary events. Notice that the variable name in the following event handlers has been redefined to better reflect the nature of the transaction.

```
Sub S1SourceEvent(ByRef ArrivingCustomer As Transaction)
  F1Facility.NextFacility ArrivingCustomer, 1, Q1Queue
End Sub

Sub F1FacilityEvent(ByRef ServicedCustomer As Transaction)
  F1Facility.Disengage ServicedCustomer
  Set ServicedCustomer = Nothing
  If Q1Queue.Length > 0 _
  Then
    F1Facility.EngageAndService Q1Queue.Depart, 1
  End If
End Sub
```

The following are a subroutine call statement and the called subroutine. There are no parameters passed and variables in the sub are all global variables. Notice that because *CheckPeeler* does not have "*Event*" appended, it is either a user-defined subroutine called by a primary event subroutine or another user-defined subroutine.

```
Call CheckPeeler
```

```
Sub CheckPeeler()
  If BeltQueue.View(1).A(EndPoint) = EndOfBelt _
  Then  'Load log into peeler
    LoadingPeeler = True
    If PeelerFacility.NumberOfIdleServers > 0 _
    Then
       LoadPeelerDelay.Start NewTransaction
    End If
  End If
End Sub
```

In the following example, the calling sub passes two parameters. All three options for calling the sub are presented. Because *SampleWoReplacement* is a user-defined sub, the first option presented is the preferred format. Notice that the called sub *SampleWoReplacement* also has defined local variables using *Dim*.

```
Call SampleWoReplacement(JobList, SampledList)
SampleWoReplacement JobList, SampledList
SampleWoReplacement OriginalValue :=JobList, _
                    SampledValue   := SampledList
```

SampleWoReplacement is a general subroutine that receives an array of values and randomly reorders the array.

```
Sub SampleWoReplacement(ByRef OriginalValue() As Variant, _
                        ByRef SampledValue() As Variant)
Dim _
   Index As Integer, _
   Jndex As Integer, _
   VectorLength As Integer
 VectorLength = UBound(OriginalValue())
 For Index = VectorLength To 1 Step -1
   Jndex = UniformDiscrete(1, Index)
   SampledValue(Index) = OriginalValue(Jndex)
   OriginalValue(Jndex) = OriginalValue(Index)
 Next Index
End Sub
```

9.9.2 Functions

As shown in the format for functions, the only difference between a *Sub* and a *function* sub is that sub is replaced by function and a value on the basis of the function's data type is returned.

```
Function WorkerOnJob(ByVal CallIndex As Integer) As Boolean
   Dim _
      Jndex As Integer
   WorkerOnJob = False
   For Jndex = 1 To JobSiteFacility.NumberOfBusyServers
      If NewCallsQueue.View(CallIndex).A(CustomerID) = _
         JobSiteFacility.View(Jndex).A(CustomerID) _
      Then 'worker already on site
         WorkerOnJob = True
         AssignedCallsQueue.Join NewCallsQueue.Detach(CallIndex)
         Exit Function
      End If
   Next Jndex
End Function
```

Figure 9.10 Sample VBA function procedure.

```
[Private | Public] Static Function name (arglist) As datatype
Statements
End Function
```

The return parameter is any VB or user-defined data type. Therefore, DEEDS node and transaction references are also valid return parameters.

In Figure 9.10 for example, the function searches for a worker on a job site and returns either *True* or *False*, depending on whether the worker was found on site. Because of the similarities between subs and functions, only one example function is presented.

9.10 Arrays

An array is a set of sequentially indexed elements with the same data type. Each array element has a unique index number. Changes to one element of an array do not affect the other elements. Arrays are also declared using *Dim, Static, Private*, or *Public* statements. The difference between arrays and other variables is that the size of the array must be specified. An array whose size is changed during program execution is a dynamic array. Arrays are extremely important in DEEDS because they provide the flexibility needed to generalize simulation models for large complex systems. The array format commonly used is

```
Dim | Static | Private | Public variablename (1 To ArraySize1, [1 To
ArraySize2]) As datatype
```

The conventional definition is that *ArraySize1* is the number of rows and *ArraySize2* is the number of columns.

For dynamic arrays the format is

```
Dim | Static | Private | Public variablename () As datatype
```

During program execution the following statement dynamically allocates the array:

```
ReDim variablename (1 To ArraySize1, [1 To ArraySize2])
```

Typical data declarations of arrays in a DEEDS model are shown in Figure 9.11. For each facility there is a queue, so *NMachines* is equal to number of instances in the *VB Collection InFacilities*.

```
Private _
  SegmentQueue(1 To 8) As Queue, _
  DropOffCenterQueue(1 To 5) As Queue, _
  PickUpCenterQueue(1 To 5) As Queue, _
  SegmentFacility(1 To 8) As Facility, _
  CenterFacility(1 To 5) As Facility, _
  JobSequence(1 To 4, 1 To 3) As Integer, _
  SegmentPath(1 To 19, 1 To 6) As Variant
```

Figure 9.11   Sample VBA array definitions.

```
NMachines = InFacilities.Count
ReDim Q(1 To NMachines)
ReDim F(1 To NMachines)
For Index = 1 To NMachines
  Set Q(Index) = Instance(CStr(Index), InQueues)
  Set F(Index) = Instance(CStr(Index), InFacilities)
Next Index
```

Figure 9.12   Sample VBA array redefinitions.

After dynamically allocating the *Q* and *F* array, the array is initialized in the *For Next* control structure. *CStr(Index)* indicates that the queue and facility names in spreadsheet *InitialModel* are integer numbers from 1 to *NMachines*.

Dynamic arrays for instances of class *Queue* and *Facility* are defined as

```
Private _
  Q() As Queue, _
  F() As Facility
```

and then as shown in Figure 9.12, initialized in sub *Initial*.

The following is an example of initializing an array of model parameters, *CutSets*, from spreadsheet *UserInitialed*. The number of cut sets (rows) and number of components (columns) are referenced by cell B4. The source data reference (i.e., row 1 and column 1) is also referenced by cell B4.

B	C	D	E	F	G	H
Cut sets	A	B	C	D	E	F
1	1	0	0	0	0	0
2	0	1	1	0	0	0
3	0	0	1	1	0	0
4	0	0	0	1	1	1

The following program segment determines the number of *NumCutSets* and *NumComponents*.

```
With Worksheets("UserInitialed").Range("B4")
  NumCutSets = _
      Range(.Offset(1, 0), .End(xlDown)).Rows.Count
  NumComponents = _
      Range(.Offset(0, 1), .End(xlToRight)).Columns.Count
End With
```

In the second segment, the outside and inside counters are *Index* and *Jndex*, respectively. The initial counter values are defaulted to 1, and the step increment for both loops is defaulted to 1. The upper limit for the outside loop is *NumCutSets*, and the inside loop will be executed *NumComponents* times. For each *Index* value, there are *NumComponents Jndex* values.

```
NumberOfNodes = Worksheets("UserInitialed").Range("B4")
With Worksheets("UserInitialed").Range("B4")
   For Index = 1 To NumCutSets
      For Jndex = 1 To NumComponents
         CutSets(Index, Jndex) = .Offset(Index, Jndex)
      Next Jndex
   Next Index
End With
```

Notice that the statement inside the loops has unique locations on the spreadsheet and in *CutSets* that are dependent on values of *Index* and *Jndex*. The *With - End* feature is an abbreviated notation to simplify references to variables and subroutines. For example, inside the *With - End*,

```
.Offset(I,J)
```

is the same as

```
Worksheets("UserInitialed").Range("B4").Offset(I,J)
```

The right-hand side of the *CutSets* assignment uses an offset from the reference cell. For example, using the spreadsheet data shown, row 1, column 1 (value = 1) has row offset of 1 and column offset of 1 from the reference cell.

9.11 Summary

This chapter discusses the essential features of VBA necessary to develop simulation models. Data types and variables are introduced and then variables are combined to form expressions and expressions are used for assignment statements and conditional expressions.

In all of the examples, visual separation is achieved by indenting lines and inserting the line VBA continuation character "_". Also, users are encouraged to decompose "excessively" complex assignment statements to several simplified statements to facilitate model development and testing. Also notice the *Option Explicit* statement in *EventsManager*. This is a VBA directive that all variables used in the program must be defined. Obviously, the user must choose this option. In a sense, the user is directing the VBA environment: "Remind me to use good programming techniques." In the next chapter, the *ProgramManager* is introduced and used to develop a complete simulation model. In addition, how to run the simulation program is demonstrated.

There were several program segments that introduced features of DEEDS. In the next chapter, the capability of DEEDS is expanded and VBA and Excel functions are introduced as additional programming tools available to the modeler. Also, VB control structures are used to compare and make inferences about system parameters. Finally, all of these features are bundled into procedures and functions that are basic building blocks for simulation models.

Chapter 10

User Interface

10.1 Introduction

ProgramManager, the user interface to design environment for event-driven simulation (DEEDS), enables users to navigate the development environment to perform data entry, run the simulation program, and view simulation output. *ProgramManager* also develops application-specific Visual Basic for Applications (VBA) code for data definitions, sub *Initial*, and sub *Parser*. Although using *ProgramManager* is optional, it has the advantage, especially for the novice user, of reducing the potential for modeling errors.

10.2 Overview of *ProgramManager*

There are a total of 11 spreadsheets in a DEEDS application. In simple models many of the sheets are not used. As shown in Chapter 14, for real applications it is possible that all 11 pages are used. Data specifications for network nodes and statistical variables are on the *InitialModel* spreadsheet. User-defined probability distributions and tables have separate spreadsheets.

The simulator reads data describing sources, queues, facilities, delays, statistical variables, probability density functions (pdfs), and tables from input spreadsheets. Depending on the model, there may also be user-defined initialization data in the spreadsheet. After performing the simulation, the simulator writes summary information on all network nodes and statistical variables, as well as any additional user-defined output.

Moving in and out of *ProgramManager* is a simple matter. For example, suppose a user is reviewing summary statistics for facilities and wants to examine a queue or user-defined statistic associated with that facility. A local approach would be to scroll through spreadsheets searching for the queues or user-defined statistics output spreadsheet. An alternative would be to reenter *ProgramManager* and then exit to the desired spreadsheet from the output spreadsheet options on the *Initial Model* page.

Tabs for six of the nine pages in *ProgramManager* are visible in Figure 10.1. The *Information* page summarizes features of *ProgramManager*. Beginning with the *Source Nodes* page, there are seven pages for data entry, one for each type of network node and one each for user-defined

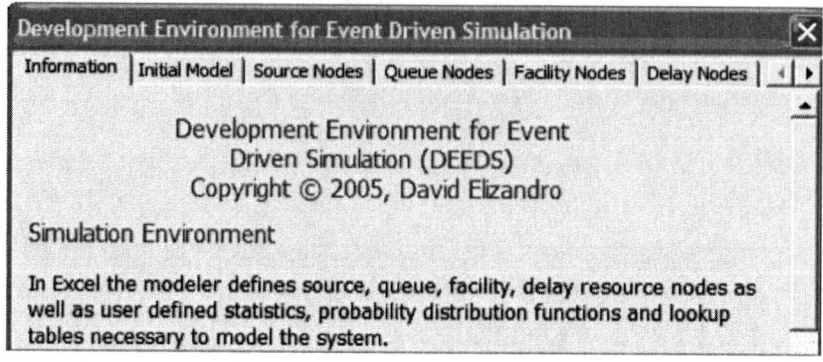

Figure 10.1 DEEDS *ProgramManager*.

probability functions, lookup tables, and statistical variables. Each of the data entry pages is presented in this chapter.

When using *ProgramManager* for data entry, all network nodes, probability functions, and tables are saved to the spreadsheet in alphabetical order, on the basis of resource name. Summary node statistics are presented in the same order that nodes are listed on the spreadsheet. Notice that there is an optional description for each data entry page in *ProgramManager*. However, for large applications, it is more than just a good idea to use this documentation tool.

With experience, the user may find it easier to bypass *ProgramManager*. However, the *Initial Model* page has several important features that actually become more valuable for complex models. *TraceReport*, VBA code generator for data definition as well as for subs *Parser* and *Initial*, and replication of experiments options are on the *Initial Model* page of *ProgramManager*.

Using the basic queueing model the following sections describe data entry features for network nodes. Recall that the interarrival rate for customers has the exponential distribution with mean of 10 time units. Service time is also exponentially distributed with mean of 7 time units per customer. Arriving customers wait in a queue when the server is busy. One source, one queue, and one single-server facility are used to model this system.

10.3 Source Nodes

As shown in Figure 10.2, source parameters are arriving transaction priority, when the first arrival occurs, how many arrivals will occur, and distribution of the arrival rate. Also shown in Figure 10.2 are source node values. These happen to be the initial values when adding a source node to the system. After all fields have been entered, the user chooses the *Add a Source* option. An editor verifies that the source description satisfies minimum conditions (i.e., no missing or invalid parameters) before adding the node to the spreadsheet. For example, *ProgramManager* may return the *Source Nodes* page with current information and prompt the user for an omitted arrival distribution.

The *Source* field is a drop-down menu with a list of source nodes currently on the spreadsheet. *ProgramManager* presents current node information for the node selected. The user may choose to edit or delete the node. The *Delete a Source* option is straightforward. The node in the *Source* field is deleted.

Source node information for an existing node may be edited. However, node names cannot be changed. To achieve this type of change, the user must do a node deletion and then a node

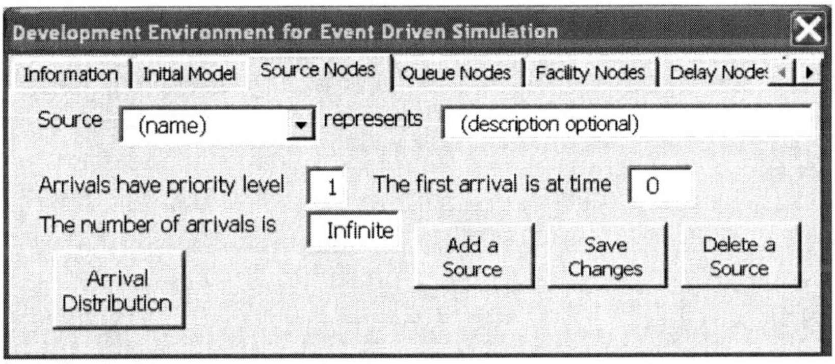

Figure 10.2 *ProgramManager Source Nodes* **page.**

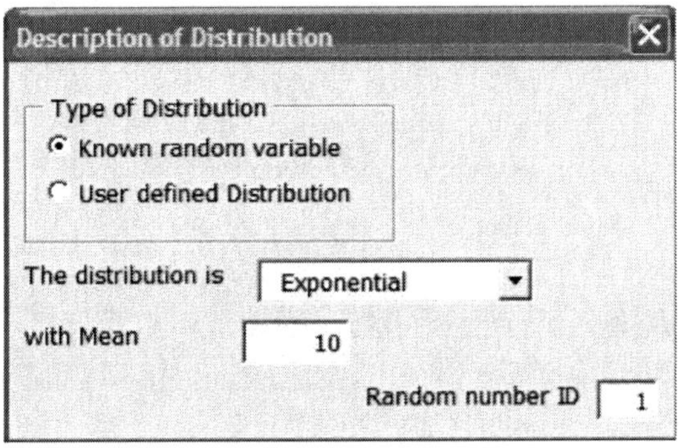

Figure 10.3 *ProgramManager Description of Distribution* **menu.**

addition. Other changes are subjected to the same edits as those for a node addition. It is important that a source description is saved to the spreadsheet before changing pages in *ProgramManager*. Otherwise, all of the information is lost.

The *Arrival Distribution* option presents the *Description of Distribution* menu (Figure 10.3) that also has a drop-down menu for selecting from well-known distributions presented in Chapter 4. Depending on the distribution, the user is prompted for distribution parameters and a random number generator. The importance of the random number sequence in experimental design is discussed in later chapters. For our source, *S1*, the interarrival rate is 10 time units from an exponential distribution. Notice that the user may indicate the preference for a user-defined distribution. In that case a drop-down menu of user-defined distributions is presented. Data entry for user-defined distributions is presented in Section 10.9. It is important that user-defined distributions are entered first so that they are an arrival rate option.

Enter the following additional examples:

- One hundred special customers arrive according to a triangular distribution with parameters 5, 10, and 15. These transactions have priority 2 and the first arrival is at simulation time 0.
- Ten new jobs arrive every 20 time units. These transactions have priority 3 and the first arrival is at simulation time 100.

Notice that the list of sources is maintained alphabetically by name.

Name	Priority	Limit	Offset	Distribution	Parm#1	Parm#2	Parm#3	RnID
NewJobs	0	10	100	Constant	20			1
S1	0	Infinite	0	Exponential	10	0	0	1
SpecialCustomers	0	Infinite	0	Triangle	5	15	10	1

After deleting *NewJobs* and *SpecialCustomers*, the remaining entry is S1 for the basic model.

10.4 Queue Nodes

Queue parameters are shown in Figure 10.4. Our basic model has a first-in, first-out (FIFO) discipline queue with infinite capacity. However, as shown in Section 11.4.2, there are class *Queue* procedures that enable the user to alter the manner in which transactions are added to the queue.

All data entry features described in *Source Nodes* page are applicable to *Queue Nodes*. Specifically, add, save, and delete options are identical. An example of the above node saved to the spreadsheet is shown below.

Name	Discipline	Capacity
Q1	FIFO	Infinite

10.5 Facility Nodes

Facility node parameters are in Figure 10.5. The default number of servers in the facility is one. Servers are identical, so the distribution is the same for each server.

Again, data entry features for *Source Nodes* and *Queue Nodes* are applicable to *Facility Nodes*. Entering *ServiceDistribution* is identical to entering arrival rates for sources. After choosing an exponential service time with mean of 10 time units, the above facility information saved to the spreadsheet is shown below.

Name	Nservers	Distribution	Parm#1	Parm#2	Parm#3	RnID
F1	1	Exponential	10	0	0	1

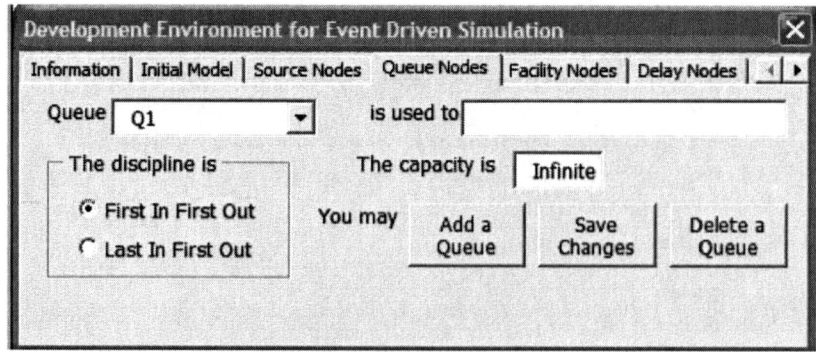

Figure 10.4 *ProgramManager* queue page.

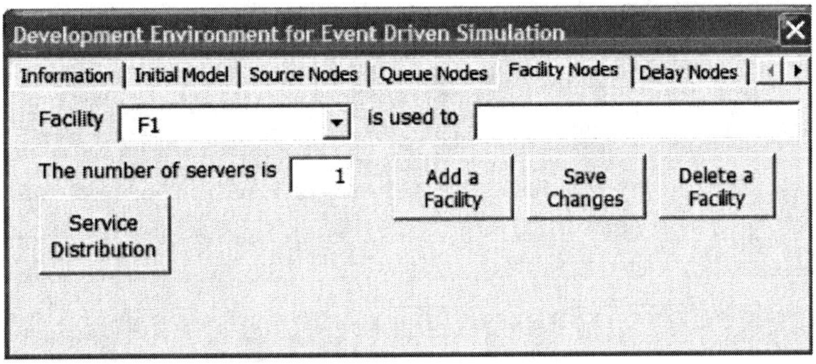

Figure 10.5 *ProgramManager* **facility page.**

Notice in the *Description of Distribution* menu for the facility that an option available to facilities and delays is *Encoded Distribution*. This option indicates that the service time or delay for the node is a parameter in the VB program. The use of this option is explained in Chapter 11.

10.6 Delay Nodes

As shown in Figure 10.6, a delay node has two parameters, name and delay time. The distribution is the same for each transaction entering the delay. Steps for data entry on delay nodes are identical to those for source and facility nodes. If there are questions, refer to the description on facility nodes in the previous section. After specifying a delay to be a constant of 30 time units, the above delay node is saved to the spreadsheet as shown below.

Name	Distribution	Parm#1	Parm#2	Parm#3	RnID
CureProcess	Constant	30	0	0	1

10.7 Initial Model

Default values of *Initial Model* options are shown in Figure 10.7.

The *Run Length* is 1000 time units. The *Transient Time* is 100 time units and there is one observation. However, the user changes these parameters to be consistent with the objectives of the

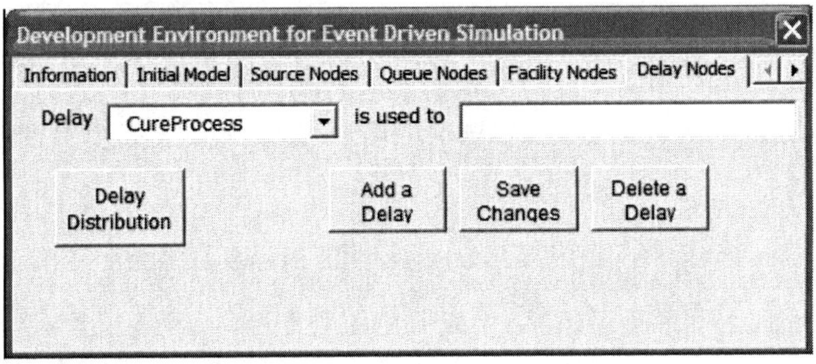

Figure 10.6 *ProgramManager* **delay page.**

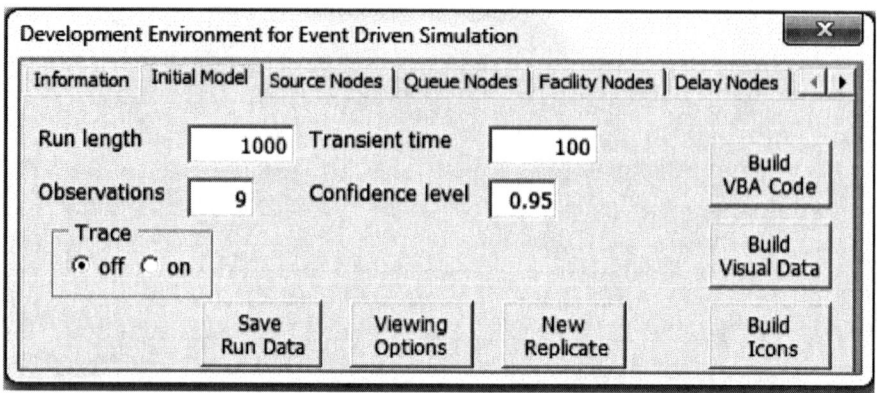

Figure 10.7 *ProgramManager Initial Model* **page.**

simulation. For example, a *Run Length* of 2000 time units, a *Transient Time* of 500 time units, and setting *Observations* to 1 enable the user to perform a replication experiment, described in Section 6.4.2. In a similar manner, a *Run Length* of 10,000 time units, a *Transient Time* of 1000 time units, and 10 *Observations* enable the user to perform the subinterval replication method discussed in Section 6.4.1. *Transient Time* for the model is also known as the model warm-up period. With *Save Run Data* the parameters are saved to the spreadsheet.

When *Number of Observations* is greater than one, *ProgramManager* prompts the user for a *Level of Confidence,* which is used to determine confidence intervals on statistical performance measures. The following is added to the *Initial Model* page.

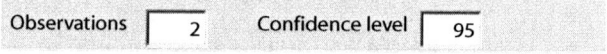

Examples of saved run data are shown below.

	A	B
1	Run Length	1000
2	Transient Time	0
3	Number of Observations	1
4	Level of Confidence	
5	Trace Report	N
6	**Run**	Start time
7	**Simulation**	Stop time

	A	B
1	Run Length	10000
2	Transient Time	100
3	Number of Observations	10
4	Level of Confidence	95.0
5	Trace Report	N
6	**Run**	Start time
7	**Simulation**	Stop time

For subsequent subinterval experiments, *New Replicate*, with an appropriate replication number, archives the current output spreadsheets and creates a new set of output sheets for the next replication. Use of the *Build Visual Data* and *Build Icons* options are described in Chapter 14.

Notice that the user did not request a trace report. The following is added to the *Initial Page* when the trace is requested. Additional discussion on the use of the *TraceReport* is presented in Section 12.7.2.

As described in Section 10.7.1, *Build VBA Code* produces application-specific VB code for the simulation program.

10.7.1 *Build VBA Code*

To facilitate coding the model, the *Build VBA Code* option of *ProgramManager* produces application-specific code to

1. Declare and initialize network nodes, pdfs, and tables
2. Define primary events and primary event subroutine calls (sub *Parser*)
3. Define primary event subroutine shells

The code is copied from the *Parser* spreadsheet to the *EventsManager* module.

In the simple queueing model the two primary events are an arrival and a service completion. Therefore, two primary event subroutines are needed. In the program the three nodes—a source, a queue, and a facility—must be declared and initialized. *Parser* must have two types of primary events and a subroutine call for each. Recall from Section 8.1.4 that the node reference and type of primary event are node name concatenated with node type. The corresponding primary event sub is node reference concatenated with "Event." From column A of spreadsheet *Parser*, references for each node are defined, as shown in Figure 10.8.

Other user-defined global variables in the model should be included under "*User defined input.*"

In column B of the *Parser* spreadsheet is sub *Initial*. As shown in Figure 10.9, it initializes each node's reference by matching each source, queue, and facility node to its corresponding *Instance*

```
Option Explicit
Private _
    InSources As New Collection, _
    InQueues As New Collection, _
    InFacilities As New Collection, _
    InDelays As New Collection, _
    InStatistics As New Collection, _
    InPDFs As New Collection, _
    InTables As New Collection, _
    PrimaryEventList As New Collection, _
    Time As Single, _
    S1Source As Source, _
    Q1Queue As Queue, _
    F1Facility As Facility, _

    ' User defined input
```

Figure 10.8 DEEDS data definitions from *Parser* spreadsheet.

```
Public Sub Initial()
    GetResources InSources, InQueues, InFacilities, _
        InDelays , InStatistics, InPDFs, InTables
    Set PrimaryEventList = ViewPEL()
    Set S1Source = Instance("S1", InSources)
    Set Q1Queue = Instance("Q1", InQueues)
    Set F1Facility = Instance("F1", InFacilities)

    ' User initialized input

End Sub
```

Figure 10.9 DEEDS resource initializations from *Parser* spreadsheet.

in the collection of sources, queues, or facilities. Notice that an area in sub *Initial* is reserved for initialization of other user-defined variables that were previously defined variables.

In column C are *Parser* and *ShutDown* subroutines shown in Figure 10.10. For our basic model there are no changes to *Parser* and *Initial*. Also, no user-defined variables are needed. However, in Chapters 18–25 are examples where both are edited to generalize the model. These models typically have arrays of queues and facilities that have similar or identical functionality. Consistent with the naming convention described in Chapter 8, column D presents primary event subroutine shells, shown in Figure 10.11, to be coded for the respective primary event.

Also from Chapter 8, the primary event details are given in Figure 10.12.

```
Public Sub Parser(ByRef CurrentEvent As Transaction)
  Dim EventType _
    As String
  EventType = CurrentEvent.EventType
  Time = CurrentTime
  Select Case EventType
    Case "S1Source"
      S1SourceEvent CurrentEvent

    Case "F1Facility"
      F1FacilityEvent CurrentEvent

    Case "SteadyState"

    Case "NewObservation"

    Case Else
      SimulatorMessage "Undefined Event Type:" & EventType
    End
  End Select
End Sub

Public Sub ShutDown()
' user defined shutdown
End Sub
```

Figure 10.10 DEEDS *Parser* segment from *Parser* spreadsheet.

```
Sub S1SourceEvent(ByRef CurrentEvent as Transaction)
End Sub

Sub F1FacilityEvent(ByRef CurrentEvent as Transaction)
End Sub
```

Figure 10.11 DEEDS *PrimaryEvent* subs from *Parser* spreadsheet.

```
Sub S1SourceEvent(ByRef ArrivingCustomer As Transaction)
  F1Facility.NextFacility ArrivingCustomer, 1, Q1Queue
End Sub
Sub F1FacilityEvent(ByRef ServicedCustomer As Transaction)
  F1Facility.Disengage ServicedCustomer
  Set ServicedCustomer = Nothing
  If Q1Queue.Length > 0 _
  Then
    F1Facility.EngageAndService Q1Queue.Depart, 1
  End If
End Sub
```

Figure 10.12 Detailed *PrimaryEvent* subs.

Wait until entering the Visual Basic Editor (VBE) to code the primary event subs. Understanding all of the programs is not important at this time. The purpose of the effort is simply to become familiar with the development process. Chapter 11 begins the explanation of the language syntax. As shown in Figure 10.13 (a), access to the VBE from Excel is selecting T̲ools → M̲acro → V̲isual Basic Editor. Access to the VBE in Excel 2007, as shown in Figure 10.13 (b), is via the *Developer* tab.

Figure 10.14 shows the VBA Project for the basic network simulation model. The simulation program resides in the *EventsManager* module. In the *Simulation* module is a program *Controller* that administers communications between DEEDS and *EventsManager*.

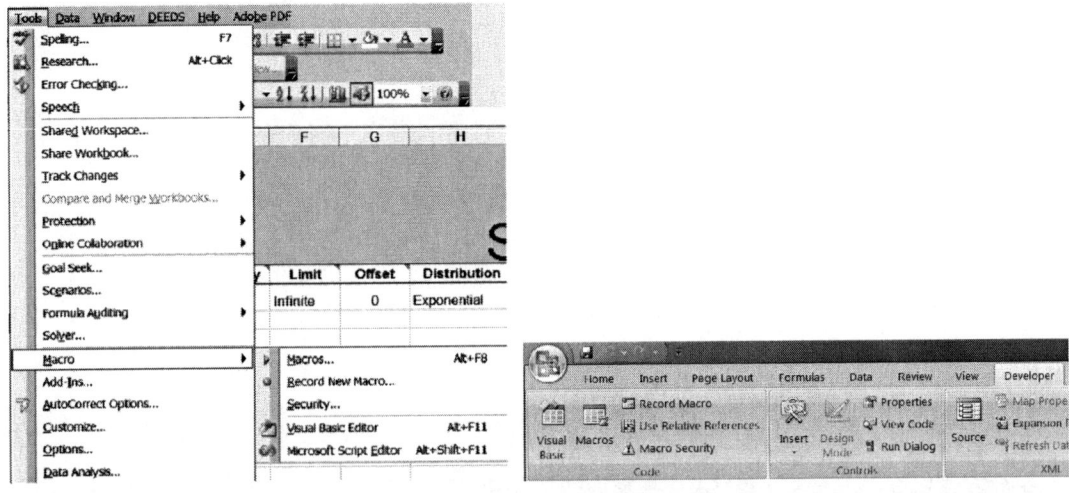

Figure 10.13 **(a) Access to the Visual Basic environment in Excel 2003. (b) Access to Visual Basic environment in Excel 2007.**

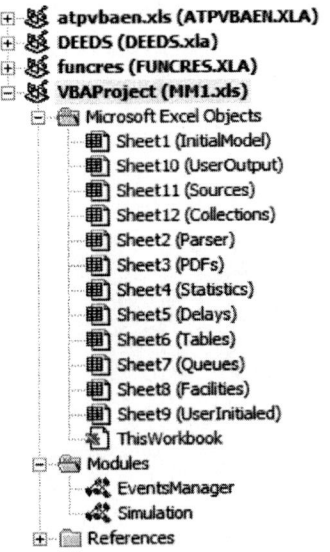

Figure 10.14 **VBE Project environment.**

10.7.2 Program Execution

After the simulation model has successfully compiled, run the program using the *Start Simulation* option shown in Figure 10.15. A progress report indicates when the simulation has completed. In some cases the simulator may trap program errors and terminate the program after displaying an error message in *SimulatorMessages*. Untrapped errors cause an execution error that returns control to VBA. The highlighted line of code has the error and the user has the option of debugging or ending the program. Details on use of the VBA Debugger are presented later in Chapter 12.

In early stages of program development, and during program testing and validation, the user spends much more time outside *ProgramManager* because coding and testing are in the VBE environment. Although additional details will be added to the development process, for now the stages of model development are as follows:

1. Enter node description information (source, queue, etc.) into the spreadsheet *InitialModel*.
2. Enter user-defined initialization data into the spreadsheet *UserInitialed*.
3. Build VBA code.

Now in the VBE environment,

1. Copy generated code to the *EventsManager* module.
2. Develop user-defined data definitions in VB.
3. Develop VB code for user-defined data in sub *Initial*.
4. For each primary event sub, convert related secondary events to VB code.
5. Develop VB code for user-defined output in sub *ShutDown*.
6. Correct compile errors.
7. Run the simulation program.
8. Resolve execution errors using *Trace, SimulatorMessages*, and the VBA Debugger.
9. View model output.

The execution step may also be carried out interactively for debugging by strategically placing break points in the VBA program. Of course, model development steps are not, as shown, a sequential process.

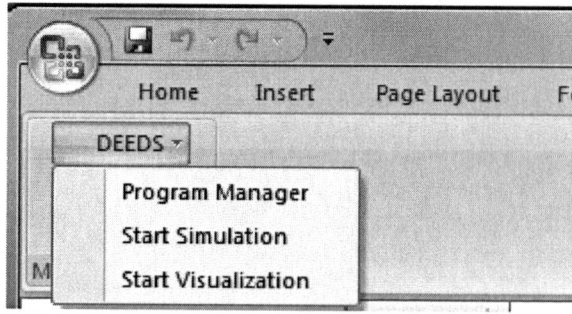

Figure 10.15 DEEDS menu options.

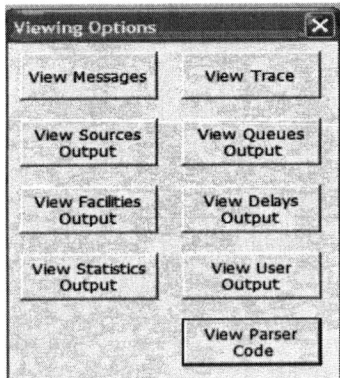

Figure 10.16 *ProgramManager* **viewing options.**

10.7.3 Viewing Options

By using the *ProgramManager* option shown in Figure 10.15, and *Viewing Options* on the *Initial Model* page of *ProgramManager* (Figure 10.7) the user can easily move between simulation results. Now is a good time to experiment with moving between spreadsheets and returning to *ProgramManager*.

Shown in Figure 10.16 are *Viewing Options*, which are spreadsheet destinations. With the exception of the *Parser* spreadsheet, each presents summary information on performance measures in the model. The top row of spreadsheet destinations exits *ProgramManager* to model validation and testing tools. The next three rows exit to summary statistics of network nodes and user-defined statistics. *View Parser Code* exits to the spreadsheet with the VBA-generated code.

10.8 Model Development

The assembly line operation problem in Example 10.1 provides another opportunity to strengthen understanding of the DEEDS modeling environment.

Example 10.1

Consider a vehicle assembly line operation where parallel stations install the left and right doors in the door installation area. The operations are such that the left and right door operations must begin simultaneously. Therefore, the operator who finishes first must wait for the other operator to finish before beginning work on the next vehicle. Vehicle interarrivals to the door installation area have a continuous uniform distribution between 10 and 15 time units. Time to install each door also has a uniform continuous distribution. However, the left door is between 8 and 12 time units while the right door is between 7 and 12 time units. The assembly line is continuous flow. Therefore, delays that result from both operators not being ready to begin installation when the vehicle arrives suggest that the line speed is too fast.

In Figure 10.17, a source, queue, and two facilities are used to model this environment. The queue is a virtual queue used to assess the line speed. Node data entered into the *InitialModel* spreadsheet produce the entries in source, queue, and facility shown in Figure 10.18. *ProgramManager* produced the data definitions and *Initial* subroutine shown in Figures 10.19 and 10.20, respectively.

A user-defined integer variable, *NumberBusyServers*, is used to assess the status of both workstations. Permissible values of *NumberBusyServers* are 0, 1, and 2. *NumberBusyServers* is included with data definitions from the spreadsheet *UserInitial*. Defining *NumberBusyServers* as *Private*

Figure 10.17 Node representation for parallel assembly operations.

Name	Priority	Limit	Offset	Distribution	Parm#1	Parm#2	Parm#3	RnID
Vehicle	1	Infinite	0	UniformContinuous	10	15	0	1

Name	Discipline	Capacity
Line	FIFO	1

Name	Nservers	Distribution	Parm#1	Parm#2	Parm#3	RnID
LeftDoor	1	UniformContinuous	8	12	0	1
RightDoor	1	UniformContinuous	7	12	0	1

Figure 10.18 Parallel operations—*InitialModel* spreadsheet.

```
Option Explicit
Private _
    InSources As New Collection, _
    InQueues As New Collection, _
    InFacilities As New Collection, _
    InDelays As New Collection, _
    InStatistics As New Collection, _
    InPDFs As New Collection, _
    InTables As New Collection, _
    PrimaryEventList As New Collection, _
    Time As Single, _
    VehicleSource As Source, _
    LineQueue As Queue, _
    LeftDoorFacility As Facility, _
    RightDoorFacility As Facility, _


    ' User defined input
```

Figure 10.19 Parallel operations—resource definition segment.

```
Public Sub Initial()
    GetResources InSources, InQueues, InFacilities, _
            InDelays , InStatistics, InPDFs, InTables
    Set PrimaryEventList = ViewPEL()
    Set VehicleSource = Instance("Vehicle", InSources)
    Set LineQueue = Instance("Line", InQueues)
    Set LeftDoorFacility = Instance("LeftDoor", InFacilities)
    Set RightDoorFacility = Instance("RightDoor", InFacilities)

    ' User initialized input

End Sub
```

Figure 10.20 Parallel operations—initial resources segment.

```
Public Sub Parser(ByRef CurrentEvent As Transaction)
    Dim EventType _
        As String
    EventType = CurrentEvent.EventType
    Time = CurrentTime
    Select Case EventType
        Case "VehicleSource"
            VehicleSourceEvent CurrentEvent

        Case "LeftDoorFacility"
            LeftDoorFacilityEvent CurrentEvent

        Case "RightDoorFacility"
            RightDoorFacilityEvent CurrentEvent

        Case "SteadyState"

        Case "NewObservation"

        Case Else
            SimulatorMessage "Undefined Event Type:" & EventType
        End
    End Select
End Sub

Public Sub ShutDown()
    ' user defined shutdown
End Sub
```

Figure 10.21 Parallel operations—*Parser* segment.

enables the variable to retain its value when the module is inactive. The scope (i.e., which subroutines have access to the variable) of *NumberBusyServers* is the entire simulation program.

An arrival and two service completions are primary events for this model. Therefore, there are three corresponding primary events subroutines. Sub *Parser* for this model is in Figure 10.21.

The *Parser* subroutine for this model has a structure identical to the previous example. That is, event types are node name concatenated with node type and subroutine names are event type concatenated with "Event." Of course, the functionality is the same. That is, *Parser* determines which subroutine has responsibility for the primary event and notifies it to "process" the transaction.

The three primary events are in Figure 10.22. As shown in subroutine *VehicleSourceEvent*, when a transaction representing a vehicle arrives in the installation area, it joins *LineQueue*. Details of subroutine *BeginNextAssembly* are presented after the following discussion on the workstations.

For *RightDoorFacilityEvent* and *LeftDoorFacilityEvent* the transaction represents an installed door. Activities in the two service completion subroutines are very similar because both have workstation synchronization requirements. The difference is the embedded service time for each station.

Upon completion of a door (the facility disengages the transaction), the system is notified that the facility is idle by reducing *NumberBusyServers* by 1. Notice that both workstation subroutines use sub *BeginNextAssembly*.

BeginNextAssembly is used to initiate work on a vehicle. This sub is common to the arrival and both service completion subroutines. If both workstations are idle (*NumberBusyServers* = *0*) and a vehicle is in the queue, the transaction departs the queue and work begins on the next vehicle. Therefore, the *NextVehicle* transaction represents either an arrival or a vehicle that has waited in the queue. The original and copy of the vehicle transaction engage the right and left workstations simultaneously. Assigning *NumberBusyServers* the value 2 communicates to the

```
                    ' PRIMARY EVENT HANDLERS
Sub VehicleSourceEvent(ByRef ArrivingVehicle As transaction)
  LineQueue.Join ArrivingVehicle
  BeginNextAssembly
End Sub
Sub LeftDoorFacilityEvent(ByRef FinishedDoor As transaction)
  LeftDoorFacility.Disengage FinishedDoor
  NumberBusyServers = NumberBusyServers - 1
  BeginNextAssembly
End Sub
Sub RightDoorFacilityEvent(ByRef FinishedDoor As transaction)
  RightDoorFacility.Disengage FinishedDoor
  NumberBusyServers = NumberBusyServers - 1
  BeginNextAssembly
End Sub
Sub BeginNextAssembly()
  Dim _
    NextVehicle As transaction
  If NumberBusyServers = 0 And LineQueue.Length > 0 _
  Then
    Set NextVehicle = LineQueue.Depart
    RightDoorFacility.EngageAndService NextVehicle, 1
    LeftDoorFacility.EngageAndService NextVehicle.Copy(), 1
    NumberBusyServers = 2
  End If
End Sub
```

Figure 10.22 Parallel operations—VBA model segment.

system that both facilities are again busy. When the simulation begins, both stations are idle and *NumberBusyServers* is zero.

Of course, in this simple model some redundant code could have been included in the three primary event subroutines. However, as models increase in complexity, additional subroutines and functions enable the user to design modules for simulation of primary and related secondary events that are much easier to test and maintain.

A technique for generalizing models, as presented in Chapter 8, is to use subroutine *Initial* to read system parameters from a spreadsheet. Then, rather than changing the simulation program, the user simply changes parameter values on the spreadsheet. A good test for understanding of the basic features of the *ProgramManager* is to run the model for 100 time units with no transient time and one observation. There are seven pages of *ProgramManager*. Five of them have been presented.

10.9 Statistical Variables

Additional performance measures may be defined as time-based or observation-based statistics. Time-based variables are automatically recorded by the simulator, and observation-based variables are recorded according to simulation program instructions. See Section 5.4 for a review of statistical variables.

For example, in a production application, the user wants information on cycle time and inventory level. *CycleTime* is defined as an observation-based statistic and *InventoryLevel* as a time-based statistic. These performance metrics shown below in the *InitialModel* spreadsheet are defined using the *Statistical Variables* page shown in Figure 10.23.

Example 10.2

Consider an assembly line operation where an operator performs an assembly operation and then an inspection. If the part passes inspection, the operator removes a part from the operations queue and begins another assembly. Of course, if the queue is empty, the operator becomes idle. If the part fails the inspection, it is immediately reworked. In essence, the part engages the operator until it passes inspection. This system is modeled with a source, queue, facility, and delay node. Also, a statistical variable for the time a transaction is in the system is included in the model. The network of the system is shown in Figure 10.24.

After entering the model description, the *InitialModel* spreadsheet has source, queue, facility, delay, and user-defined statistical variable information, as shown in Figure 10.25.

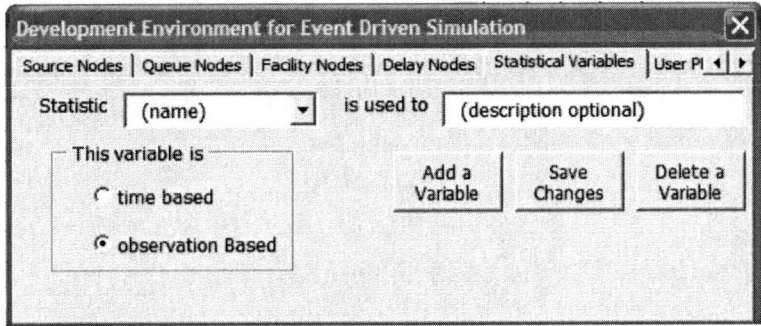

Figure 10.23 *ProgramManager Statistical Variables* **page.**

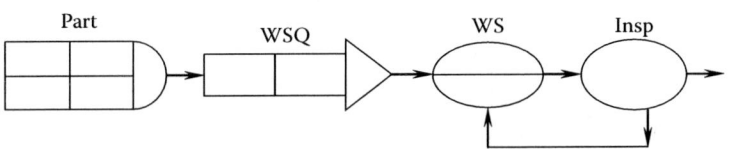

Figure 10.24 **Node representation of sequential operations.**

Name	Priority	Limit	Offset	Distribution	Parm#1	Parm#2	Parm#3	RnID
Parts	1	Infinite	0	Exponential	2	0	0	1

Name	Discipline	Capacity
Job	FIFO	Infinite

Name	Nservers	Distribution	Parm#1	Parm#2	Parm#3	RnID
WS	1	Exponential	1			1

Name	Distribution	Parm#1	Parm#2	Parm#3	RnID
Insp	Constant	1.5			1

Name	Type
CycleTime	O

Figure 10.25 **Sequential operations** *InitialModel* **spreadsheet.**

Visual Basic code produced by *ProgramManager* for data declarations and initialization is shown in Figures 10.26 and 10.27. Notice that reference variables for *InspDelay* and *CycleTime* are automatically included. Sub *Parser* and sub *ShutDown* are shown in Figure 10.28. Sub *ShutDown* is not edited in this model. However, it is essential that it be included in the model. Otherwise, an execution error occurs. Detailed primary events for the model are shown in Figure 10.29. In sub *PartsSourceEvents* the arriving part joins the queue when another part is being processed or inspected. In sub *WSFaciltityEvent* the part is immediately inspected upon completion of processing. In this model 30% of parts fail inspection. A part that passes inspection exits the system after it disengages the facility and its time in the system is collected. If there is a part in the queue, it is scheduled for service. If the part fails inspection, it is immediately scheduled for service again. It may be helpful to develop the inspection process program and run the model for 100 time units with no transient time and one observation.

The last two pages of *ProgramManager* are for user-defined pdfs and user-defined function tables. Both are presented below. It is important that when a model includes user-defined distributions and tables, they must be defined before entering node descriptions. Otherwise, the options in the drop-down menu for user-defined pdfs and tables will be empty.

```
Option Explicit
Private _
    InSources As New Collection, _
    InQueues As New Collection, _
    InFacilities As New Collection, _
    InDelays As New Collection, _
    InStatistics As New Collection, _
    InPDFs As New Collection, _
    InTables As New Collection, _
    PrimaryEventList As New Collection, _
    Time As Single, _
    PartsSource As Source, _
    JobQueue As Queue, _
    WSFacility As Facility, _
    InspDelay As Delay, _
    CycleTimeStatistic As Statistic, _

    ' User defined input
```

Figure 10.26 Sequential operations—resource definition segment.

```
Public Sub Initial()
    GetResources InSources, InQueues, InFacilities, _
            InDelays , InStatistics, InPDFs, InTables
    Set PrimaryEventList = ViewPEL()
    Set PartsSource = Instance("Parts", InSources)
    Set JobQueue = Instance("Job", InQueues)
    Set WSFacility = Instance("WS", InFacilities)
    Set InspDelay = Instance("Insp", InDelays)
    Set CycleTimeStatistic = Instance("CycleTime", InStatistics)

    ' User initialized input


End Sub
```

Figure 10.27 Sequential operations—initial resources segment.

```
Public Sub Parser(ByRef CurrentEvent As Transaction)
   Dim EventType _
     As String
   EventType = CurrentEvent.EventType
   Time = CurrentTime
   Select Case EventType
     Case "PartsSource"
        PartsSourceEvent CurrentEvent

     Case "WSFacility"
        WSFacilityEvent CurrentEvent

     Case "InspDelay"
        InspDelayEvent CurrentEvent

     Case "SteadyState"

     Case "NewObservation"

     Case Else
        SimulatorMessage "Undefined Event Type:" & EventType
        End
   End Select
End Sub

Public Sub ShutDown()
   ' user defined shutdown
End Sub
```

Figure 10.28   Sequential operations—*Parser* segment.

```
Sub PartsSourceEvent(ByRef Arrival As Transaction)
  If WSFacility.Length + InspDelay.Length = 0 _
  Then
    WSFacility.EngageAndService Arrival, 1
  Else
    JobQueue.Join Arrival
  End If
End Sub
Sub WSFacilityEvent(ByRef CompletedPart As Transaction)
  InspDelay.Start CompletedPart
End Sub
Sub InspDelayEvent(ByRef InspectedPart As Transaction)
  InspDelay.Finish InspectedPart
  If StatProcs.RandomNumber(1) > 0.7 _
  Then    ' passed inspection
    WSFacility.Disengage InspectedPart
    CycleTimeStatistic.Collect InspectedPart.SystemTime
    Set InspectedPart = Nothing
    If JobQueue.Length > 0 _
    Then
      WSFacility.EngageAndService JobQueue.Depart, 1
    End If
  Else    ' failed inspection
    WSFacility.Service InspectedPart, 1
  End If
End Sub
```

Figure 10.29   Sequential operations—VBA model segment.

10.10 User-Defined Probability Functions

Before delving into this section, if may be helpful to briefly review material on discrete and continuous distributions in Section 5.3. The data entry pages for probability distributions are shown in Figures 10.30 and 10.31. Begin by choosing an existing distribution from the drop-down menu or creating a new distribution by entering a new distribution name. Again, these functions must be available to DEEDS when entering source, facility, and delay nodes that use these distributions.

For a new distribution, specify the number of cells and type of distribution (continuous or discrete) and then choose the *Create Pdf* option.

The blank table in Figure 10.31 is presented for data entry. With x_m as the value of the mth random value in the discrete pdf and $p(x_m)$ the probability associated with random value x_m, the general format for both discrete and piecewise continuous functions is the same. Notice that values of x_m must be listed in ascending order. After entering the data, the table is closed and the distribution is saved to the spreadsheet by selecting *Add a Distribution*.

For an existing distribution, changes allowed are type of distribution, number of cells, and cell contents. In this case choose *View/Edit Pdf*, then complete changes to the table, close the table,

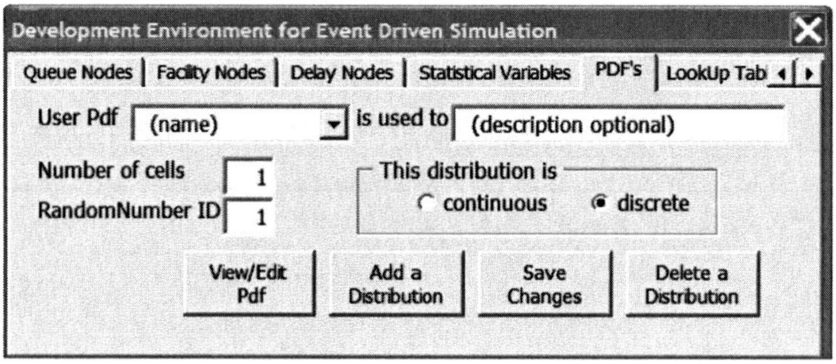

Figure 10.30 *ProgramManager* user-defined PDF.

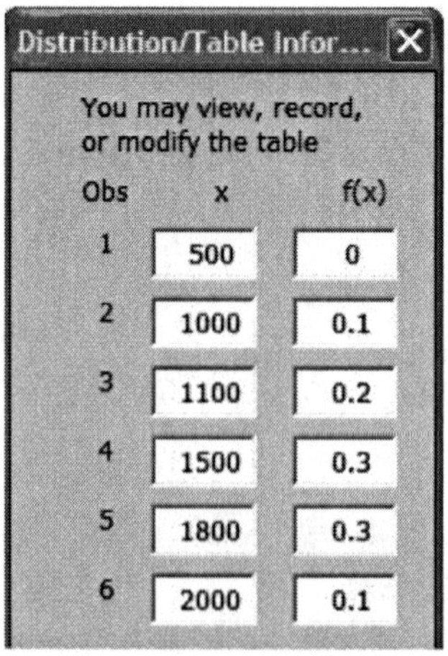

Figure 10.31 *ProgramManager* PDF table.

Name	ServiceTime	
Type	Continuous	
RnID	1	
Comment		
Values	*x*	*p(x)*
	500	0
	1000	0.1
	1100	0.2
	1500	0.3
	1800	0.3
	2000	0.1

Figure 10.32   DEEDS statistical variable spreadsheet.

and save the changes. Figure 10.32 shows an example of a *Continuous* distribution saved on the *PDFs* spreadsheet. The PDF table has room for 15 pairs of *x* and *p(x)*. If there are more terms, the distribution must be entered directly into the *PDF* spreadsheet.

10.11 User-Defined Tables

A VBA function queries a user-defined function. The simulation program calls the VBA function with an *x* parameter and the function returns a value of $f(x)$. Steps for creating user-defined tables are the same as for probability functions. The *UserTables* page is shown in Figure 10.33.

In Figure 10.34, table lookup functions have a format similar to the discrete pdf. With x_m as the *m*th value of the independent variable in the table and $f(x_m)$ as the dependent value associated with x_m, the format for both discrete and piecewise continuous tables are the same. Figure 10.35 is an example of a continuous and discrete table definitions saved to the *UserTables* spreadsheet.

10.12 Program Execution—Expanded

The program execution described in Section 10.6.2 is a simple example to get started. The following description is the unabridged version of program execution. The primary difference is the result of user-defined pdfs and tables. As mentioned in Section 10.3, it is important for pdfs and tables to be created before the *InitialModel* spreadsheet is populated (nodes). Of course, all

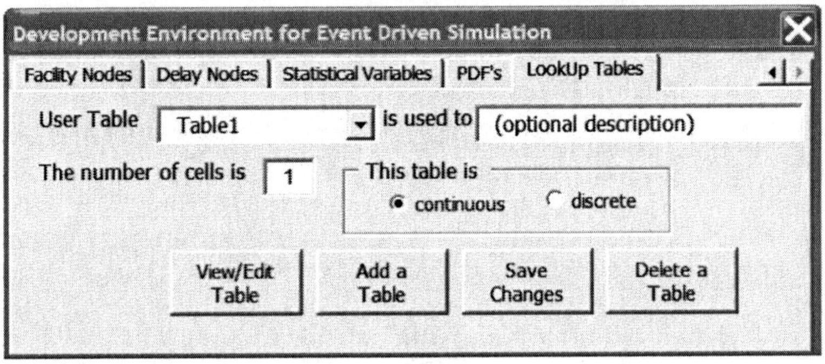

Figure 10.33   *ProgramManager* table page.

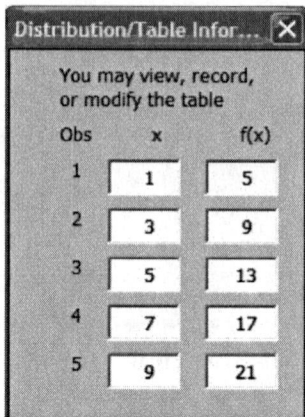

Figure 10.34 *ProgramManager* **table.**

	A	B	C	D	E
1	Name	Table1		Table2	
2	Type	Discrete		Continuous	
3	Comment				
4	Values	*x*	*f(x)*	*x*	*f(x)*
5		1	5	1	5
6		3	9	3	9
7		5	13	5	13
8		7	17	7	17
9		9	21	9	21

Figure 10.35 **DEEDS** *Table* **spreadsheet.**

network nodes, pdfs, and tables must be on their respective spreadsheet before using the *Build VBA Code* option. After preparing the Excel worksheets for a model, the model development steps are as follows:

1. Enter *ProgramManager*.
2. Enter user-defined pdfs and tables data in spreadsheets *PDFs* and *tables*, respectively.
3. Enter node description information (source, queue, etc.) into the spreadsheet *InitialModel*.
4. Enter statistical variables descriptions into the spreadsheet *InitialModel*.
5. Enter user-defined initialization data into the spreadsheet *UserDefined*.
6. Request *ProgramManager* to generate application-specific VBA code.
7. Exit *ProgramManager*.

Details of steps 2, 3, 4, and 5 are application specific. For example, if there are no tables or user-defined pdfs, step 4 is omitted. Now enter the VBA environment and do the following:

1. Copy *ProgramManager*-generated code to the *EventsManager* module.
2. Complete application-specific data definitions.
3. Complete application-specific code for sub *Initial*.

4. For each primary event sub, convert related secondary events to VBA code.
5. Develop VBA code for user-requested information in sub *ShutDown*.
6. Correct any compile errors.
7. Return to Excel and *ProgramManager*.

In a similar manner, details of the above steps are also application specific. Now from *ProgramManager*

1. Run the simulation program.
2. Resolve execution errors using *Trace, SimulatorMessages*, and the VBA Debugger.
3. View modeling result.

Finally, it is important that to be familiar with materials in Chapters 12 and 13 on user output, interpreting simulation results, and the potential problems caused by using the results incorrectly.

10.13 Summary

This chapter presented *ProgramManager*, the user interface to DEEDS. The interface enables users to navigate the development environment and assists with entering descriptive network information into the spreadsheets, running the simulation program, and viewing simulation results. Also shown is how to use *ProgramManager* to assist with developing application-specific VBA code for the model. Procedures for running a program and viewing simulation results were also presented.

Chapter 11

Modeling Procedures

11.1 Introduction

Beginning with Chapter 7, slightly different views of the design environment for event-driven simulation (DEEDS) modeling environment have been presented. In Chapter 9 essentials of Visual Basic (VB) programming for modeling in DEEDS were presented. This chapter focuses on DEEDS tools for developing models. Simply stated, DEEDS is a suite of VB procedures that manage instances of the DEEDS classes, communicate with the DEEDS simulator, and provide information for modeling complex decisions. In addition to DEEDS procedures, the modeler also has access to Excel and Visual Basic for Applications (VBA) procedures.

11.2 VBA Procedures

From definitions in Chapter 9, a VB procedure is a general term that refers to VB subs and functions. When working with VB *classes*, subs and functions have a slightly different terminology that is very helpful. Object-oriented programming terminology includes objects, properties, and methods. An *object* is an instance of a class. For example, there are often several facilities in a model. Each facility is an instance of class *Facility*. The number of facility servers is a facility *property*. Whether the server is idle or busy is a property. A *method* changes the state of the facility. For example, a server becomes busy as a result of a method that changed the server status from idle to busy.

There are similar properties and methods for classes *Source, Queue, Delay, Statistic, PDF, Table,* and *Transaction*. A transaction routed between nodes alters the state of the system. Other examples include changing transaction attributes, node capacities, and node distributions. It may be helpful to think of objects (instances of a class) as nouns, properties as adjectives that describe the object, and methods as verbs that act on an object.

The language syntax for procedures and methods is similar because in a sense a VB module may be viewed as an object. The following two examples demonstrate the syntax as well as the similarity:

```
Q1Queue.JoinFront Arrival
```

169

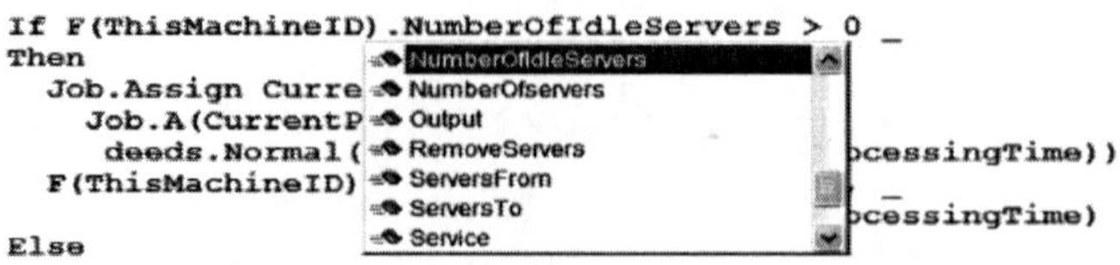

```
If F(ThisMachineID).NumberOfIdleServers > 0 _
Then                    NumberOfIdleServers
    Job.Assign Curre    NumberOfservers
        Job.A(CurrentP   Output
          deeds.Normal(  RemoveServers            cessingTime))
    F(ThisMachineID)    ServersFrom
                        ServersTo               cessingTime)
    Else                Service
```

Figure 11.1 VBA procedures for the *Facility* class.

In the above example *Q1Queue* is an object, an instance of class *Queue*. The method is *JoinFront* and *Arrival* is an instance of class *Transaction*, placed to the front of *Q1Queue*.

```
Distribution.Normal 20,5,2
```

Normal is a function in module *Distribution*.

Figure 11.1 demonstrates useful features of the VBA development environment. The programmer has access to a menu of subs and functions available to the class or module being referenced. As demonstrated by examples in this and the following chapters, another advantage of modeling in the Excel/VBA environment is that many special-purpose VBA and Excel procedures may be included in the simulation program.

11.3 Simulator Procedures

The *Simulator* module has numerous subs and functions. Table 11.1 presents several functions available to the modeler.

The format for a function call is presented in Chapter 9. However, it is recommended that the following expanded version be used while learning DEEDS features.

Variable = **Simulator.***name* [*arglist*]

The type of *Variable* must match the return type of the function and *arglist* is optional. In the following example, attribute 1 of transaction *Vehicle* is assigned the value of *CurrentTime*.

```
Vehicle.Assign 1, Simulator.CurrentTime
```

In the next example, to obtain 50 outcomes of a statistical variable after the fourth simulation observation, the following program segment is included in sub *Parser*.

```
Case "New Observation"
If Simulator.CurrentObservation > 3_
Then
```

The example below should be familiar. The *SimulatorInstance* function is used to initialize an object variable to reference an instance of class *Source* named *S1*.

```
Set S1Source = Simulator.Instance("S1", InSources)
```

Table 11.1   DEEDS Simulator Procedures

Name	Parameters	Description
CurrentTime		Returns single simulator *ClockTime*
CurrentObservation		Returns Integer *ObservationNumber*
IndexOf	*Name, CollectionOfObjects*	Returns Integer *Index* to instance with name equal to *Name*
Instance	*Name, CollectionOfObjects*	Returns *ObjectReference* to instance with name equal to *Name*
NewTransaction		Returns new Transaction initialized with CreationTime and SerialNumber
ViewPEL		Returns Collection reference to the PEL

The function finds an object in the Collection *InSources* and returns an object variable for that instance. If the instance is not found, the function ends the simulation with an error message indicating the name of the node that was not found. *InSources* is simply a *CollectionOfObjects*. *CollectionOfObjects* may be a Collection of *Sources, Queues,* and so on. *InSources.Count* is the number of sources in the Collection. Additional information on Collections may be found using *ObjectBrowser* to search the VBA library.

Normally, transactions enter the model through the source node. However, in some situations, it may be convenient to create a transaction. The following statements create a new transaction that immediately enters node *MaintenanceDelay*.

```
Set Maintenance = Simulator.NewTransaction
MaintenanceDelay.Start Maintenance
```

NewTransaction returns an instance of *Transaction*. Therefore, *Maintenance* is a transaction reference variable and *Set* must be used in the assignment statement. The new transaction is assigned a *SerialNumber* and a *CreationTime* equal to the simulation time that the function was called. As the transaction enters the delay, the *EventType* for *MaintenanceDelay* is assigned to the transaction. Transaction attributes for new transactions are defaulted to zero.

There are additional *Simulator* procedures presented in Chapter 12 that are primarily for testing and validating the simulation program. In Chapter 16 there are other *Simulator* procedures that increase the flexibility needed to model complex decisions and routing transactions.

11.4   DEEDS Classes

Each *Source, Queue, Facility,* and *Delay* class has methods to monitor and alter states of the respective network nodes. Similarly, there are methods for assessing and changing the state of *Transactions*. DEEDS also has *Statistic, PDF,* and *Table* classes to manage instances of these objects. Instances of *Statistic* collect user-defined observation- and time-based statistical variables as described in Section 5.4. *PDF* and *Table* instances are user-defined data tables to evaluate decisions. The format for function or sub calls is also the same as presented in Chapter 9. However, because class objects are used the following syntax is required for functions:

Variable = DEEDSClassObject.name [arglist]

where *name* is the function name and *DEEDSClassObject* is a reference variable for the class instance. If *Variable* is a class reference, the *Set* assignment statement must be used as shown below.

Set Variable = DEEDSClassObject.name [arglist]

In the following example are two function calls. *Q1Queue.Length* returns a variable equal to the length of *Q1Queue*. *Q1Queue.Depart* returns a reference variable for the transaction departing *Q1Queue*.

```
If Q1Queue.Length > 0 _
Then
    Set NextTransaction = Q1Queue.Depart
End If
```

For subs the following format must be used for class methods:

DEEDSClassObject.name [arglist]

11.4.1 Source

The *Simulator* automatically schedules arrivals from each source on the basis of an interarrival rate distribution, limit count, offset, and priority specified in spreadsheet *InitialModel*. Table 11.2 shows *Source* functions that assess the state of a source.

An active source has a transaction scheduled on the primary events list (PEL). A source becomes inactive when the number of arrivals equals the *Limit* specified in the *InitialModel* spreadsheet or

Table 11.2   DEEDS Functions for Class *Source*

Name	Parameters	Description
Active		Returns Boolean status of *Source* (*True* → not suspended)
EventType		Returns String with *EventType*
Name		Returns String with name
Priority		Returns Byte with source priority

Table 11.3   DEEDS Subs for Class *Source*

Name	Parameters	Description
NewDistribution	Name, P1, P2, P3	Next arrival scheduled from this distribution
NewPriority	Priority	Next arrival scheduled has this *Priority*
Restart		Schedule arrival at current time
ScheduleNextArrival		Next arrival scheduled
Suspend		Arrival removed from PEL

the simulation program suspends the source. Function *Active* returns *True* when the source node has a transaction scheduled on the PEL. Table 11.3 contains *Source* subs that change either source parameters or states of transactions associated with the source.

The model may include time-dependent arrival rates. For example, restaurant arrivals increase during the middle of the day. To modify the model environment the program must monitor simulation time and change the arrival distribution to reflect the time of day.

NewDistribution changes the interarrival distribution during program execution. The first parameter, *Name*, is a VB String. Other parameters are Variant data. Depending on the distribution chosen, up to three parameters may be needed to define the distribution. For example, the exponential has only one parameter and the triangular distribution has three. A description of common distributions and their parameters is in Table 11.15.

The following example changes the interarrival rate for *S1Source* to a discrete uniform distribution when queue *Q1Queue* has more than five transactions.

```
If Q1Queue.Length > 5 _
Then
    S1Source.NewDistribution "UniformDiscrete", 5, 10
End If
```

The lower and upper limits for the distribution are 5 and 10, respectively. In contrast, the following example has two character strings that change the interarrival distribution to a user-defined distribution. The first string indicates that the distribution is user defined and the second is the name of the user-defined distribution.

```
S1Source.NewDistribution "PDF", "Failures"
```

The queue discipline is either first-in, first-out (FIFO) or last-in, first-out (LIFO). However, transactions with different priority are queued according to priority. In other words, transactions with priority level 1 join the queue in front of transactions with priority level 2.

Transactions enter the model with the *Priority* specified for the source in spreadsheet *InitialModel*. *NewPriority* resets the source priority. The next arrival scheduled from that source enters the model with the new priority. *Priority* evaluates to an integer value greater than 0 and less than 128. The following subroutine call changes the priority of arriving transactions from source node *S1Source* to 1:

```
S1Source.NewPriority 1
```

The priority may be reduced by 1 with the following:

```
S1Source.NewPriority S1Source.Priority - 1
```

Suspend removes the scheduled arrival from the PEL for the indicated source node and the source becomes inactive. The program may restart source arrivals with *Restart* or *ScheduleNextArrival*. The *Restart* procedure schedules an arrival at *CurrentTime*. In contrast, *ScheduleNextArrival* schedules an arrival at a randomly generated interarrival time of the source node plus *CurrentTime*. In either case, the source node is again active. In the following example, *S1Source* is suspended:

```
S1Source.Suspend
```

Elsewhere in the program it is restarted:

```
If Not S1Source.Active _
Then
    S1Source.Restart
End If
```

When the number of source node arrivals equals the *Limit* specified in spreadsheet *InitialModel*, *ScheduleNextArrival* ignores additional requests for arrivals. However, with *Restart* the simulator ignores the limit specification and resets *Limit* to an infinite number of transactions. It then becomes the responsibility of the model to control the number of arrivals with the method *Suspend*.

A *Restart* or *ScheduleNextArrival* call, when an arrival is on the PEL, will cause the simulator to terminate with an appropriate *SimulatorMessage* because only one transaction per source is allowed on the PEL.

11.4.2 Queue

In Chapter 8 *Join* and *Depart* methods were used to demonstrate simple transaction routing. However, there are several other class methods that increase program flexibility by enabling the model to alter the state of a queue.

Before discussing other queue methods, it is helpful to remember that queued transactions are maintained in a VB Collection. At the front of the queue, *Q1Queue*, is *Q1Queue.item (1)* and at the end is *Q1Queue.item (Q1Queue.Count)*. Queue disciplines recognized by DEEDS are FIFO and LIFO. A FIFO join places the transaction at location *Q1Queue.item (Q1Queue. Count+1)*. For LIFO, a join places the transaction at location *Q1Queue.item (1)*. Therefore, *Q1Queue.Depart* always affects *item (1)* in the Collection. Several *Queue* functions are given in Table 11.4.

An example the average queue length during the current observation, for an object variable reference, *Q1Queue*, is

```
Q1Queue.AverageLq()
```

Notice that with the exception of *Depart* and *Detach*, these functions monitor the state of a queue. *Depart* and *Detach* alter the state of the queue by removing a transaction from the queue. In both cases, a transaction reference variable is *Set* equal to the transaction removed from the queue.

The *QueueIndex* parameter for *Detach* and *View* is an index from the front of the queue. *View* simply returns a *Transaction* reference to a transaction in the queue. The position of the transaction in the queue is not altered. An error condition occurs when *QueueIndex* exceeds *Length*.

It is important to remember that *Depart* removes from the front of the queue and the queue discipline determines which transaction is placed at the front of the queue. An error condition results from an attempt to remove a transaction from an empty queue. The program is terminated with an appropriate simulator error message indicating an attempt to remove from an empty queue and the time that the error occurred.

Table 11.4   DEEDS Functions for Class Queue

Name	Parameters	Description
AverageLq		Returns a Single of average length since last observation
AverageWq		Returns a Single of average waiting time since last observation
Capacity		Returns an Integer of capacity
Depart*		Returns a transaction removed from the front of the queue
Detach*	QueueIndex	Returns a transaction removed from the indexed location
Discipline		Returns a String with discipline LIFO or FIFO
Full		Returns a Boolean *True* when full and *False* otherwise
Length		Returns an Integer queue length
Location	AttributeIndex, Relational operator, Comparison value	Returns the index of the first transaction that satisfies the condition
Name		Returns a String with a name
View	QueueIndex	Returns a reference to the indexed transaction

*Changes the queue state.

For *Detach*, the program indicates which transaction is to be detached from the queue. Normally, the transaction to be detached is dependent on an attribute value. *View* allows specific transaction access to evaluate transaction attributes. In the following example, *Q1Queue* is searched for the first transaction with attribute 1 equal to "Platinum." If there is a match, that transaction becomes the next *Customer* detached from the queue.

```
Found = False
For Index = 1 to Q1Queue.Length
   Set Customer = Q1Queue.View (Index)
   If Customer.A(1) = "Platinum" _
   Then
      Found = True
      Exit For
   End If
Next Index
If Found = True _
Then
   Q1Queue.Detach Index
Else
   Set Customer = Nothing ' no match
End If
```

Subs for class *Queue* are in Table 11.5. These subs alter either the queue parameters from the *InitialModel* spreadsheet or the state of the queue by inserting or removing transactions from the queue.

Table 11.5   DEEDS Subs for Class Queue

Name	Parameters	Description
Join	Transaction	Transaction joins according to the queue discipline
JoinAfter	JoinIndex, Transaction	Transaction placed behind indexed transaction
JoinFront	Transaction	Transaction placed in front
JoinOnHighAttribute	AttributeIndex, Transaction	Transactions ordered ascending on attribute value
JoinOnLowAttribute	AttributeIndex, Transaction	Transaction ordered descending on attribute value
JoinOnPriority	Transaction	Transaction placed after other transactions with same priority
NewCapacity	Capacity	Resets capacity
NewDiscipline	Discipline	Resets discipline and reorders transactions on the basis of new discipline
Remove	NumberRemoved, RemoveIndex	Removes a range of transactions in the queue from the system

The first six class methods insert transactions into the queue. These methods override the queue discipline. An attempt to insert a transaction when the queue is *Full* results in the transaction being destroyed. As shown in Chapter 12, the number of destroyed transactions is standard output for queues.

Join simply queues the transaction on the basis of the queue discipline, which is either LIFO or FIFO. For example, in the following, a customer joins *Q1Queue* when the server *F1Facility* is busy.

```
If F1Facility.NumberOfIdleServers > 0 _
Then
    ' schedule for service
else
    Q1Queue.Join Customer
End If
```

In the following example, *JoinFront* transaction *Arrival* becomes *item(1)* in *Q1Queue*.

```
Q1Queue.JoinFront Arrival
```

JoinFront has the same effect as the LIFO queue discipline. *JoinAfter* inserts a transaction behind a specific indexed transaction in the queue. *Index* must be an integer greater than 0 and cannot exceed queue *Length*. For example,

```
Q1Queue.JoinAfter 5, Arrival
```

inserts the arrival after the fifth transaction in *Q1Queue*. The simulation program terminates with a simulator error message when there are an insufficient number of transactions in the queue. We can place a transaction in front of a specific transaction using a *JoinAfter* with *Index* decremented by 1. To insert an arrival into the queue in front of transaction 3, use

```
Q1Queue.JoinAfter 2, Arrival
```

JoinOnHighAttribute and *JoinOnLowAttribute* use attribute values to determine the relative position in the queue. Transactions are ranked on ascending (descending) attribute value for *AttributeLow* (*AttributeHigh*). Remember when the queue is ordered on attribute value, it is important for the queue discipline to be FIFO to ensure removing from the front of the queue. *Index* must be a valid *Attribute* index. For example,

```
Q1Queue.JoinOnHighAttribute 2, Arrival
```

inserts transaction *Arrival* in *Q1Queue* on the basis of the value of attribute 2. *JoinOnPriority* has the effect of FIFO within transaction priority. For example, transactions with priority level 1 are in front of transactions with priority level 2. Within a priority level, the discipline is FIFO. An example is as follows:

```
Q1Queue.JoinOnPriority Arrival
```

NewCapacity changes the queue capacity. When the new queue capacity is less than the current queue length, DEEDS allows the queue to exceed *Full* until sufficient departures occur and the queue *Length* is equal to *NewCapacity*. However, as described later in this section, the model may use method *Remove* to force transactions out of the queue when the new capacity becomes less than current queue *Length*. To reduce the capacity of queue *Q1* by 1, use the following:

```
Q1Queue.NewCapacity Q1Queue.Capacity - 1
```

To change the queue discipline to FIFO for queue *Q1Queue*, use the following:

```
Q1Queue.NewDiscipline "FIFO"
```

When the model changes the queue discipline, it is incumbent on the user to reorder the queue by moving the transactions to a temporary queue, changing the queue discipline, and then returning transactions to the original queue.

In contrast to the *Depart* and *Detach* functions that enable transactions taken from the queue to be routed elsewhere in the model, *Remove* causes the transactions to exit the system. For example, the following causes transactions 2, 3, and 4 in *Q1Queue* to exit the model:

```
Q1Queue.Remove 3, 2
```

The value 3 is the number of transactions removed and 2 is the starting location in the queue. The queue *Length* must be at least 4. Otherwise, DEEDS terminates the program with an appropriate error message.

11.4.3 Facility

In Chapters 9 and 10 *EngageAndService* and *Disengage* were used to demonstrate simple facility methods. However, there are several other methods that increase program flexibility by enabling the model to monitor and alter the state of a facility. Facility functions are shown in Table 11.6.

Notice in the description that function *Interrupt* alters the state of the facility by removing a transaction from the facility. Because the return data type is a *Transaction* object, a *Set* assignment must be used to create an object reference to the transaction and remove the transaction from the facility. Other functions are for assessing the state of the facility.

Identical to viewing a queued transaction, the *WIPIndex* parameter for *View* is an index into the facility. *View* returns a *Transaction* reference to a transaction in the facility. The position of the transaction in the facility is not altered. *WIPIndex* must not exceed the *Length* of the facility. Otherwise, the program is terminated with an appropriate simulator error message indicating which facility was involved and the time that the error occurred. Methods for class *Facility* are shown in Table 11.7. These subroutines either alter the facility parameters in the *InitialModel* spreadsheet or alter the state of the facility by inserting or removing transactions from the facility.

Before examples of *Engage, EngageAndService, Service*, and *NextFacility* are described, a brief overview of features for each method is presented. *Engage* allocates the requested number of servers to a transaction. The transaction must occupy at least one server until disengaged from the facility.

Table 11.6   DEEDS Functions for Class Facility

Name	Parameters	Description
AvgUtilization		Returns Single with average server utilization since last observations
EventType		Returns String with the *EventType*
Interrupt*	*WIPIndex*	Returns transaction removed from the facility and server(s) become idle
Length		Returns Integer number of transactions in the facility
Name		Returns String with the facility name
Location	*AttributeIndex*, Relational operator, Comparison value	Returns index of the first transaction that satisfies the condition
NumberOfBusyServers		Returns Integer number of busy servers
NumberOfIdleServers		Returns Integer number of idle servers
NumberOfServers		Returns Integer number of servers
View	*WIPIndex*	Returns transaction for viewing

*Changes the facility state.

However, while engaged, the transaction may *ReceiveServers* and *ReturnServers*. It is possible that the transaction may never schedule a service completion in that facility. For example, a transaction that occupies a facility represents a customer using a shopping cart. Rather than receiving service in the facility, a series of delays models the customer's shopping pattern. In this situation, the transaction is on the internal facility *WIP List* but moves on and off the PEL because of the delays.

EngageAndService is the more traditional view of service in a facility. The transaction is allocated servers and goes directly into service (i.e., schedules a service completion on the PEL). A workstation is an example of a component engaging a facility and immediately beginning service. The default service time is from spreadsheet *InitialModel*. The *ServiceTime* option of *EngageAndService* enables the user to specify a service time that is transaction specific.

Service schedules a service completion in the facility. From the definition, the transaction must have previously acquired servers. As in *EngageAndService*, *ServiceTime* is a transaction-specific option. The transaction servers may have been acquired by *Engage*, or it may have previously completed an *EngageAndService* and is now ready for another *Service*. *NextFacility* schedules

Table 11.7 DEEDS Subs for Class Facility

Name	Parameters	Description
Disengage	Transaction	Transaction disengages server(s) and facility
Engage	Transaction, NumberServers	Transaction engages server(s) but not scheduled for service
EngageAndService	Transaction, NumberServers, [MyServiceTime]	Transaction engages server(s) and scheduled on PEL
NewDistribution	Name, P1, P2, P3	Next arrival scheduled from this distribution
NewNumberOfServers	NumberOfServers	Resets *NumberOfServers*
NextFacility	Transaction, NumberServers, Queue	The transaction engages the facility or joins the queue when servers are not available
ReceiveServers	Transaction, NumberRequested	Transaction receives additional server(s)
ReturnServers	Transaction, NumberReturned	Transaction returns server(s) and maintains its position in the facility
Service	Transaction, [MyServiceTime]	Transaction scheduled on PEL

the transaction for service when sufficient servers are available; otherwise, the transaction joins the queue.

The most common example of a transaction, *Customer*, acquiring a server and receiving service is shown in the following example:

```
F1Facility.NextFacility Customer, 1, Q1Queue
```

The above statement is the same as

```
If F1Facility.NumberOfIdleServers > 0 _
Then
    F1Facility.EngageAndService Customer, 1
Else
    Q1Queue.Join Customer
End If
```

A server in facility *F1Facility* engages *Customer*. The *Simulator* schedules a service completion event for facility *F1Facility* at time equal to *CurrentTime* + *ServiceTime*. If no *F1Facility* server is available, then *Customer* joins *Q1Queue*. If the program attempts to have the facility engage a customer without verifying that servers are available, the program is terminated with an appropriate simulator error message indicating which facility was involved and the time when the error occurred. When two servers are needed, the above would become

```
F1Facility.NextFacility Customer, 2, Q1Queue
```

The next example demonstrates the *Engage* and *Service* procedures respectively. Assume a bank teller services two customer lines. For modeling purposes, the idle teller resides in a teller queue. When the teller queue is empty the teller is servicing one of the two windows. An arriving transaction engages an empty window or joins the queue. If the arrival engages the facility (window) and the teller queue is empty, the customer must wait to begin service. The program segment to model the customer arrival to line 1 is as follows:

```
If F1Facility.NumberOfIdleServers > 0 _
Then
  F1Facility.Engage Customer, 1
  If TellerQueue.Length > 0 _
  Then
    TellerQueue.Remove 1,1
    F1Facility.Service Customer
  End If
Else
  Q1Queue.Join Customer
End If
```

The customer in *F1Facility* gains control of the teller by removing the teller from the teller queue. Otherwise, the customer in *F1Facility* must wait. As shown in the following, when the customer in *F2Facility* is finished, the teller services the customer in *F1Facility* or returns to *TellerQueue*.

```
If F1Facility.NumberBusyServers > 0 _
Then
  Set Teller = Nothing
  Set Customer = F1Facility.View(1)
  F1Facility.Service Customer
Else
  TellerQueue.Join Teller
End IF
```

When *F1Facility.NumberBusyServers* > *0*, a customer in *F1Facility* is waiting for service. To identify that transaction, *View (Index)* returns a transaction reference variable. The reference variable indicates to *F1Facility* which customer to service. A transaction joins *TellerQueue* when *F1Facility.NumberBusyServers* = *0*, to indicate that the teller is idle.

During the course of the simulation, a transaction may acquire additional servers or return unneeded servers. Each facility has an internal list of transactions that have engaged its servers. From that internal list, the facility also knows how many servers are engaged by each transaction. An invalid attempt by a transaction to *ReceiveServers* that are not available or *ReturnServers* that are not engaged will cause the program to terminate with an appropriate error message.

In the following example, assume a component engages two *F1Facility* servers for service and then returns a server before additional service in the same facility. After the second service, the transaction exits the facility. Attribute 1 equals 2 when the first service begins and is decremented after each service completion. The following example processes the service completion event for *F1Facility*:

```
Customer.Assign 1, Customer.A(1) - 1
If Customer.A(1) = 1 _
Then
  F1Facility.ReturnServer Customer, 1
  F1Facility.Service Customer
Else
  F1Facility.Disengage Customer
  ' route to next destination
End If
```

Notice that sub *ReturnServer* must include an identifier for the transaction that returns servers because each transaction has a counter for the number of servers it occupies. *Disengage* is also introduced in the above example. After the second service, the customer disengages the facility and all servers are returned because the facility knows how many servers the transaction controls. A variation of the example is to receive another server after the first service; the program must verify that another server is available before the *ReceiveServers* request.

Function *Interrupt* removes a selected transaction from the facility. The program must identify which transaction will be interrupted and a destination for the interrupted transaction. DEEDS automatically returns all servers of interrupted transactions to idle. A program segment to shut down workstation *F1Facility* for maintenance and move the interrupted component to the front of *Q1Queue* where other components are waiting to be serviced by *F1Facility* is as follows:

```
Set Component = F1Facility.Interrupt(1)
Q1Queue.Front Component
```

The above can be compactly written as

```
Q1Queue.Front F1Facility.Interrupt(1)
```

The following program segment interrupts all transactions in the facility:

```
Index = F1Facility.Length
For Jndex = 1 to Index
  Q1Queue.Front F1Facility.Interrupt(F1Facility.Length())
Next Jndex
```

Notice that the interrupted transaction always corresponds to the last transaction in the facility, and therefore transactions are queued in the same order that they engaged *F1Facility*.

Preemption is another common modeling situation. Simply stated, preemption occurs when a transaction in service is forced to relinquish its server(s) and is replaced by another transaction. A preemption example is a telephone customer put on hold because of the arrival of a higher-priority call. Preemption is modeled in DEEDS by a combination of *Interrupt* and either *Engage* or *EngageAndService*. The preempted transaction is interrupted and the preempting transaction engages the facility.

The model must ensure that a sufficient number of servers are available before engaging the preempted facility. Before preemption occurs, the program determines which transaction is to be preempted. Recall that *View* provides access to a transaction in the facility. In the following program segment a transaction with attribute 1 value of "A" in *F1Facility* is preempted. Assuming that the transaction in service occupies only one server and the preempting transaction also needs only one server, the program must find the transaction in the work-in-progress (WIP) collection of the facility and then the transaction is preempted.

```
Found = False
For WIPIndex = 1 to F1.Length
  Set CustomerInService = F1Facility.View(WIPIndex)
  If CustomerInService.A(1) = "A" _
  Then
     Found = True
     Exit For
  End If
Next WIPIndex
If Found = True _
Then
   Q1Queue.Front F1Facility.Interrupt(WIPIndex)
   F1Facility.EngageAndService NewCustomer, 1
End If
```

The preempted transaction is placed to the front of *Q1Queue* and, assuming that *F1Facility* receives transactions from *Q1Queue*, then the preempted transaction will be the next transaction to begin service.

Suppose in the previous example that the preempted transaction has more than one server and the preempting transaction needs three servers and the index, *WIPIndex*, into the facility for the transaction to be preempted was previously determined. The program compares the number of servers needed by the preempting transaction with the number of idle servers and the servers controlled by the transaction to be preempted using the following:

```
If F1Facility.NumberOfIdleServers + _
   F1Facility.WIP(WIPIndex).NumberOfServers > 2 _
Then
   Q1Queue.Front F1Facility.Interrupt(WIPIndex)
   F1Facility.EngageAndService NewCustomer, 1
Else
   ' Route NewCustomer to alternate destination
End If
```

F1Facility.WIP(WIPIndex).NumberOfServers returns the number of servers engaged by the transaction to be preempted. *NewNumberOfServers* changes the number of servers in the facility. For example, the following changes the number of servers in *F1Facility* to 3:

```
F1Facility.NewNumberOfServers 3
```

NumberOfNewServers may reduce the number of servers below the number currently engaged. Transactions keep their servers; however, disengaged servers are removed from the system until the total number of busy servers is equal to the number of servers in the facility. The modeler may choose to immediately remove servers using the function *Interrupt* with a criterion for an interruption. In the following example, the number of facility F1 servers is decremented by 1:

```
F1Facility.NumberOfNewServers F1Facility.NumberOfServers - 1
```

The previous description of *NewDistribution* for sources in Section 11.4.1 is also applicable to facilities.

11.4.4 Delay

Delays may be thought of as an infinite capacity facility. The functions in Table 11.8 are *Delay* class methods. With the exception of *Interrupt*, these functions assess the state of the delay node. Notice that the delay node *Interrupt* has features identical to a facility node *Interrupt*. The subroutines in Table 11.9 are methods that alter a delay node definition or affect transactions using the node. The only parameter for delays in spreadsheet *InitialModel* is the distribution. Changes to a delay distribution are identical to changes for source and facility nodes.

Start initiates the transaction delay. When the delay starts, the transaction is placed, similar to facility, in the Collection of delay node transactions. When the transaction completes the delay, DEEDS removes the transaction from the PEL and delay node Collection. Normally, delay time is from the distribution in the *InitialModel* spreadsheet. However, *Start* also includes an optional delay time parameter.

A program segment to *Start* a transaction delay for a cure process is

```
CureProcess.Start Component
```

The following example includes the optional delay time:

```
CureProcess.Start Component 10
```

Table 11.8 DEEDS Functions for Class Delay

Name	Parameters	Description
EventType		Returns String with *EventType*
Length		Returns Integer number of transactions in the *Delay*
Location	*AttributeIndex*, Relational operator, Comparison value	Returns index of the first transaction that satisfies the condition
Interrupt*	*DelayIndex*	Returns transaction removed from the *Delay*
Name		Returns String with *Delay* name
View	*DelayIndex*	

*Changes the delay state.

Table 11.9   DEEDS Subs for Class Delay

Name	Parameters	Description
NewDistribution	*Name, P1, P2, P3, P4*	Next delay scheduled from this distribution
Start	*Transaction*, [MyDelay]	Transaction scheduled on PEL

11.4.5 Transaction

Recall from Figure 7.4 that transactions move between lists that represent facilities, queues, delays, and the PEL. A single transaction may be on several lists. For example, a transaction on a delay node list is also on the PEL. Also, when a transaction engages a facility but is not scheduled for a service completion, it is on the facility list but not on the PEL. When a transaction is scheduled for service completion, it is on the PEL and the internal facility list.

Recall from Figure 5.2 that the state of the system changes as a primary event causes a transaction to move between these lists. As a minimum, the state of a transaction's origin and destination node (list) are altered. Changes to transaction parameters also affect the state of the system.

The transaction state is often a criterion for decisions that affect state changes to nodes, variables, and other transactions. Examples in Chapter 16 demonstrate the use of transaction attributes as the basis for synchronizing and assembling transactions. The functions in Table 11.10 are *Transaction* class methods. With the exception of *Copy*, these functions assess the state of a transaction. The *NewTransaction* procedure discussed in Section 11.3 creates a new transaction. In some situations, it may be easier to start with a copy of a transaction and change a few attributes. DEEDS has the function *Copy* that returns a duplicate of the specified transaction, with the exception of *SerialNumber* and *CreationTime*. These transaction parameters are unique. Because *Copy* returns a transaction reference, the *Set* assignment must be used. For example, after the following program segment, there are duplicate transactions, *ComponentA* and *ComponentB*.

```
Set ComponentB = ComponentA.Copy
```

Similar to the *NewTransaction* function, *ComponentB* is assigned a *SerialNumber* and a *CreationTime* equal to the time the statement was executed. All other parameters and attributes are the same as those in *ComponentA*.

TransitTime is the time for the transaction to travel between two program segments. Assume that in a previous segment the transaction attribute was assigned *CurrentTime*. *TransitTime* will return the difference in *CurrentTime* and the value of the specified attribute. For example, *CurrentTime* was recorded in attribute 1 when a vehicle chassis is placed on an assembly line. When the vehicle transaction comes off the assembly line, the following expression is the length of time on the assembly line:

```
CycleTime = Vehicle.Transit 1
```

SystemTime is similar to *TransitTime*. However, *SystemTime* is lapsed time since the transaction entered the system. Because each transaction has its creation time, there are no *SystemTime*

Table 11.10 DEEDS Functions for Class Transaction

Name	Parameters	Description
A	*AttributeIndex*	Returns Variant value of attribute
Copy*		Returns copy of transaction with different *SerialNumber* and *CreationTime*
CreationTime		Returns time the transaction entered the model
EventType		Returns String of *EventType*
Priority		Returns Byte of transaction *Priority*
SourceID		Returns String of *Source*
SystemTime		Returns *CurrentTime – CreationTime*
TimeStamp		Returns last time the transaction was associated with a primary event
TransID		Returns Integer transaction number for *Source*
TransitTime	*AttributeIndex*	Returns *CurrentTime – ValueOfAttribute* (AttributeIndex)

*Changes the transaction state.

Table 11.11 DEEDS Subs for Class Transaction

Name	Parameters	Description
Assign	*AttributeIndex, AttributeValue*	Assigns attribute the attribute value
DimAttributes	*Number*	Creates *Number* attributes
NewEventType	*EventType*	Resets transaction *EventType*
NewPriority	*Priority*	Resets transaction priority
NewTimeStamp	*TimeStamp*	Resets transaction time stamp

parameters. For example, if the vehicle chassis went onto the assembly line immediately upon arrival into the system, the following expression is also the vehicle's transit time since entering the system.

```
CycleTime = Vehicle.SystemTime
```

The methods in Table 11.11 alter a transaction.

The default number of transaction attributes is 10. However, that be increased using *DimAttributes*. For example, the following statement changes the number of *Customer* attributes to 15.

```
Customer.DimAttributes 15
```

Because the attributes are the Variant data type, any value can be saved in an attribute. The flexibility of Variant data requires that the user be extremely careful to use attributes consistently. For example, if the program sums over attribute 1 for several transactions, then the attribute value must be defaulted to 0 or explicitly assigned a numeric value. Any other data type causes a run-time error condition.

Several examples of assigning transaction attributes are as follows. *Orders* are received from sources *S1* and *S2*. The order quantity is dependent on the source. The following program segment determines an order quantity on the basis of the transaction source and saves the order quantity in attribute 2.

```
If Order.SourceID = "S1Source" _
Then
    Order.Assign 2, UniformDiscrete 5, 10, 1
Else
    Order.Assign 2, Poisson 4, 1
End If
```

Order is a transaction reference variable. As will be discussed in Section 11.5, the value 1 refers to the first random number stream.

Vehicles arriving at a car wash choose between the premium and deluxe service. However, 40% choose the premium service that includes cleaning the interior. The following program segment randomly assigns the service type to arriving vehicles in attribute 3:

```
If Random(1) < 0.40 _
Then
    Vehicle.Assign 3, "Premium"
Else
    Vehicle.Assign 3, "Deluxe"
End If
```

Because attributes are Variant data types, the following are also valid assignment statements for transaction *ArrivingCustomer*:

```
ArrivingCustomer.Assign 1, 27.3
ArrivingCustomer.Assign 2, "Red"
```

Attribute 1 is assigned a value of 27.3 and attribute 2 is assigned the value "Red." In the following example, attribute 1 of *Customer* is randomly assigned a color:

```
Select Case Random(1)
    Case < 0.25
        Customer.Assign 1, "Red"
    Case 0.25 to 0.49
        Customer.Assign 1, "Yellow"
    Case 0.5 to 0.74
        Customer.Assign 1, "Blue"
    Case>= 0.75
        Customer.Assign 1, "Violet"
    Else
End Select
```

Transactions are assigned a priority when they originate from a source. *NewPriority* enables the model to alter the transaction priority without affecting the priority associated with the source. Permissible values for priority are 1–127. For example,

```
Customer.NewPriority 1
```

changes the priority of *Customer* to 1.

As described in Section 9.5, for more complex models it may improve model documentation to use VB constants to reference transaction attributes. Using the following job attributes definition,

```
' job attributes
Private Const _
  ID As Integer = 1, _
  Destination As Integer = 2
```

examples of references to attributes are

```
Clerk.Assign Destination, "Warehouse"
```

```
If Clerk.A(Destination) = "Warehouse" _
Then
  ' statements
End If
```

In both examples, *Destination* references attribute 2. Attributes may also be used for indirect addressing of nodes. For example, in the following program statement,

```
F(Clerk.A(ID)).Service Clerk, ProcessingTime
```

F() is an array of facilities and the transaction, *Clerk*, is serviced by the facility with array position given by the *Clerk*'s value of attribute 1. If *Clerk*'s attribute 1 is 3, *Clerk* is serviced by *F(3)*.

As discussed in Section 8.1.2, each transaction on the PEL has an *EventType* that uniquely maps the transaction to a source, facility, or delay primary event. Network node procedures have internal controls that assign the *EventType* and *TimeStamp*. Therefore, *NewEventType* and *NewTimeStamp* cannot be used to affect a transaction that resides in a facility or delay.

These methods enable the modeler to define an *EventType* and associated *TimeStamp* for an event that may not be conveniently represented with a facility or delay node. The modeler must define an event handler for the associated primary event, modify *Parser* to recognize the *EventType*, and schedule the transaction on the *PEL*.

The following sub inserts the transaction on the PEL:

Simulator.InsertOnPEL Transaction

Only advanced users should consider these options. In Chapter 24 is an example of complex modeling situations that use these features. The point is that users must be very careful when using these features because misuse can cause unpredictable results that may be difficult to detect and correct.

11.4.6 Statistic

Statistic is a DEEDS class for measuring user-defined statistical performance measures. Recall from Section 5.4 that statistical variables are either observation-based or time-based. In the first case, the sample average equals the sum of observations divided by the number of observations, and in the second, the average is obtained by dividing the (time-weighted) area under the curve by the time base. Statistical variables must be defined in the *InitialModel* spreadsheet. Table 11.12 shows class methods characterized as time-based, observation-based, or both.

It is incumbent on the modeler to ensure that the program collects statistical data. *Collect* records data, as requested by the model. For an observation-based statistic, *Collect* actually records an observation. For a time-based statistic, *Collect* changes the current value of a time-based variable. The simulator records the observation during the data collection phase of the simulation described in Figure 5.2.

As described in the Section 11.4.5, *TransitTime* and *SystemTime* are performance measures for transactions. Suppose *CycleTime* is defined in the vehicle assembly line example from the previous section to be a statistical variable. The assignment statement

```
CycleTime = Vehicle.Transit 1
```

becomes

```
CycleTime.Collect Vehicle.Transit 1
```

where *CycleTime* is defined as an object reference for an observation-based statistic. An example of a general expression for an observation-based statistical variable is

```
CycleTimeStatistic.Collect Vehicle.A(1) - CurrentTime
```

An example of updating a time-based statistical variable is

```
InventoryLevelStatistic.Collect InventoryLevel
```

where *InventoryLevel* has been defined in the *InitialModel* spreadsheet as a time-based statistical variable. In this example, the modeler is notifying *Simulator* that *InventoryLevel* has changed. Notice the name of a statistical variable reference is simply the statistic name appended to *Statistic*.

Table 11.12   DEEDS Procedures for Class Statistics

Statistical Variable	Observation Based	Time Based
Average*	X	X
BetweenArrivals	X	
Collect	X	X
Name*	X	X
List(Trials)	X	

*Denotes a function.

BetweenArrivals is a measure of the lapsed time between successive arrivals of transactions at the point in the network (program segment) where *BetweenArrivals* is computed. For example, to record statistical information on system balks, the following sub is called each time a balk occurs:

```
CustomerBalks.BetweenArrivals
```

CustomerBalks must be defined in the *InitialModel* spreadsheet as an observation-based statistic.

The *List* procedure produces observations of statistical data on queues and user-defined statistics in a form that is easily accessible by Excel's *Histogram* program. The output consists of 10 equal-size *Bins* for grouping the data and a maximum of *Trials* observations. The advantage of using Excel's *Histogram* program is that the user can experiment with *Bins* to find the "best" histogram and then edit the histogram for a professional presentation.

The following segment produces a *CycleTime* observation list in the *UserStatistics* spreadsheet. The first observation may be collected after *ObservationNumber* 1 begins and continues for 100 trials during the observation.

```
If ObservationNumber > 0 _

Then
```

The accumulation of statistical variable data is independent of the *List* subroutine. The subroutine simply allows for recording observations in the spreadsheet.

In addition to observation-based and time-based, there are end-of-run statistics collected only once at the termination of the simulation. This type of variable is useful for, among other things, collecting cumulative counts as well as percentages. In Chapter 12, end-of-run statistics are defined and written to the *UserOutput* spreadsheet from the *ShutDown* procedure.

11.4.7 PDF

Presented in Section 11.5 are functions for several well-known probability distributions described in Chapter 4 as mathematical functions. As discussed in Section 4.2.2, empirical probability distributions are developed from histograms to represent either a discrete or piecewise continuous random variable. The concept is illustrated in Figures 11.2 and 11.3.

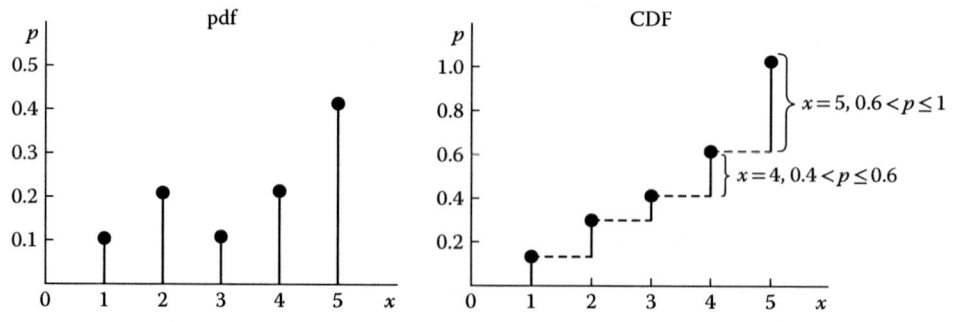

Figure 11.2 Example of discrete random variable.

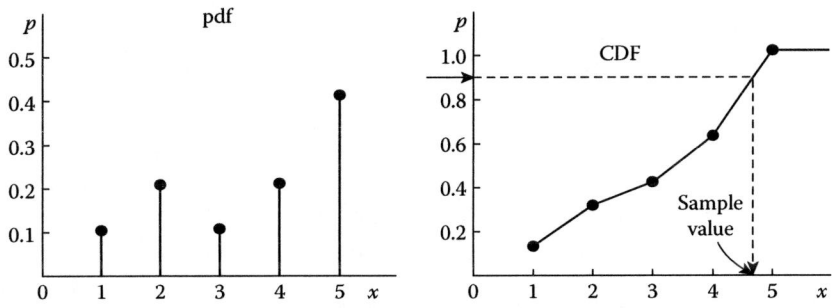

Figure 11.3   Example of piecewise continuous random variable.

Notice in the discrete pdf, the discrete value of the random variable is associated with a range of values for the (0, 1) random number. In the piecewise continuous case, a sample value is determined by a single value of the (0, 1) random number using linear interpolation.

The piecewise continuous random variable may yield incorrect results unless p is defined for the entire range (0, 1). As shown, the function is referenced about the point (0, 0) so that interpolation will be possible anywhere in the range $0 \leq p \leq 1$. However, the value of the random variable associated with $p = 0$ need not be 0. The only method for class *PDF* is the function given in Table 11.13.

RnID is an optional parameter and the default is *RnID* = 1. As shown below, these user-defined distributions are defined in the *PDF* spreadsheet. With x_m as the value of the *m*th random value in the pdf, $p(x_m)$ is the probability associated with random value x_m. The general format for both discrete and piecewise continuous functions is as shown below. The following distribution is used to assign a job type to arriving transactions:

Name	JobType	
Type	Discrete	
RnID	1	
Comment		
Values	*x*	*p(x)*
	1	0.2
	2	0.3
	3	0.4
	4	0.1

The reference variable name for a pdf is the distribution name from the *PDF* spreadsheet appended to "PDF." A valid reference for the above distribution is

```
Customer.Assign JobType, JobTypePDF.RV
Customer.Assign JobType, JobtypePDF.RV(2)
```

Table 11.13   DEEDS Function for Class PDF

Name	Parameters	Description
RV	*[RnID]*	Returns Single value of *PDF*

Table 11.14   DEEDS Function for Class Table

Name	Parameters	Description
F	TableIndex	Returns Single value of *f(x)*

Both statements sample from PDF *JobType* and *Assign* transaction *Customer*'s attribute *JobType*. In the first case *RnID* is defaulted to 1 and in the other, *RnID* is 2. Note that whether the pdf is continuous or discrete is dependent on the *PDF* spreadsheet specification.

11.4.8 Table

Similar to the idea of a PDF, a table is defined for a mathematical relationship that does not conform to a closed-form algebraic expression. The only method for class *Table* is the function in Table 11.14.

For a discrete table, if *TableIndex* does not equal a defined value for the independent variable, an error occurs and the program is terminated. When evaluating the function using linear interpolation, *TableIndex* must be between the minimum and maximum values of the independent variable.

A table lookup function has a format similar to the discrete pdf. x_m is the *m*th value of the independent variable in the table and $f(x_m)$ is the dependent value associated with x_m. The format for both discrete and piecewise continuous tables is as shown in the table definition below. The following is from the *Table* spreadsheet:

Name	InCraneCorridor	
Type	Discrete	
Comment	In Corridor Crane Map	
Values	*x*	*f(x)*
	1	15
	2	19
	3	23

The reference variable name is the table name from the *Tables* spreadsheet appended to "Table." A valid reference for the above table is

```
Pallet.Assign Destination, InCraneCorridorTable.F(1)
Pallet.Assign Destination, InCraneCorridorTable.F(3)
Pallet.Assign Destination, InCraneCorridorTable.F(4)
```

Examples 1 and 2 return 15 and 23, respectively. The third example results in an error message because *TableIndex* is outside the defined range of the independent variable.

When the table is defined as continuous, the method performs a linear interpolation of a piecewise continuous function. The piecewise continuous function may also yield incorrect results unless the entire range of *x* is well defined. As shown, the function is defined using $(f(x), x)$, so that interpolation will be possible anywhere in the range of *x*.

11.5 Distribution Functions

DEEDS provides several mathematical distributions commonly used in discrete event simulation modeling. Because these procedures are functions, they can be used as variables in assignment statements. Table 11.15 presents the distributions and their respective function parameters. It may

Table 11.15 Probability Distribution Functions in DEEDS

Variable	*Description*[a,b]
Exponential (P1, [RnID])	A sample from *Exponential* pdf with mean P1 given *RnID*
Poisson (Pl, [RnID])	A sample from *Poisson* pdf with mean P1 given *RnID*
Beta (Pl, P2, [RnID])	A sample from *Beta* pdf with parameters P1 and P2 given *RnID*
Gamma (P1, P2, [RnID])	A sample from *Gamma* pdf with parameters P1 and P2 given *RnID*
Lognormal (P1, P2, [RnID])	A sample from *Lognormal* pdf with mean P1 and standard deviation P2 given *RnID*
Normal (Pl, P2, [RnID])	A sample from *Normal* pdf with mean P1 and standard deviation P2 given *RnID*
UniformContinuous (Pl, P2, [RnID])	A sample from *UniformContinuous* pdf in the interval P1–P2 given *RnID*
UniformDiscrete (Pl, P2, [RnID])	A sample from *UniformDiscrete* pdf in the interval P1–P2 given *RnID*
Weibull (P1, P2, [RnID])	A sample from *Weibull* pdf with parameters P1 and P2 given *RnID*
Triangular (Pl, P2, P3, [RnID])	A sample from *Triangular* pdf in the interval P1–P3 with mode P2 given *RnID*
Random ([RnID])	(0, 1) Pseudo-random number using random stream *RnID*

[a] Parameters P1, P2, and P3 are Single or Integer constants or variables.
[b] The parameter RnID must assume one of the integer values ±l, ±2, ..., ±50 corresponding to DEEDS' 50 random streams. A negative RnID generates antithetic random (0, 1) values and an error when RnID is outside the specified range. The default value of RnID is 1.

help to reference Chapter 4 for details on the parameters for each distribution. These functions are in module *Distribution. RnID* is an optional parameter and the default is 1.

Examples of expressions from statistical functions are as follows:

```
Distribution.Normal 20,5,2
Normal 20,5
```

In both examples, the mean is 20 and standard deviation is 5. In the first case, the program specifically has *RnID* equal to 2; in the other example, the default *RnID* is 1. Including *Distribution* in the program enables the user to select from a menu of distributions. However, as discussed in Section 11.3, *Distribution* is an optional reference.

11.6 Visual Basic Functions

VBA has a variety of functions that can be incorporated into the simulation model. To view the classes and functions within each class, switch from Excel to the Visual Basic Editor (VBE). In VBE select View → Object Browser → VBA, as shown in Figure 11.4. The left column presents

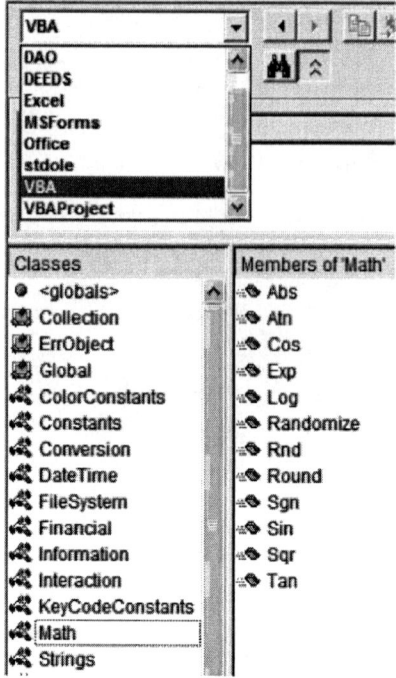

Figure 11.4   VBA functions available to DEEDS.

VBA classes. To the right are members of each class. For example, member functions of class *Math* are shown in the right column.

An example of a *Math* function in an assignment statement is as follows:

```
y = Math.Sqr(x)
```

Notice that *Conversion, Financial,* and *Strings* are examples of other VBA classes with functions that may be useful in simulation modeling.

11.7 Excel Worksheet Functions

In a manner similar to finding VBA functions, from the View Object Browser select Excel, as shown in Figure 11.5. Scroll down to *WorksheetFunction*. There are over 100 Excel functions available for use in a simulation program.

An example of a *WorksheetFunction* in an assignment statement is

```
z = WorksheetFunction.Max(x, y)
```

Several problems and their models that demonstrate programming techniques discussed in this chapter are now presented.

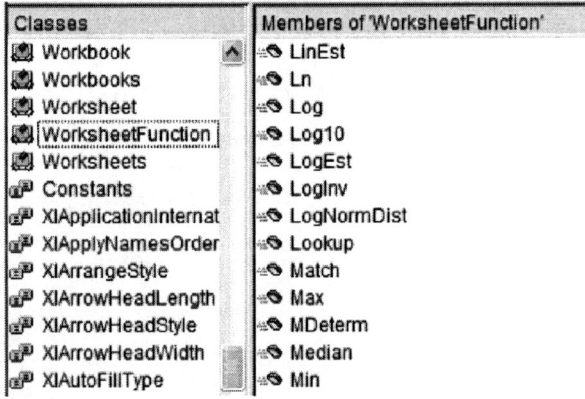

Figure 11.5   Excel worksheet functions available to DEEDS.

Name	Priority	Limit	Offset	Distribution	Parm#1	Parm#2	Parm#3	RnID
System	0	Infinite	0	Exponential	10	0	0	1

Name	Discipline	Capacity	Name	Nservers	Distribution	Parm#1	Parm#2	Parm#3	RnID
WashBay	FIFO	5	WashBay	1	UniformContinuous	10	15	0	1

Name	Type
CycleTime	O
TimeBetweenBalks	O

Figure 11.6   Car wash resource definitions.

Example 11.1

Cars arrive at a one-bay car wash every *Exponential* (10) minutes. Arriving cars wait in a queue that can accommodate five waiting cars. If the queue is full, arriving cars balk the system. The time to wash a car is uniformly continuous between 10 and 15 minutes. Performance metrics are statistics on the time a car spends in the system until it is washed and the time interval between successive balks.

The *InitialModel* spreadsheet data for this problem is in Figure 11.6. Notice that *CycleTime* and *TimeBetweenBalks* are observation statistics. The primary event subroutines are based on VBA code from *ProgramManager*.

There are two primary events, an arrival from *System* and service completion in *WashBay*. In sub *SystemSourceEvent* an *ArrivingVehicle* goes directly into service when the wash bay is idle. When the wash bay is busy, *ArrivingVehicle* joins the queue or balks the system if the queue is full. Time between balks is collected when *ArrivingVehicle* balks the system.

In Figure 11.7 the procedure for service completions is sub *WashBayFacilityEvent*. After a vehicle service is completed in the wash bay, *FinishedVehicle* disengages the server and exits the system after the statistical variable *CycleTime* has been collected. If there is a waiting transaction in *WashBayQueue*, it departs the queue, engages the facility, and is scheduled for service completion. Otherwise, the wash bay is idle.

Example 11.2

Buses arrive at a station every *Exponential* (120) seconds. The number of passengers on each bus is between 20 and 50, uniformly distributed. Between two and six passengers, uniformly distributed, leave the bus. Passengers arrive at the bus stop every *Exponential* (20) seconds. Each passenger leaves

```
Sub SystemSourceEvent(ByRef ArrivingVehicle As Transaction)
  If WashBayFacility.NumberOfIdleServers > 0 _
  Then
     WashBayFacility.EngageAndService ArrivingVehicle, 1
  Else
    If WashBayQueue.Full _
    Then
       TimeBetweenBalksStatistic.TimeBetweenArrivals
       Set ArrivingVehicle = Nothing
    Else
       WashBayQueue.Join ArrivingVehicle
    End If
  End If
End Sub
Sub WashBayFacilityEvent(ByRef FinishedVehicle As Transaction)
  WashBayFacility.Disengage FinishedVehicle
  CycleTimeStatistic.Collect FinishedVehicle.SystemTime
  Set FinishedVehicle = Nothing
  If WashBayQueue.Length > 0 _
  Then
     WashBayFacility.EngageAndService WashBayQueue.Depart, 1
  End If
End Sub
```

Figure 11.7   Car wash model procedures.

Name	Priority	Limit	Offset	Distribution	Parm#1	Parm#2	Parm#3	RnID
Bus	0	Infinite	0	Exponential	120	0	0	1
Passenger	0	Infinite	0	Exponential	20	0	0	1

Name	Discipline	Capacity	Name	Distribution	Parm#1	Parm#2	Parm#3	RnID
BusStop	LIFO	Infinite	Load	UniformContinuous	1	3	0	1
			Unload	UniformContinuous	2	3	0	1

Figure 11.8   Bus stop resource definitions.

the bus according to a continuous uniform distribution between two and three minutes. To board the bus the time is uniformly continuous between one and three minutes. The *InitialModel* spreadsheet data for sources, queue, and delays for this problem are in Figure 11.8.

In Figure 11.9, sub *BusSourceEvent*, a bus stops only when the bus stop is idle. When the bus arrives, the number of passengers on the bus and how many will leave the bus at the station are determined. Because passengers exit the bus one at a time, only the first departure is scheduled.

In sub *PassengerSourceEvent*, arriving passengers join the queue and wait to either board the bus or be removed from the system.

Sub *UnloadDelayEvent* in Figure 11.10 removes the person from the bus and schedules the next bus departure until there are no more departures. When there are no more departures, *UnloadDelayEvent* schedules the first person to board the bus, assuming that there are persons waiting to board the bus. Because there are no transactions waiting to exit the bus, a transaction is created to schedule a bus departure using *Simulator* function *NewTransaction*.

Sub *LoadDelayEvent*, also in Figure 11.10, boards the passenger and schedules boarding passengers until the bus is full or no more passengers are waiting. In either case, the bus exits the bus stop. As the bus departs the stop, all passengers that were unable to board the bus are tabulated and removed from the system.

To generalize the bus problem, the bus capacity is from the *UserInitialed* spreadsheet and summary data is from the *UserOutput* spreadsheet. Sub *ShutDown* and the related portion of sub *Initial* are shown in Figure 11.11.

```
Sub BusSourceEvent(ByRef ArrivingBus As Transaction)
  If BusOccupiesStop _
  Then
    NumberOfByPassBuses = NumberOfByPassBuses + 1
    Exit Sub
  End If
  BusOccupiesStop = True    ' available for unloading/loading
  NumberOnBus = Distribution.UniformDiscrete(20, BusCapacity, 1)
  NumberDepartingBus = Distribution.UniformDiscrete(2, 6, 1)
End Sub
Sub PassengerSourceEvent(ByRef ArrivingPassenger As Transaction)
    ' simply join queue
    '   1. first to bus stop and no bus
    '   2. bus available and someone boarding
  BusStopQueue.Join ArrivingPassenger
End Sub
```

Figure 11.9 Bus stop—*BusSourceEvent*.

```
Sub UnloadDelayEvent(ByRef DepartedPassenger As Transaction)
  UnloadDelay.Finish DepartedPassenger
  NumberDepartingBus = NumberDepartingBus - 1
  If NumberDepartingBus > 0 _
  Then
     UnloadDelay.Start DepartedPassenger
     Exit Sub
  End If
  Set DepartedPassenger = Nothing
  If BusStopQueue.Length > 0 _
  Then        ' begin boarding
    LoadDelay.Start BusStopQueue.Depart
    Exit Sub
  End If
  BusOccupiesStop = False ' no one waiting for bus
End Sub
Sub LoadDelayEvent(ByRef LoadedPassenger As Transaction)
  Dim _
    Index As Integer
  LoadDelay.Finish LoadedPassenger
  Set LoadedPassenger = Nothing
  If NumberOnBus = BusCapacity _
  Then
    BusOccupiesStop = False
    NumberOfAngryCustomers = BusStopQueue.Length
    BusStopQueue.Remove 1, NumberOfAngryCustomers
    Exit Sub
  End If
      ' bus not full
  If BusStopQueue.Length > 0 _
  Then
    LoadDelay.Start BusStopQueue.Depart
    Exit Sub
  End If
  BusOccupiesStop = False ' no more to board; bus departs
End Sub
```

Figure 11.10 Bus stop—*LoadDelayEvent* and *UnloadDelayEvent*.

```
' user defined initialization

BusCapacity = Worksheets("UserInitialed").Range("B1").Value
NumberOfAngryCustomers = 0
NumberOfByPassBuses = 0
BusOccupiesStop = False    ' bus not available when model begins
End Sub
Public Sub ShutDown()
  Worksheets("UserOutput").Range("A:Z").ClearContents
  Worksheets("UserOutput").Range("A1").Value = _
                              "NumberOfAngryCustomers"
  Worksheets("UserOutput").Range("B1").Value = _
                              NumberOfAngryCustomers
  Worksheets("UserOutput").Range("A2").Value = _
                              "NumberOfByPassBuses"
  Worksheets("UserOutput").Range("B2").Value = _
                              NumberOfByPassBuses
End Sub
```

Figure 11.11 Bus stop—initialization and sub *ShutDown*.

Example 11.3

Jobs arrive for processing at a machine every *Exponential* (15) minutes. Processing time is *Exponential* (10) minutes. Every eight hours the machine is shut down for maintenance. A job in the machine process is allowed to finish before maintenance begins. Time for maintenance is uniformly continuous between 15 and 20 minutes, and jobs continue to arrive during maintenance. The *InitialModel* spreadsheet data for source, facility, delay, and queue for this problem are in Figure 11.12. Notice that *CycleTime* is an observation statistic. Primary event subroutines are also based on VBA code from *ProgramManager*.

In addition to primary events for arriving and serviced jobs, there are primary events for the production and maintenance delay. A portion of sub *Initial* in Figure 11.13 shows that the model is initialized for the production cycle. An alternative to using sub *Initial* to initialize a delay node is to include a source node with a single arrival. However, that requires an extra primary event sub.

In Figure 11.14, jobs arrive in sub *S1SourceEvent* and are scheduled for service if the machine is idle and there is no maintenance activity. Otherwise, the arrival joins the machine queue.

In sub *WorkstationFacilityEvent*, also in Figure 11.14, the finished part disengages the workstation and exits the system after *SystemTime* has been collected. If the machine is blocked and waiting for maintenance, then maintenance begins; otherwise, normal operations continue. The invocation of sub *ProductionDelayEvent* in Figure 11.15 occurs at the end of the production cycle. If the

Name	Priority	Limit	Offset	Distribution	Parm#1	Parm#2	Parm#3	RnID
S1	0	Infinite	0	Exponential	15	0	0	1

Name	Nservers	Distribution	Parm#1	Parm#2	Parm#3	RnID
Workstation	1	Exponential	10	0	0	1

Name	Distribution	Parm#1	Parm#2	Parm#3	RnID
Maintenance	UniformContinuous	15	20	0	1
Production	Constant	480	0	0	1

Name	Discipline	Capacity
Workstation	LIFO	Infinite

Name	Type
CycleTime	O

Figure 11.12 Maintenance resource definitions.

```
      ' User initialized input
      'Intialize maintenance schedule
   ProductionDelay.Start NewTransaction 'ProductionCycleTimer
   InMaintenanceMode = False
   InShutDownMode = False  ' complete part before shutdown
End Sub
```

Figure 11.13 Maintenance—S1 Source and Workstation Facility Event.

```
Sub S1SourceEvent(ByRef RawMaterial As Transaction)
  If WorkstationFacility.NumberOfIdleServers > 0 And _
                                 Not InMaintenanceMode _
  Then
    WorkstationFacility.EngageAndService RawMaterial, 1
  Else
    WorkstationQueue.Join RawMaterial
  End If
End Sub
Sub WorkStationFacilityEvent(ByRef FinishedPart As Transaction)
  WorkstationFacility.Disengage FinishedPart
  CycleTimeStatistic.Collect FinishedPart.SystemTime
  Set FinishedPart = Nothing   ' exit the system
  If InShutDownMode _
  Then  ' start maintenance
    MaintenanceDelay.Start NewTransaction
    InMaintenanceMode = True
    InShutDownMode = False
    Exit Sub
  End If
    ' Continue operations
  If WorkstationQueue.Length() > 0 _
  Then
    WorkstationFacility.EngageAndService _
                              WorkstationQueue.Depart, 1
  End If
End Sub
```

Figure 11.14 Maintenance—*ProductionDelayEvent*.

```
Sub ProductionDelayEvent(ByRef ProductionCycleTimer As Transaction)
  Set ProductionCycleTimer = Nothing
  If WorkstationFacility.NumberOfbusyServers = 0 _
  Then
    InMaintenanceMode = True
    MaintenanceDelay.Start NewTransaction ' Maintenance
  Else
    InShutDownMode = True
  End If
End Sub
Sub MaintenanceDelayEvent(ByRef MaintenanceActivity As Transaciton)
  MaintenanceDelay.Finish MaintenanceActivity
  Set MaintenanceActivity = Nothing
  InMaintenanceMode = False
  ProductionDelay.Start NewTransaction 'ProductionCycleTimer
  If WorkstationQueue.Length() > 0 _
  Then
    WorkstationFacility.EngageAndService WorkstationQueue.Depart, 1
  End If
End Sub
```

Figure 11.15 Maintenance—*ProductionDelayEvent* and *MaintenanceDelayEvent*.

machine is idle, maintenance begins. Otherwise, the machine is in shutdown mode waiting for the current job to finish so that maintenance can begin.

Sub *MaintenanceDelayEvent*, also in Figure 11.15, is invoked at the end of the maintenance activities. Another production cycle delay is scheduled and if there is a job waiting for service, the machine is restarted.

11.8 Summary

Presented in this chapter is the flexibility of the DEEDS modeling environment. Details include DEEDS methods for altering states of sources, queues, facilities, and delays, as well as transactions. Other DEEDS methods initialize network nodes and assess the states of the nodes. Still others enable the user to model routing transactions and control the simulation. Also presented are features of Excel and VBA that enhance the capability of developing complex models. In Chapter 12 are additional *Simulator* procedures to assist with program testing and validation. In Chapter 16 there are additional DEEDS subroutines to assist with complex transaction routing schemes.

Problems

11.1. For the following statistical variable definitions:
   a. *Customer.SystemTime*.
      Suppose *Customer* entered the model at *CurrentTime* = 22. Determine the corresponding value of *SystemTime* at *CurrentTime* = 34.
   b. *Span.BetweenArrivals*.
      Suppose that the times when *Span* is computed are 10, 32, 45, and 57. Compute the resulting average value of *Span*.
   c. *InventoryLevel* is a time-based variable.
      Suppose that *InventoryLevel* is 0, 10, 20, and 10 at time = 10, 15, 18, and 22, respectively. Compute the average for *InventoryLevel*.
11.2. For the following VBA data declarations,

```
Private _
   CustomerTypeTable As Table, _
   ServiceTimePDF As PDF, _
   Customer As Transaction
      ` CustomerType Table has service time by customer id
      ` attribute 1 has customer id
```

indicate errors, if any, in each of the following mathematical expressions:

   a. ServiceTime = Exponential (10)
   b. ServiceTime = ServiceTimePDF
   c. ServiceTime = ServiceTimePDF(1)
   d. ServiceTime = Exponential (CustomerTypeTable(Customer.A(1)))
   e. ServiceTime = Exponential (CustomerTypeTable(Customer.A(1))*60)
   f. ServiceTime = Exponential (CustomerTypeTable(Customer.A(1))*60,1)

11.3. Describe the source node in *InitialModel* and the corresponding sub in *EventsManager* for each of the following:

   a. Customers start to arrive at a bank five minutes after opening time. The interarrival time is *Exponential* (4) minutes. An arriving customer always prefers the teller with the shortest lane.

   b. Jobs arrive from a production line every *Exponential* (2) minutes. The first 10 jobs are usually defective and must be discarded.

   c. A bus arrives with 30 tourists at a rest area every *Exponential* (5) hours. Tourists leave the bus to be served lunch in the restaurant.

11.4. TV sets arrive at a two-inspector station for testing. The time between arrivals is *UniformContinuous* (3.5, 7.5) minutes. The inspection time per TV set is *Exponential* (9) minutes. On average, 85% of the sets pass inspection. The remaining 15% are routed to an adjustment station with a single operator. Adjustment time per TV set is *UniformContinuous* (8, 12) minutes. After adjustments are made, sets are routed back to the inspection station to be retested. Determine the total time a set spends in the system.

11.5. In Problem 11.4, show how one can determine the average number of times a given job is adjusted.

11.6. In Problem 11.4, suppose that a job is not allowed more than two adjustments. That is, a job needing a third adjustment is discarded. Determine the number of discarded jobs in a given simulated period.

11.7. For the following model:

Name	Priority	Limit	Offset	Distribution	Parm#1
S1		Infinite	10	Constant	4

Name	Discipline	Capacity
Q1	FIFO	2

Name	Nservers	Distribution	Parm#1
F1	2	Constant	3

```
Sub S1SourceEvent (ByRef Arrival As Transaction)
    F1Facility.NextFacility Arrival, 1, Q1Queue
End Sub
```

```
Sub F1FacilityEvent (ByRef FinishedJob As Transaction)
    Set FinishedJob = Nothing ' exit the system
    If Q1Queue.Length > 0 _
    Then
        F1Facility.EngageAndService Q1Queue.Depart, 1
    End If
End Sub
```

   a. Determine the maximum value of *Q1Queue.Length* + *F1Facility.Length* that may occur during simulation.

   b. When will the first transaction be created from *S1*?

   c. When will the first transaction leave *F1*?

   d. When will the second transaction leave *F1*?

11.8. Cars arrive every *Exponential* (18) minutes at a car-wash facility that also offers vacuum cleaning. It takes *Exponential* (12) minutes to wash and *Exponential* (15) minutes to vacuum clean. When a car arrives, it can go to either wash or vacuum cleaning, depending on which queue is shorter. After completing the first activity (wash or clean), the car must go to the remaining activity (clean or wash). Assuming infinite queue sizes, determine the average time a car spends in washing and in cleaning (two separate variables), as well as the time it spends in the entire facility (a third variable).

11.9. A car-wash facility includes four bays and has a waiting space that can accommodate six cars. It takes *Exponential* (10) minutes to finish a car wash. After a car is washed, it is cleaned by one of three vacuum cleaners. The waiting space in front of the vacuum cleaners can accommodate a maximum of four cars. It takes *Exponential* (6) minutes to finish the cleaning. Cars arrive every *Exponential* (5) minutes and will drive around the block only once if both the washing and vacuum-cleaning facilities are full. If upon return, the two facilities are still full, the car will drive away. Otherwise, the car will go to the facility having an empty space. If spaces are available in both facilities, the car will go to washing first. If a car is vacuum cleaned first, it will attempt washing next. Again, the driver is willing to go around the block once if no space is available. However, a car that does not find a space upon completing washing will simply leave without being vacuum cleaned. Compute the average time between balking before and after washing as categorized by those that arrive fresh from source and those that come from vacuum cleaning. Also, determine the average time in the system for those cars that go through both washing and vacuum cleaning.

11.10. A shop consists of two types of identical machining centers. Each center includes a total of two machines. Jobs arrive at the shop randomly every *Exponential* (3) hours. A job must be processed in both machining centers in a serial fashion. The processing times are *UniformContinuous* (8, 10) and *Normal* (16, 2) hours, respectively. The in-process inventory between the two centers must never exceed ten jobs. If the space is full, a job is not allowed to leave machining center 1 until a space becomes available. Determine the percentage of time machining center 1 is in a blocked state because the in-process buffer is full.

11.11. Consider a single-queue model where the facility with one server utilizes one unit of a given resource. The resource will stay in the facility as long as the queue is not empty. When the queue becomes empty, maintenance is performed on the resource. It takes *Exponential* (10) minutes to carry out the maintenance task. Assume that customers arrive at the facility every *Exponential* (15) minutes. The service time per customer is *Exponential* (12) minutes. Determine the average utilization of the resource as well as the percentage of time the facility is blocked because of the unavailability of the resource.

11.12. Jobs arrive every *Exponential* (2) minutes for processing on a single machine. The processing time is *Exponential* (1.2) minutes. After a job is completed, it must be inspected. Thirty percent of the jobs fail inspection and are reworked immediately. All three operations (process, inspection, and rework) are performed on the single machine. It takes *Exponential* (1) minutes to do the rework. Compute the total time a job spends in the system.

11.13. Rework Problem 11.12 assuming that the inspection per job is 30 seconds.

11.14. Jobs arrive in batches of 10 items each. The interarrival time is *Exponential* (2) hours. The machine shop contains two milling machines and one drill press. About 30% of the items require drilling before being processed on the milling machine. Drilling time per item is *UniformContinuous* (10, 15) minutes. The milling time is *Exponential* (15) minutes for items that do not require drilling, and *UniformContinuous* (15, 20) for items that do. Determine the utilization of the drill press and the milling machines as well as the average time an item stays in the system. All items are processed on a FIFO basis.

11.15. Jobs arrive from two separate sources. The first source creates jobs every *Exponential* (10) minutes, whereas the second source generates them every *Exponential* (12) minutes. A job may be processed on any one of the two machines. The first machine takes *UniformContinuous* (5, 8) minutes, and the second one takes *UniformContinuous* (10, 15) minutes. Arriving jobs always prefer machine 1 over machine 2, but will go to machine 2 if it is the only one available. When either of the two machines finishes its load, it always gets its next job from the longer queue. Determine the utilization of each machine and the time it takes to process each job (separately).

11.16. A cafeteria provides a salad bar and a sandwich bar. Customers arrive every *Exponential* (10) minutes. About 60% of them will visit both bars. The remaining 40% will choose the salad bar only. Each bar can accommodate a (moving) line of 15 customers. Collected data indicate that when the cafeteria is busy, a customer will leave the salad bar every *UniformContinuous* (2, 3) minutes. Similarly, a customer will leave the sandwich bar every *UniformContinuous* (4, 6) minutes. The salad bar is located ahead of the sandwich bar, and the space between the two can accommodate a line of four customers. If this space is full, a customer wishing to visit the sandwich bar must wait until a space becomes available. This action will also block all other customers in the salad bar line. Determine the average time needed for a customer to be served.

11.17. Jobs arrive at a machine every *Exponential* (20) minutes. It takes *Exponential* (16) minutes to complete the processing of each job. Every eight hours the tools of the machine must be replaced for resharpening. Replacement time is *Exponential* (5) minutes, after which time the machine may start processing again. Occasionally, every *Exponential* (48) hours, the machine will break down. Repair time is estimated at *Exponential* (1) hour. Estimate the net utilization of the machine.

11.18. The initial inventory level of an item is 200 units. Orders for the item arrive every *Exponential* (0.5) days. The size of the order is *Poisson* (5) units. Orders are processed on a FIFO basis. Those that cannot be filled immediately are usually backlogged. Every four weeks the stock level is reviewed. If the inventory position (on-hand minus backlogged) drops below 50 units, a replenishment order that brings the inventory to 200 units is placed. Delivery of the stock usually takes place a week later. Determine the average stock level, the average stock replenishment size, and the average inventory position.

11.19. Assume for Problem 11.18 that the inventory position is reviewed continuously. If it drops below 50 units, the stock is replenished immediately up to 200 units.

11.20. Two types of messages, I and II, arrive at a transmitting station every *Exponential* (2) and *UniformContinuous* (1, 2) minutes, respectively. The station has two transmitting lines. Type I messages have higher priority for using the lines but may not preempt another message that is already in progress. It takes *UniformContinuous* (0.5, 0.8) minutes to transmit a Type I message and *UniformContinuous* (0.6, 0.9) minutes to transmit a Type II message. Determine the net utilization of the lines.

11.21. Cars arrive for repair at a garage every *Exponential* (3) hours. The supervisor assigns jobs to the shop's two mechanics on a rotational basis. It takes *UniformContinuous* (1, 5) hours to do a repair job. About 20% of the time, after a repair job is completed, one mechanic may need the help of the other mechanic for a period of *UniformContinuous* (10, 30) minutes. In this case, the mechanic, if busy, will interrupt the repair job to consult with the other mechanic, after which time the original repair job may be resumed. Determine the percentage of time each repairman spends consulting with the other.

11.22. A shop is operated by two salespersons. Customers arriving every *Exponential* (20) minutes will be helped by the salesperson that may be free at the time. It takes *UniformContinuous* (10, 20) minutes to help a customer decide on a purchase. The salesperson then goes to the cash register to collect money from the customer before returning to help another customer. It takes about five minutes to complete this transaction. However, if the cash register happens to be busy, the salesperson must wait until it becomes available. Determine the percentage of time the salespersons stay busy and stay idle.

11.23. Jobs arrive at a single-machine shop in three priority classes, I, II, and III, every *UniformContinuous* (2, 3), *UniformContinuous* (3, 4), and *UniformContinuous* (5, 7) hours, respectively. Jobs of Type III have the lowest priority for processing and may be preempted by higher-priority jobs. Similarly, Type I jobs may preempt Type II jobs. Assuming that the respective processing times for the three types are *UniformContinuous* (1, 2), *UniformContinuous* (5, 5), and *UniformContinuous* (6, 8) hours, respectively, determine the net utilization of the machine by each job type.

11.24. Customers arrive at a barbershop with three barbers every *Exponential* (20) minutes. A haircut takes *UniformContinuous* (20, 30) minutes. There is a waiting area for eight customers. If a customer arrives and finds the place full, he will drive around the block for about *UniformContinuous* (10, 15) minutes before returning to the shop to try again. Determine the average time from the moment a customer arrives for the first time until he gets a haircut.

11.25. Rework Problem 11.24 assuming that only once will the customer return to the shop.

11.26. Telephone orders are received at a mail-order facility every *Exponential* (10) minutes. Telephone service is in operation 24 hours a day. Orders are filled by six clerks. It takes *Exponential* (30) minutes to have an order ready for shipping. The clerks work one eight-hour shift per day. However, a clerk may not leave an order unfinished at the end of the shift. Determine the average number of orders that remain unfinished at the end of each shift as well as the average time needed to complete these orders.

11.27. Assume in Problem 11.26 that the clerks will always leave at the end of a shift even if an order is not ready. Upon return the next morning, they start from where they stopped the day before. Determine the average number of orders that remain unfinished at the end of a day and the average time each of these orders takes to be completed.

11.28. Four trucks haul waste material from a collection center to the dumping grounds. Waste material is piled by a bulldozer at the rate of one pile every *Exponential* (4) minutes. It takes three piles to make one truckload. Two mechanical loaders, each operated by an operator, are used to load the trucks. Loading times for the two loaders are *Exponential* (15) and *Exponential* (12) minutes, respectively. The operator must spend five minutes between loads preparing the loader before attending to the next truck. The trip to the dumping grounds takes *Normal* (20, 3) minutes. The return trip for empty trucks takes *UniformContinuous* (18, 3) minutes. The unloading time per truck is *UniformContinuous* (2, 8) minutes. Compute the time period from the moment a waste pile is brought by the bulldozer until it reaches the dumping grounds.

11.29. Rework Problem 11.28 assuming that every 24 hours each truck has preventive maintenance and that the maintenance lasts *Exponential* (60) minutes.

11.30. Assume a model includes queues *Q1* and *Q2* and a facility *F1*. *Q1* is ordered on lowest value of attribute 2 and F2 has two servers. The contents of the three nodes are given below

Node	A(1)	A(2)	A(3)
Q1	1	2	4
	5	4	6
	3	7	2
	4	8	-3
Q2	1	5	9
	4	4	10
	3	5	2
	2	7	1
	5	8	4
F1	2	4	3
	5	1	6

Show the contents of the three files in each of the following (independent) cases:

a. Q2.JoinAfter 2, Q1.Detach(2)
b. Q1.JoinOnLowAttribute(2) = Q2.Detach(Q2.Length)
c. Q2.Join = Q1.Detach(Q1.Length)
d. Q1.Remove 1, Q1.Length
   Q2.JoinAfter 2, Q1.Detach(Q1.Length)
e. Do While (Q1.Length > 0)
      Q2.JoinAfter 1, Q1.Detach(Q1.Length)
   Loop
   Q2.Remove 2,1
f. Do While (Q1.Length > 0)
      Q2.JoinAfter 1, Q1.Detach(Q1.Length)
   Loop
   Q2.Remove 1,2
g. Do While (Q2.Length > 0)
      Q1.JoinOnLowAttribute 2, Q2.Depart
   Loop
   Q2.Remove 1,2
h. Set NewJob = Copy.F1.View(2)
   NewJob.Assign 2, 99
   Q1.JoinOnLowAttribute(2) NewJob

11.31. Jobs arrive at a machine every *UniformContinuous* (2, 6) minutes. It takes *Exponential* (4) minutes to process each job. Rush jobs arrive every *Exponential* (45) minutes and must preempt any regular job that might be in progress. However, a rush job may not preempt another rush job. It takes *Exponential* (5) minutes to process a rush job. When a job is interrupted, the operator spends about one minute removing it from the machine. Interrupted jobs are placed at the head of the regular jobs queue. Determine the average time regular and rush jobs spend in the system.

11.32. Two machines are used to sharpen tools. Machine 1 takes *Exponential* (6) minutes and machine 2 takes *Exponential* (8) minutes per tool. Operators prefer to take their tools to machine 1, but will use machine 2 if it is the only one available. It is estimated that tools arrive at the machines every *Exponential* (5) minutes. A special class of tools is given higher priority for processing and will interrupt machine 1 if both machines are busy. These tools arrive every *Exponential* (15) minutes. An interrupted tool must be placed at the head of the queue for machine 1, and its processing will start from the point where the tool sharpening was interrupted. Determine the average number of times a regular job is interrupted on machine 1.

11.33. Jobs arrive at a machining center every *Exponential* (10) minutes. The center has three machines. All jobs must go through machine 1, where they are processed for *Exponential* (3) minutes. As the jobs leave machine 1, about 30% of them will go to machine 2. The remaining 70% will be completed on machine 3. The processing times for the two machines are *Exponential* (5) and *Exponential* (3) minutes, respectively. Every *Exponential* (120) minutes, either machine 2 or machine 3 will break down with equal probability. In this case, the job on the broken machine will be discarded and the machine will be repaired in *Exponential* (40) minutes. During this repair period, all jobs leaving machine 1 will be diverted to the other (unbroken) machine. At the instant the broken machine is repaired, all jobs whose route was changed because of breakdown must now be rerouted to their originally intended machine. Determine the average number of diverted jobs during each machine's breakdown.

11.34. Consider the operation of a small manufacturing company. Every *UniformContinuous* (1, 1.5) minutes, one unit of product I and two units of product II are produced. Orders arrive every *Exponential* (5) minutes. About 40% of the orders request product I. The remaining 60% request product II. The sizes of the orders are decided according to the following distributions:

Product	Order Size (Units)	Probability
I	4	0.1
	5	0.4
	6	0.3
	7	0.2
II	5	0.3
	7	0.35
	9	0.15
	10	0.15
	12	0.05

It takes (1.5 + order size * 0.9) minutes to package an order. Backlog is allowed for up to 12 orders only. When the limit is reached, newly arriving orders are sent elsewhere. Determine the percentage of lost orders, the average time between lost orders, the total time needed to fill an order, the average stock level for each product, and the average number of orders waiting to be filled.

11.35. Orders in a mail-order facility are received by three telephone operators. The interarrival time for orders by three operators is *Exponential* (4.5), *Exponential* (3), and *Exponential* (2) minutes, respectively. Upon receipt, the operator enters the order information directly into a computer. All orders from the three operators are put in one file and filled on a FIFO basis. Orders come in four different types that require different packaging times. The following table summarizes the percentages of occurrence for each order type and its associated processing time:

Order Type	Probability	Packaging Time (Minutes)
1	0.2	UniformContinuous (4, 6)
2	0.4	UniformContinuous (2, 3)
3	0.3	UniformContinuous (3, 5)
4	0.1	UniformContinuous (7, 11)

In addition to packaging time, preparation of order shipping papers depends on the source from which the order is received. It is estimated that this task takes three, two, and one minute for orders received by operators 1, 2, and 3, respectively. Determine the average cycle time for orders processed by each operator. (Note: This problem is ideal for applying arrays of sources and statistics.)

11.36. Develop a single-queue model twice in one simulation session under the following conditions: In run 1, the interarrival time at the source is *Exponential* (10) minutes. In run 2, on the other hand, the interarrival time is *UniformContinuous* (15, 20) minutes. The service time distribution in both runs is triangle (5, 7, 10) minutes.

11.37. A commercial savings and loan opens at 8:00 a.m. and closes at 4:00 p.m. daily. The interarrival time for customers is *Exponential* (3) minutes. It takes *Exponential* (6) minutes to serve a customer by one of the bank's two tellers. At the closing time at 4:00 p.m., all the customers that are inside the bank must be served. Estimate the time needed to finish all the customers after 4:00 p.m.

11.38. Two types of subassemblies are produced in separate departments and are assembled to produce a heavy-duty machine. Subassembly A is produced every *Exponential* (5) hours and subassembly B every *Exponential* (6) hours. The assembly facility takes *Exponential* (4) hours per assembled unit. The assembled product is then prepared for shipping. It takes *Exponential* (1) hour to finish this task. Concurrently, the preparation of shipping papers and warranties, which are prepared in a different department, requires *Exponential* (2) hours. When both the machinery and the papers are ready, the unit is sent to a shipping area. Estimate the time for each unit in the factory.

Chapter 12

Simulation Output

12.1 Introduction

Of course, the purpose of a simulation is to obtain performance metrics for the system and system components. However, before drawing conclusions about the system performance, it is important to ensure that the model performs to specification (model verification) and the model accurately reflects the system being modeled (model validation). The various features of the simulation program output are important tools for model verification and validation. In this section, how to obtain output data from the simulation model is presented.

12.2 Gathering Observations

As described in Section 6.4, common methods for gathering observations in a simulation are

- Subinterval method
- Replication method

Briefly, the subinterval method divides a single run into equal time subintervals, each representing an observation. In the replication method, each run represents an observation. Design environment for event-driven simulation (DEEDS) allows the implementation of both methods through the use of control statements that define the *Number of Observations*, the *Run Length*, and the *Transient Time*. The parameters are on the *InitialModel* spreadsheet in Figure 12.1. By definition, lapsed time for each observation is

$$(Run\ Length - Transient\ Time)/Number\ of\ Observations$$

Recall from Section 6.4, the transient time is the warm-up period for the model at the start of a run. The default values for *Run Length, Number of Observations*, and *Transient Time* are infinity, 1, and 0, respectively. In either case, *Observation* 0 output corresponds to data collected during the transient time.

209

	A	B
1	Run Length	10000
2	Transient Time	1000
3	Number of Observations	9
4	Level of Confidence	0.95
5	Trace Report	Y
6	Start time	0
7	Stop time	50

	A	B
1	Run Length	2000
2	Transient Time	1000
3	Number of Observations	1
4	Level of Confidence	
5	Trace Report	Y
6	Run Simulation Start time	0
7	Stop time	50

Figure 12.1   Control parameters for simulation model.

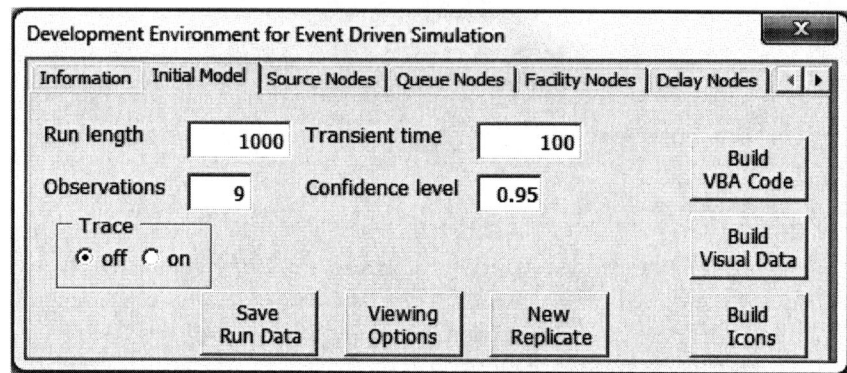

Figure 12.2   *ProgramManager—InitialModel* page.

The user selects the desired sampling method by specifying the number of observations. The first example demonstrates the subinterval method. There are nine observations of length 1000 time units, after a steady-state period of 1000 time units. The second example has a single observation of 1000 time units after a transient time of 1000 time units. Implicitly, the replication method assumes that there are changes to the system before repeating the experiment. Otherwise, the pseudo-random number generator produces the same results on each run.

Simply by invoking the *New Replicate* option shown in the *InitialModel* page of *ProgramManager* (Figure 12.2), the modeler is able to archive results of the current model and create new output spreadsheets for the next replicate. User-defined input and output spreadsheets are also archived. After revising the model parameters, the model is ready to conduct the next replication, which can be used to compare with results of other configurations. The replication tools are particularly helpful when using experimental design techniques for analysis. Notice that the subinterval method can be used within the replication.

12.3 Simulation Messages

DEEDS automatically produces *Simulator* messages that identify circumstances that may adversely affect the model validity or stability. For unusual conditions, the message is simply an advisory with a time stamp so that the programmer knows when the condition occurred. In Figure 12.3, for seven of the ten observations, there was no activity in the rework queue. Notice that this is not a program error condition because, as indicated, the simulation terminated successfully.

Table 12.3   DEEDS Facility Output Definitions

Cell	Definition
B1	Facility name
B2	Normal engagements
B3	Normal disengagements
B4	Engaged when simulation ends
B5	Interrupted engagements
B6	Unable to engage a server
B7	Start of the simulation
B8	During the simulation
B10–B13	Mean, standard deviation, upper and lower confidence interval for average busy servers
B17–C17	Transient response observation
B18–C21	Mean and standard deviation for each observation

	A	B	C
1	Facility	F1	
2	Number In	1917	
3	Number Out	1916	
4	Residing	1	
5	Interrupted	0	
6	Destroyed	0	
7	Minimum	0	
8	Maximum	1	
9		AvgBusy	Servers
10	Average	0.79	
11	StdDev	0.04	
12	LCL	0.78	
13	UCL	0.80	
14			
15	Observation	Avg	StdDev
16	0	0.82	0.38
17	1	0.85	0.35
18	2	0.70	0.46
19	3	0.84	0.37
20	4	0.77	0.42
21	5	0.78	0.41

Figure 12.7   DEEDS facility output.

Table 12.4   DEEDS Delay Output Definitions

Cell	Definition
B1	Delay name
B2	Starts
B3	Finishes
B4	Currently in delay
B5	Interrupted starts
B6	Minimum during the simulation
B7	Maximum during the simulation

	A	B	C
1	Delay	Storm	Tug
2	Number In	174	2278
3	Number Out	174	2270
4	Residing	0	0
5	Interrupted	0	8
6	Minimum	0	0
7	Maximum	1	1

Figure 12.8   DEEDS delay output.

12.6.4 Delay *Sheet*

Recall that a delay node is an "infinite" capacity facility. The transaction is automatically taken from an internal delay list by the *Simulator*. Therefore, a transaction that has finished the delay (i.e., not on the primary events list, or PEL) will not reside on the internal delay node list. Table 12.4 describes output in Figure 12.8.

12.6.5 Statistic *Sheet*

Table 12.5 describes output in Figure 12.9. For an observation-based variable, *Number In* is the actual number of occurrences. For example, average cycle time is based on the number of transactions that complete the cycle. For a time-based variable, *Number In* is the number of times the variable changes values during the simulation.

12.6.6 UserOutput *Sheet*

There are *End-of-Run* statistics collected at the termination of the simulation. These statistics are useful for, among other things, collecting cumulative counts as well as percentages. *End-of-Run* statistics are defined in sub *ShutDown* and written to the *UserOutput* spreadsheet. The user has complete control over the *UserOutput* spreadsheet. Typically, the user prepares a report outline and then writes results to the outline in Figure 12.10.

Table 12.5   DEEDS Statistic Output Definitions

Cell	Definition
B1	Statistic name and type
B2	Observations
B3	Lowest observation value
B4	Highest observation value
B6–B9	Mean, standard deviation, upper and lower confidence interval for statistic
B12–C12	Transient response observation
B13–C17	Mean and standard deviation for each observation

	A	B	C
1	Statistic	(O) SystemTime	
2	Number In	12776	
3	Min Value	5.00	
4	Max Value	51.88	
5			
6	Average	9.37	
7	StdDev	0.63	
8	LCL	9.20	
9	UCL	9.55	
10			
11	Observation	Mean	S
12	0	9.69	4.50
13	1	10.57	5.41
14	2	8.76	3.61
15	3	9.48	3.92
16	4	9.23	4.21
17	5	8.44	3.07

Figure 12.9   DEEDS statistic output.

	A	B
1		
2	**Summary Interrupt Information**	
3	Type of Interrupt	
4	Bay to Port	1
5	Port to Bay	4

Figure 12.10   DEEDS user-defined output.

```
Public Sub ShutDown()
  ' user defined shutdown
  Sheets("UserOutput").Range("B4:Z200").ClearContents
  With Worksheets("UserOutput").Range("B4")
    .Offset(0, 0) = PortToBayInterrupts
    .Offset(1, 0) = BayToPortInterrupts
  End With
End Sub
```

Figure 12.11   DEEDS—sub *Shutdown*.

As indicated in Figure 12.8, the standard delay node output details the number of interrupts that occur during the simulation. However, in the following example, we use the *ShutDown* subroutine to obtain additional information on the type of interrupt that occurred. In Figure 12.11, the report outline describes and formats data received from sub *ShutDown*. At the end of the simulation, sub *ShutDown* in Figure 12.11 completes the report. Notice that the *ClearContents* cleans around the report template leaving the first column and first three rows as formatted by the user. Beginning with Chapter 18, there are several other example programs that demonstrate *End-of-Run* statistics as well as other user types of defined output.

12.7 Model Verification

Table 12.6 presents *Simulator* methods available to the user as model verification tools. These methods enable the user to obtain additional details about simulator-maintained lists. *ViewPEL* is an interactive function for program testing and the others are subs that produce program output. The format for these methods is the same as those presented in Section 9.9. *ViewPEL* is discussed in Section 12.8.

12.7.1 User-Defined Simulator Messages

Section 12.3 presented the standard output of *SimulatorMessage*. However, the simulator also enables the modeler to define special conditions and use the *SimulatorMessage* sub to indicate that the condition(s) occurred. *SimulatorMessage* includes the user message and appends the time when the event occurred to the modeler's string message. The format is

```
If ConditionOccurs _
Then
  SimulatorMessage "Condition Description"
End If
```

By including *End*, the following segment terminates the program when the condition is encountered:

```
If ConditionOccurs _
Then
  SimulatorMessage "Condition Description"
  End
End If
```

Table 12.6   DEEDS Procedures for Simulator Lists Output

Name	Parameters	Description
PrintNodes	*NodeObjects*	Prints listed nodes on *Collections* spreadsheet
PrintPEL		Prints PEL on *Collections* spreadsheet
SimulatorMessage	*Message (Transaction)*	Displays *Message* and *TimeStamp* in *SimulatorMessages* area of *InitialModel* spreadsheet
TraceReport	*Message (Transaction)*	Displays *Message* and *TimeStamp* in *TraceReport* area of *InitialModel* spreadsheet
ViewPEL		Returns *Collection* reference to the PEL

12.7.2 Trace Report

DEEDS produces an easy-to-read trace report, shown in Figure 12.12, that enumerates the sequence of primary events as they occur during program execution for a user-specified time interval. The request on the *InitialModel* spreadsheet in Figure 12.13 is a trace report for 50 time units beginning at time period 0. A portion of the report is shown in Figure 12.13 for the time between 0 and 5.

When specifying a *TraceReport* time interval, it is very easy to produce a large amount of useless information. To examine the model environment before an error occurred, for best results, a trace should begin "immediately" before the time of the questionable event. A standard trace report includes the time when the event occurred, the event type and its associated node, a brief description on the node status, and transaction information that includes the serial number and values of the first two attributes. When more detailed information on the sequence of events is needed, the programmer can strategically incorporate event descriptions into the trace report by inserting user-defined sub calls to *TraceReport* into the program. The *TraceReport* subroutine automatically appends the time when the event occurred to a String argument. The format for the user-defined subroutine call is

```
TraceReport "Event Description"
```

12.7.3 Collection Report

Another important DEEDS feature is the ability to include transaction lists for the PEL, as well as queue, facility, and delay nodes as part of the simulation output. The sub for the PEL is

```
Simulator.PrintPEL()
```

For network nodes, the sub is

```
Simulator.PrintNodes NodeObjects separated by comma
```

where node objects are references for queues, facilities, and delays. In Figure 12.14 these procedures write the transaction lists to the *Collections* spreadsheet. Remember that source node

	A	B
1	Run Length	10000
2	Transient Time	1000
3	Number of Observations	9
4	Level of Confidence	0.95
5	Trace Report	Y
6	Start time	0
7	Stop time	50

Figure 12.12   DEEDS *TraceReport* request.

Trace Report

Time	Event
0.000	S1 arrival; First arrival at 0.000
0.000	S1 arrival; next arrival at 0.583
0.000	Engage F6 with 1 Server, Busy servers = 1
	SourceID = S1 SN = 2 A(1) = ExactChange A(2) = 6
0.000	Begin Service in: F6; Completion at: 5.703
	SourceID = S1 SN = 2 A(1) = ExactChange A(2) = 6
0.583	S1 arrival; next arrival at 3.793
0.583	Engage F1 with 1 Server, Busy servers = 1
	SourceID = S1 SN = 3 A(1) = A(2) = 1
0.583	Begin Service in: F1; Completion at: 13.277
	SourceID = S1 SN = 3 A(1) = A(2) = 1
3.793	S1 arrival; next arrival at 4.523
3.793	Engage F5 with 1 Server, Busy servers = 1
	SourceID = S1 SN = 4 A(1) = ExactChange A(2) = 5
3.793	Begin Service in: F5; Completion at: 10.391
	SourceID = S1 SN = 4 A(1) = ExactChange A(2) = 5
4.523	S1 arrival; next arrival at 6.182
4.523	Engage F2 with 1 Server, Busy servers = 1
	SourceID = S1 SN = 5 A(1) = A(2) = 2
4.523	Begin Service in: F2; Completion at: 17.085
	SourceID = S1 SN = 5 A(1) = A(2) = 2

Figure 12.13   DEEDS *TraceReport* output.

A	B	C	D	E	F	G	H
Collection Name	SerialNumber	Source	TransID	Priority	CreationTime	EventType	Time Stamp
PEL @ Time: 1000							
	111	S1	110	0	987.77	F1Facility	1003.271
	114	S1	113	0	1023.44	S1Source	1023.44
Q1 @ Time: 1000							
	113	S1	112	0	997.574	S1Source	997.574
	112	S1	111	0	992.264	S1Source	992.264
	110	S1	109	0	984.731	S1Source	984.731
	109	S1	108	0	980.798	S1Source	980.798
F1 @ Time: 1000							
	111	S1	110	0	987.77	F1Facility	1003.271

Figure 12.14   DEEDS *Collections* spreadsheet output.

transactions are in the *PEL*. Typically, transaction lists are requested as a result of a condition that occurs during the simulation or at the end of the simulation. The other case is a combination of the two, where the user wants to output the lists and terminate the model when a condition occurs.

The following statements in the *ShutDown* sub put the transactions on the PEL, the internal list of transactions in facility *F1*, and the transactions in queue *Q1* in the *Collection* spreadsheet at the end of the simulation:

```
PrintPEL
PrintNodes F1Facility, Q1Queue
```

The following requests the same information as a result of a condition in the model:

```
If Condition _
Then
   PrintPEL
   PrintNodes F1Facility, Q1Queue
End If
```

If the condition is recurring, the *Collection* spreadsheet will have lists from the last occurrence. The following stops the simulation after requesting the transaction lists:

```
If Condition _
Then
   PrintPEL
   PrintNodes , Q1Queue, F1Facility
   End
End If
```

To include statistics in the output, replace *End* with *StopSimulation*.

12.8 VBA Interactive Debugger

Standard output and *Simulator* reports enable the modeler to identify anomalous simulation results. In some cases, these reports enable the modeler to determine what caused the problem and when the problem occurred. The modeler can then use the VBA debugger to advance the program execution until the simulation clock is just before the time the questionable conditions occurred and from there step through the execution sequence to determine precisely what happens in the model. Debugger features are as follows:

- Display listing of input model statements.
- View current values of model parameters.
- Advance the simulation clock time.
- Invoke the debugger as a result of changes in the value of selected variables.
- Perform a single step trace.
- Change values of variables during program testing.

VBA provides excellent debugger documentation. In this section a summary of the debugger features is presented. Test the debugger with a simple program like the simple queueing model. Then, to obtain additional information, a good place to start is by entering *Debug* in the Index of the VBA *Help* menu. Figure 12.15 shows the debugger icons with a brief description of the most important ones for modeling in DEEDS.

Before clicking the *StartSimulation* option in the DEEDS drop-down menu, select program variables to watch during program execution and activate strategically selected break points. A left click on a program variable highlights the variable. It may then be added to the Quick Watch window. A left click in the left-hand column next to a VB statement sets the breakpoint at that line. Typically, the approach is to start the simulation from *Excel* with toggle breakpoints set or, as explained below, breaks on variables set in the Watch window.

Section 12.7.3 details how to obtain simulator lists in the *Collections* spreadsheet. The same information is available during interactive debugging. In Figure 12.16, resources initialized in the *EventsManager* module are managed by VBA Collections. To include a Collection in the *Watch List*, highlight and add it to *Quick Watch* with the add option.

Figure 12.17 shows a facility, queue, and source node for the basic queueing network program when the debugger is active. Clicking on the appropriate resource at any time while in debug mode

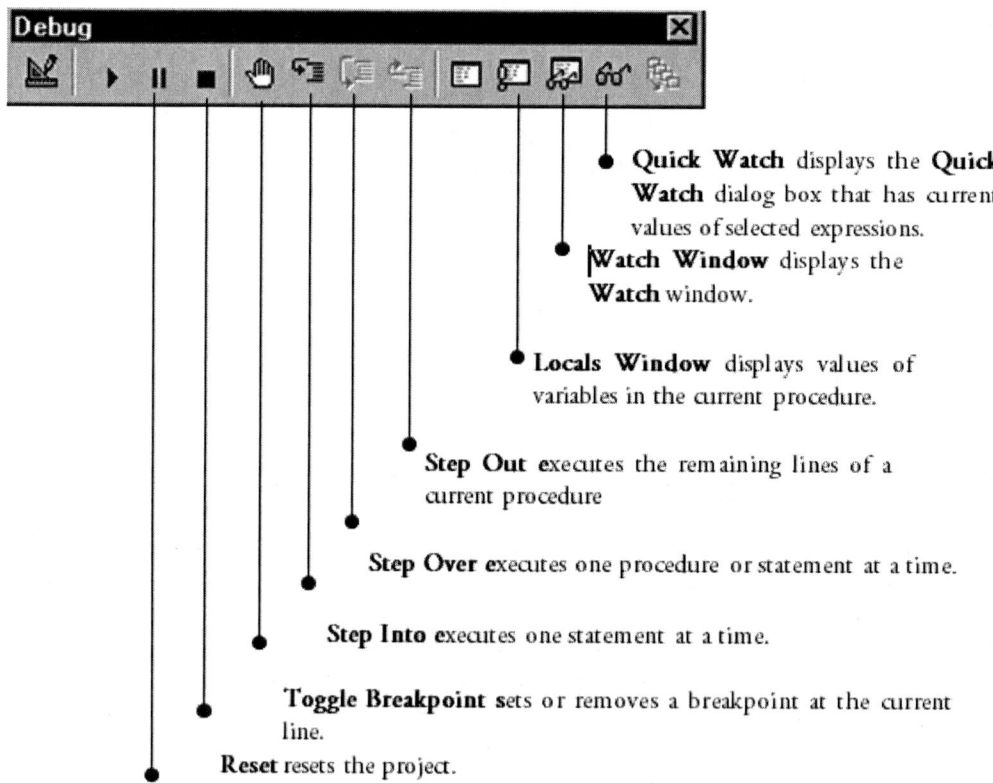

Quick Watch displays the **Quick Watch** dialog box that has current values of selected expressions.

Watch Window displays the **Watch** window.

Locals Window displays values of variables in the current procedure.

Step Out executes the remaining lines of a current procedure

Step Over executes one procedure or statement at a time.

Step Into executes one statement at a time.

Toggle Breakpoint sets or removes a breakpoint at the current line.

Reset resets the project.

Break stops execution of a program and switches to break mode.

Figure 12.15   VBA debugger options.

```
Public _
    InSetOfSources As New Collection, _
    InSetOfQueues As New Collection, _
    InSetOfFacilities As New Collection, _
    InSetOfDelays As New Collection, _
    InSetOfUserStatistics As New Collection, _
    InSetOfUserPDFs As New Collection, _
    InSetOfUserTables As New Collection, _
    CurrentTime As Single
```

Figure 12.16   VBA definition of DEEDS resources.

Watches				
Expression	Value	Type	Context	▲
👁 ⊟ InFacilities		Collection/Collection	EventsManager	
└ ⊞ Item 1		Variant/Object/Facility	EventsManager	
👁 ⊟ InQueues		Collection/Collection	EventsManager	
└ ⊞ Item 1		Variant/Object/Queue	EventsManager	
👁 ⊟ InSources		Collection/Collection	EventsManager	
└ ⊞ Item 1		Variant/Object/Source	EventsManager	▼

Figure 12.17   VBA watch list of DEEDS resources.

shows detailed information on that resource. When there is more than one node of each type, the order in their respective collection is the same as the order in the *InitialModel* sheet. By clicking on item 1 of *InQueues*, the view of the queue shown in Figure 12.18 is available. *Q* is the internal list of queued transaction. Notice there is only one transaction in the queue.

With *PrimaryEventsList* defined as a Collection and included in the *Watch List*, the following assignment statement enables the user to view the PEL:

```
PrimaryEventsList = ViewPEL()
```

In Figure 12.19 there are three transactions on the *PEL*. Item 1 and item 2 have been opened so that detailed transaction information may be examined. It may be helpful to review the transaction description in Section 8.1.2. To create a Watch break, assume an unusual condition is detected in the model at *CurrentTime* 105.2 and the modeler wants to view details of program execution after a *CurrentTime* of 100.

In Figure 12.20, *CurrentTime* is added to the list of *Quick Watch* variables; a right click on *CurrentTime* in *Edit Watch* provides the "Edit Watch" form shown. The modeler chooses the Watch Expression option and adds "100" to the expression. Now, the debugger is invoked when *CurrentTime* exceeds 100. Notice that procedure is *All Procedures*. This option ensures that *CurrentTime* is viewable throughout the simulation. Without that option, a variable is viewable only when the module in which the variable is defined is currently executing.

A view of the *F1FacilityEvent* sub in debug mode is shown in Figure 12.21. In the left column, the circle represents break points in the code and the arrow indicates the next line of code to be executed in debug mode. Clicking on Step Over causes the highlighted line of code to be executed.

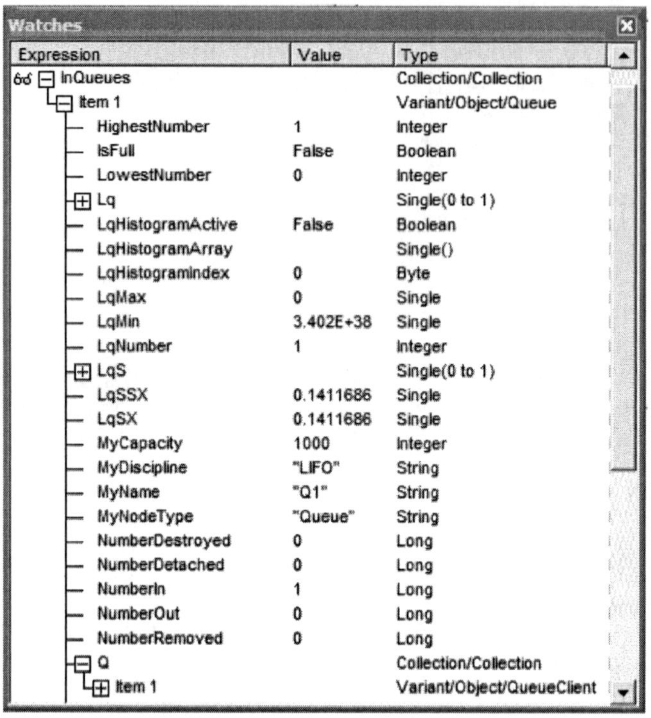

Figure 12.18 Debugger view of DEEDS resources.

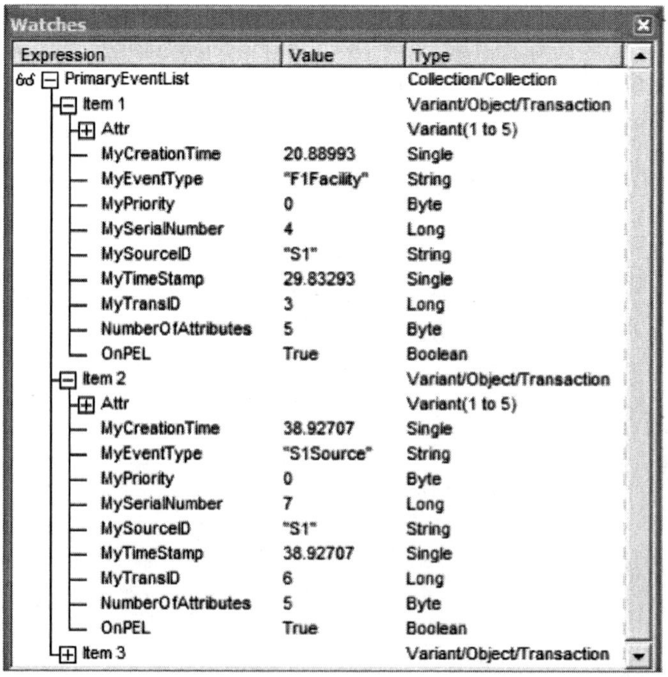

Figure 12.19 Debugger view of PEL.

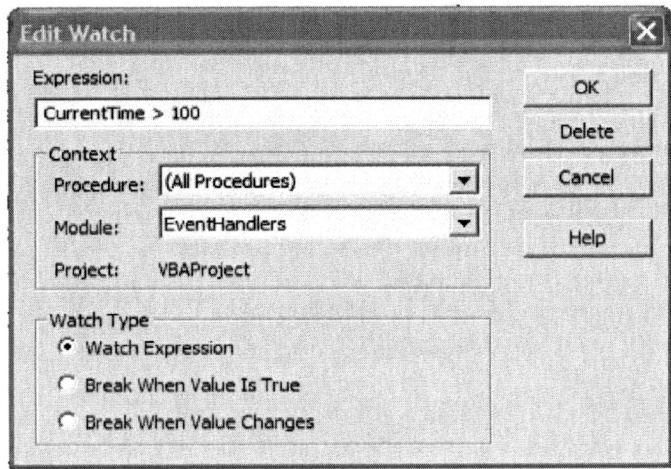

Figure 12.20   Debugger break options.

```
Sub F1FacilityEvent(ByRef ServicedCustomer As Transaction)
   F1Facility.Disengage ServicedCustomer
   Set ServicedCustomer = Nothing
   If Q1Queue.Length > 0 _
   Then
     F1Facility.EngageAndService Q1Queue.Depart, 1
   End If
End Sub
```

Figure 12.21   Simulation model in debug mode.

12.9 Summary

Of course, the purpose of a simulation is to obtain performance metrics for the system and system components. Chapters 7 and 13 are important chapters for understanding how to design a model and correctly interpret statistical program output. However, before drawing conclusions about the performance measures, it is important to ensure that the model performs to specification (model verification) and the model accurately reflects the system being modeled (model validation). The various features of the simulation program reports presented in this chapter are important tools for model verification and validation.

Chapter 13

Analysis of Simulation Results

13.1 Introduction

Chapters 7 through 12 are dedicated to the mechanics for constructing simulation models using design environment for event-driven simulation (DEEDS). In practice, model development is often equated to accomplishing the project objective. All the user must do to evaluate the simulated system is to run the model for an unspecified length of time. The results are then taken as true representations of the real system's performance. However, such an approach erroneously ignores the fact that simulation is a statistical experiment.

There are several historical reasons for the misuse of simulation output:

- Simulation induces a degree of confidence in model output because the modeler can "see" how observations are gathered during the simulation. As a result, the user may ignore the statistical aspects of simulation results.
- Simulation model development represents a major undertaking, and the fact that the model is producing (logical) results may be psychologically equivalent to completing an assignment.
- The type of knowledge needed to construct simulation models is different from that of a person knowledgeable on statistical techniques, particularly as it relates to the subtle statistical peculiarities of the simulation experiment.
- Misuse of simulation also stems from the fact that, in the past, emphasis in some of the widely used simulation languages was on building models while all but ignoring the statistical aspects of simulation.

The objectives of this chapter are to underscore the seriousness of problems caused by ignoring the statistical aspects of the simulation experiment, and to show how such problems may be avoided using DEEDS programming facilities. Although these issues were presented on a theoretical level in Chapter 6, the emphasis in this chapter is on practical procedures for handling these problems.

The chapter deals with the following statistical aspects of simulation:

- The effect of a transient state on simulation output
- Collection of independent observations
- Construction of confidence intervals
- Comparison of alternative designs

These points will be explored by using the following DEEDS example.

Example 13.1

This problem is from [1]. It deals with cars arriving at a six-booth tollgate every *Exponential* (2) seconds. Booths 4–6 are reserved for cars with exact change. However, an exact-change car may pass through any of the booths. The traffic pattern dictates that exact-change cars scan the booths right to left. Other cars scan the booths from left to right. Preference is always given to the shortest lane. The time to drive through the gate is *UniformContinuous* (5, 7) seconds for exact-change cars and *UniformContinuous* (10, 13) seconds for others. Sixty percent of the cars have the exact change. The performance metric for this study is an estimate of the average time for an arriving car to clear the gate.

The model has a single source of cars. Also included are six facilities, *F1–F6*, that represent tollbooths and each tollbooth has its respective queue, *Q1–Q6*, of waiting cars. An observation-based statistical variable, *SystemTime*, is used to tabulate the difference between the time a car exits the tollbooth and the time it arrives in the system. As indicated in spreadsheet *InitialModel*, tollbooth service times are encoded in the program. An array of queues and facilities, as shown below, are used in the model:

The arrays are initialized as follows:

```
         ' User defined input
Private _
    F(1 To 6) As Facility, _
    Q(1 To 6) As Queue
Private Const _
    Class As Integer = 1, _
    GateID As Integer = 2, _
    ServiceTime As Integer = 3
```

As shown in *Parser*, all tollbooth service completions invoke sub *ServiceCompletionEvent*.

```
' user input initialization
Dim _
    Index As Integer
For Index = 1 To InFacilities.Count
    Set Q(Index) = Instance("Q" & Index, InQueues)
    Set F(Index) = Instance("F" & Index, InFacilities)
Next Index
```

Because a car has an attribute for its tollbooth number, the car knows which facility to disengage. If the respective tollbooth has waiting cars, the next car departs the queue and is scheduled for service using its previously determined service time. The time in the system for the car that just finished service is then collected before the car exits the system.

```
Case "F1Facility", "F2Facility", _
     "F3Facility", "F4Facility", _
     "F5Facility", "F6Facility"
     ServiceCompletionEvent CurrentEvent
```

In Figure 13.1, 60% of the arrivals are randomly selected as an exact fare car and then assigned an exact fare service time. Exact fare cars scan the booths from number 6 to number 1 and join the shortest queue. Of course, if a server is idle, the car immediately begins service at that booth. Other cars scan booths from 1 to 3.

```
Sub ServiceCompletionEvent(Vehicle As Transaction)
   F(Vehicle.A(GateID)).Disengage Vehicle
   If (Q(Vehicle.A(GateID)).Length > 0) _
   Then
       F(Vehicle.A(GateID)).EngageAndService _
              Q(Vehicle.A(GateID)).Depart, 1,
   Vehicle.A(ServiceTime)
   End If
   SystemTimeStatistic.Collect Vehicle.SystemTime
End Sub
```

```
Sub S1SourceEvent(ByRef Vehicle As Transaction)
   Dim _
      Index As Integer, _
      CurrentLength As Integer
   If RandomNumber(1) < 0.4 _
   Then
      Vehicle.Assign ServiceTime, UniformContinuous(10, 13)
   Else
      Vehicle.Assign Class, "ExactChange"
      Vehicle.Assign ServiceTime, UniformContinuous(5, 7)
   End If
   CurrentLength = 999
   Select Case Vehicle.A(Class)
      Case "ExactChange"
         For Index = 6 To 1 Step  -1
            If (Q(Index).Length + F(Index).Length) < CurrentLength _
            Then
               CurrentLength = Q(Index).Length + F(Index).Length
               Vehicle.Assign GateID, Index
            End If
         Next Index
      Case Else
         For Index = 1 To 3
            If (Q(Index).Length + F(Index).Length) < CurrentLength _
            Then
               CurrentLength = Q(Index).Length + F(Index).Length
               Vehicle.Assign GateID, Index
            End If
         Next Index
   End Select
   If (F(Vehicle.A(GateID)).NumberOfIdleServers > 0) _
   Then
      F(Vehicle.A(GateID)).EngageAndService Vehicle, 1, _
                                      Vehicle.A(ServiceTime)
   Else
      Q(Vehicle.A(GateID)).Join Vehicle
   End If
End Sub
```

Figure 13.1   Tollbooth—*S1SourceEvent.*

13.2 Effect of Transient State

Transient conditions are prominent as one of the more serious problems in the analysis of simulation output. Values of performance metrics may fluctuate erratically over time while the system is in a transient state. As a result, estimates of output obtained during the transient period may be misleading. Steady state is achieved when the values of means and standard deviations of performance metrics become time-independent. It is important to point out that these conditions are necessary but not sufficient for achieving stationarity.

As presented in Chapter 6, there are many heuristics for detecting a steady-state system response. However, none of these seem easy to implement within the framework of a simulation language. The end of the transient state can be estimated by visually inspecting plots of performance metrics. In essence, a steady state begins when the plot ceases to show erratic variations. DEEDS enables the user to consider the stationary series question by automatically presenting the mean and standard deviation of the performance metrics under study as a function of time. The plots are prepared using Excel's *Chart* function that can plot any number of variables simultaneously. This is important because different variables may not reach stable values at the same time.

The impact of transient conditions is examined in this section by using the tollgate model presented in Section 13.1. The *SystemTime* and *Utilization* variables of facility $F(1)$, tollbooth 1, are used to demonstrate how the transient period is estimated. Figures 13.2 and 13.3 are plots of the mean and standard deviation of *SystemTime* and *Utilization*. There are 50 observations with a sampling interval of 400 seconds. On the basis of these plots, the transient period is assumed to be 15 observations (6000 seconds). The first steady-state observation corresponds to the first observation after the utilization of 0.81.

Two conclusions can be drawn from Figures 13.2 and 13.3:

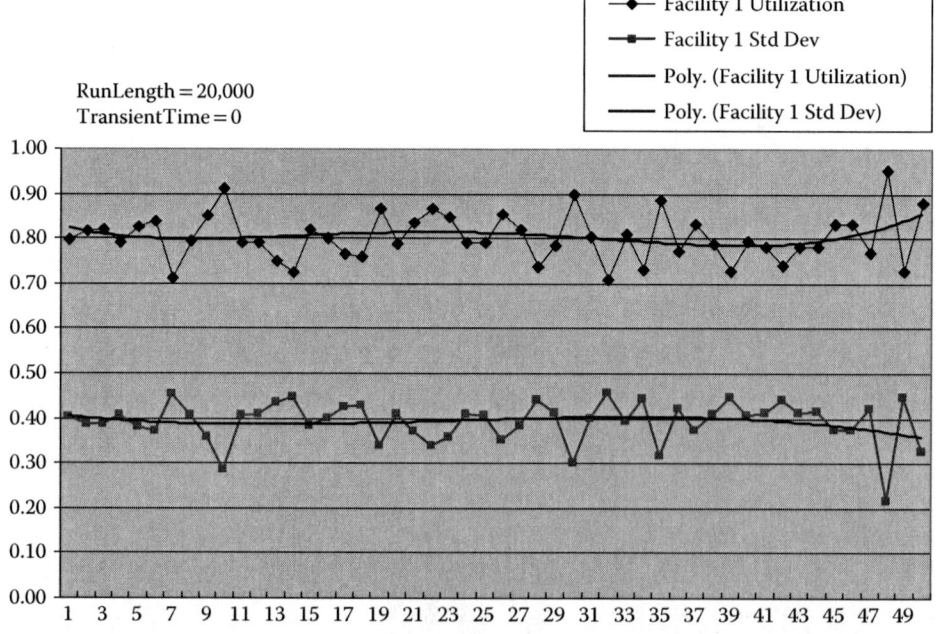

Figure 13.2   Tollbooth—Facility 1 utilization.

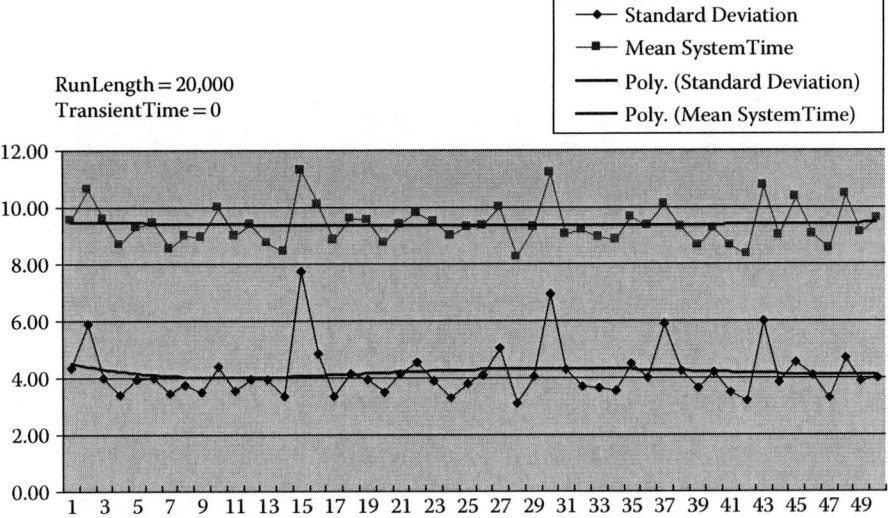

Figure 13.3 Tollbooth—transient analysis for system.

1. The mean and standard deviation of each variable appear to reach a stable value simultaneously.
2. Different variables of the same model may not reach a steady state simultaneously. For example, the *SystemTime* statistic is assumed to become stable at about $T = 6000$ seconds, whereas the *Utilization* parameter of $F(1)$ is assumed to reach stability at time $T = 4000$ seconds.

Although the conclusions above are drawn from a single experiment, experience with a large number of models supports the generality of the second conclusion. As a result, it is important that all performance parameters be examined simultaneously. Steady state is assumed to start after the longest of all transient periods has elapsed. Given that we are interested only in the two variables, steady state is estimated to start at about $T = 4000$ seconds from the start of the simulation run.

A general procedure for estimating the length of the transient period is summarized as follows:

1. Identify the performance characteristic variables to be estimated by the simulation model.
2. Use Excel's *Chart* function to graph the variation of the desired variables (and their standard deviations) as a function of simulation time.
3. By visually inspecting the resulting plot, estimate the time at which the values and standard deviations of all the variables stabilize. This time marks the approximate start of the steady state.
4. Collect the statistical observations by rerunning the model with the transient period truncated.

The seriousness of the transient conditions in biasing simulation results is demonstrated by data in Figure 13.4 that shows *SystemTime* as a function of the number of observations for time = 0–6000 seconds. The table shows that the average value of *SystemTime* varies from a minimum of 8.3 seconds to a maximum of 13.9 seconds, a considerable variation that could lead to erroneous conclusions. Also notice that the polynomial trend line for average system time varies as a function of time, an indication that the data are not stationary.

Figures 13.5 and 13.6 show steady-state plots for parameters *F1 Utilization* and *SystemTime* after the transient period is truncated. The *RunLength* is 26,000 seconds. Of that, 6,000 seconds

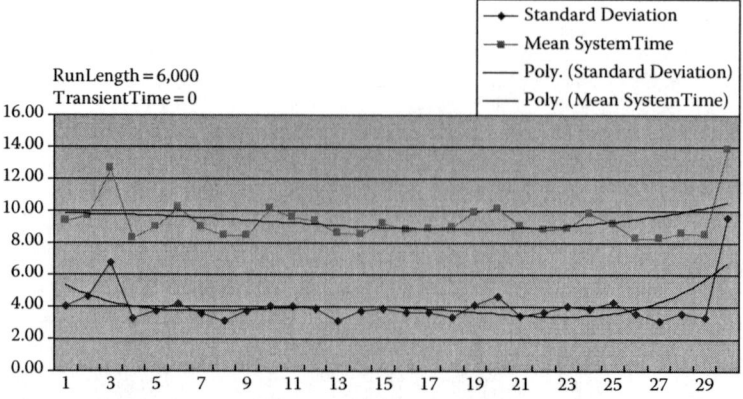

Figure 13.4   Tollbooth—*SystemTime* during transient period.

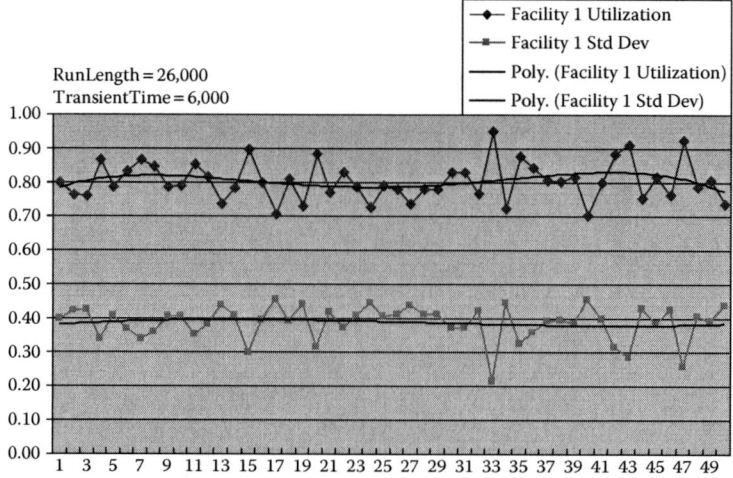

Figure 13.5   Tollbooth—steady state for *F1 Utilization*.

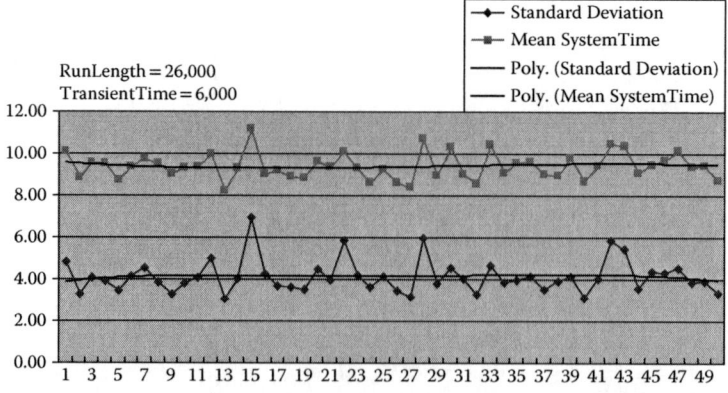

Figure 13.6   Tollbooth—steady state for *SystemTime*.

are allocated to transient time. After the elapsed transient time, 50 observations (400 seconds per observation) are recorded and the confidence interval on parameter estimates is at the 95% level. Notice that the variations exhibited in Figures 13.2 and 13.3 have been substantially dampened. The range of values for *SystemTime* is from 5.0 to 57.14 seconds. Average *SystemTime* is 9.41 seconds and the standard deviation of the *SystemTime* sample means is 0.79.

Included on each plot are polynomial trends for the parameter. The trend lines indicate that constant mean and constant variance are reasonable assumptions. However, it is also reasonable to assume that there is some autocorrelation in the data. These observations demonstrate the plausibility of the graphical procedure to detect the termination of transient conditions.

13.3 Gathering Statistical Observations

The two most common methods for gathering observations in simulation are the subinterval and replication methods. As indicated in Section 6.4, the subinterval method divides the run length (after truncating the transient period) into a number of equal subintervals (or batches) with each subinterval representing a single observation. In contrast, the replication method assumes that each observation is defined by a separate run with its transient period truncated. The subinterval method is efficient because the transient period is truncated only once, as compared with the replication method where the transient period is truncated in each run. The disadvantage of the subinterval method is that early successive observations may exhibit a degree of correlation and hence a lack of independence. This is not a problem in the replication method because each observation is associated with an independent run. In general, the subinterval method is favored over the replication method because of computational efficiency. This result is particularly plausible when generous truncation of the initial transient period is applied. Implementation of the replication method is explained in Chapter 25.

In this section, the subinterval method is applied to the tollgate operation example. It is necessary first to decide on the number of statistical observations for the experiment. Unfortunately, traditional statistical techniques are not suitable because they assume prior knowledge of the standard deviation of the variable under consideration. Generally, about 10 observations should be adequate for most simulation studies. In DEEDS, the subinterval method is implemented using the set of control statements given in Table 13.1.

The difference between *RunLength* and *TransientTime* $(a - b)$ must be sufficiently long to allow for valid data collection. If c represents the number of observations, the time period associated with each observation $(a - b)/c$ must be sufficiently long to collect pertinent information on the desired performance metrics. Specifically, as discussed in Chapter 1, the time period must be sufficiently long to allow for an adequate number of updates associated with the metric to occur. If no updates are recorded during the interval $(a - b)/c$, then the observation is probably not valid.

Table 13.1 Simulation Time Parameters

Parameter	Subinterval Method
RunLength	A
TransientTime	B
NumberOfObservations	C

To ensure that the user is aware of a questionable observation, DEEDS issues a *SimulatorMessage* similar to the following:

No Entries in Queue: queuename

A similar message is displayed for user-defined statistics. However, DEEDS does not issue this message when the number of updates during the interval $(a - b)/c$ exceeds 0. Again, biased performance metrics may occur when the number of updates in an observation is too small.

The number of updates can be verified using the following control statements in sub *Parser* under case *NewObservation*:

```
If NumberOfObservations = 1 _
Then
    StatisiticalVariable.List(NumberRequested)
Endif
```

The result is the number of requested updates on the output spreadsheet of the appropriate performance metric since steady state was achieved. Update values are listed while the actual number is less than the number requested. The user can then decide if the number of updates is sufficient. When insufficient, the problem is resolved by increasing *RunLength* or decreasing *NumberOfObservations*.

The subinterval method is demonstrated below for the tollgate model. The focus is on estimating the value of *SystemTime* only. However, other variables are treated in a similar manner. A transient period of 4000 seconds as estimated from the analysis in Section 13.2 is used. Also, the estimate is based on 10 observations and each observation spans 1000 seconds of simulation time.

The subinterval method control statements are as follows:

Parameter	Subinterval Method
RunLength	14,000
TransientTime	4,000
NumberOfObservations	10

The subinterval method has *RunLength* of 14,000 seconds, which is the length of the transient period plus the steady-state length of 10 observations ($10 \times 1,000 = 10,000$). The results are summarized in the above table.

Notice that the parameters in Table 13.2 that are based on 10 observations compare favorably with the results in Section 13.2, where the mean was 9.41 and standard deviation 0.79 for 50 observations with run length of 24,000 seconds and transient period of 4,000 seconds.

Table 13.2 System Time Estimation by Subinterval Method

Parameter	Subinterval
Mean	9.4
Standard deviation	0.73

13.4 Establishing Confidence Intervals

For the subinterval method, DEEDS automatically computes the upper and lower confidence limits of all the output variables of the model based on a user-specified confidence level. However, for reasons explained in Chapter 25, when using the replication method it is incumbent on the user to make these calculations. This section demonstrates confidence interval calculations for the replication method using standard simulation output.

Before proceeding with the presentation, it is important to emphasize that in a number of simulation languages, the output automatically provides standard deviation figures based on (possibly highly) correlated updates of a single run. These estimates are generally severely biased, and hence are not suitable for any type of statistical tests. This is precisely the reason DEEDS does not provide standard deviation estimates based on single runs. Such estimates are produced only when the subinterval method is used.

Suppose a 90% confidence interval for *SystemTime* is the design criteria for the model. As discussed in Section 4.3.1, given that $\bar{x}$ and s are the sample mean and standard deviation of a variable, a $(1 - \alpha)\%$ confidence interval is defined as

$$\bar{x} \mp \frac{s}{\sqrt{n}} t_{n-1,1-\alpha/2}$$

where n is the sample size and $t_{n,1-\alpha/2}$ is the upper $(1 - \alpha/2)$ critical point from the t-distribution with $n - 1$ degrees of freedom. Applying the formula to *SystemTime* data in Table 13.2 for $n = 10$, $\alpha = 0.1$, and $t_{9,0.95} = 1.833$, the 90% confidence interval based on the subinterval method is given as follows:

$$\bar{x} = 9.4 \quad \text{and} \quad s = 0.72$$

and the 90% confidence interval $= 9.4 \mp \dfrac{0.72}{\sqrt{10}} * 1.833 = [8.98, 9.82]$

The confidence interval indicates that the probability that the true value of *SystemTime* is between 8.98 and 9.82 seconds is 0.90. Comparable results from DEEDS are that the true value of *SystemTime* is between 9.02 and 9.78 seconds. The difference is that DEEDS uses the normal distribution to estimate the confidence interval.

13.5 Hypothesis Testing in Simulation Experiments

Hypothesis testing in simulation is demonstrated in this section. Suppose that in the tollgate model the objective is to test the effectiveness of installing a new device for the exact-change lanes that reduces the time to pass through the gate from *UniformContinuous* (5, 7) to *UniformContinuous* (4, 5) seconds. A plausible way to test the new design is to study its effect on the average time for a car to pass through the tollgate by comparing *SystemTime* before and after the new device is installed.

Let *SystemTime1* and *SystemTime2* represent the total time for a car to clear the tollgate before and after the device is installed. Then the situation can be translated into the following hypothesis test:

$$H_0 : SystemTime2 = SystemTime1$$

$$A : SystemTime2 \leq SystemTime1$$

Table 13.3 *SystemTime1* and *SystemTime2* for Tollgate Model

Variable	Number of Observations	Mean	Standard Deviation
SystemTime1	10	9.30	0.72
SystemTime2	10	8.18	0.68

If the hypothesis H_0 is not rejected, the conclusion is that the new device is not effective. If H_0 is rejected, the device is considered effective.

To test the given hypothesis, the model is run for the two stated cases: one with the time to pass the exact-change gate as *UniformContinuous* (5, 7) and the other by *UniformContinuous* (4, 5) seconds. The subinterval method is used to collect the statistical observations and the results are summarized in Table 13.3.

The associated *t*-statistic for comparing two means is computed using the formula in Section 4.3.2 as follows:

$$t_0 = \frac{8.18 - 9.30}{\sqrt{\dfrac{0.68^2}{10} + \dfrac{0.72^2}{10}}} = \frac{-1.12}{0.313} = -3.58$$

The degrees of freedom of the *t*-statistic are now calculated as

$$N = \frac{(0.72^2 + 0.68^2)^2/10}{\dfrac{(0.72^2/10)^2}{11} + \dfrac{(0.68^2/10)^2}{11}} - 2 = 22 - 2 = 20$$

For a 90% confidence level, $t_{20,0.95} = 1.725$. Because $t_0 = 2.94$ is greater than 1.725, H_0 is rejected and the conclusion is that the device is effective in reducing the time needed to pass through the gate.

13.6 Summary

The emphasis in this chapter is on the statistical aspects and peculiarities of the simulation experiment. It is important to remember that simulation output must be interpreted statistically. DEEDS is designed to provide practical programming facilities to assist the user in completing this important phase of a simulation study. In particular, DEEDS output and Excel's *Chart* function enable the user to estimate the start of the steady state. Once this is done, DEEDS enables the user to estimate global statistics of the simulated model. In addition, the global statistical summary for the subinterval method automatically computes the upper and lower confidence limits of all the output variables of the model on the basis of a user-specified confidence level.

Reference

1. Pegden, C. D., *Introduction to SIMAN*. College Station, PA: Systems Modeling Corporation, 1985.

Chapter 14

Model Visualization

14.1 Introduction

The visualization features of design environment for event-driven simulation (DEEDS) enable the modeler to graphically represent simulation model activity as a function of time. These visualization features are accessible in Excel 2007 and later versions. Visualization is an invaluable development tool for isolating problems with respect to program logic. Of course, visualization is also an important tool for identifying binding resource constraints and the operational environment when the constraints are binding. This chapter demonstrates how to build and launch the visualization program.

14.2 Model Design

Source, queue, facility, and delay nodes representing physical system resources are described in the *VisualData* spreadsheet and presented on the *VisualPallet* spreadsheet. A *visualization* program slows the simulation model and reflects node activity by changes in color of the nodes. Before creating the visualization model the *InitialModel* spreadsheet must be populated and a preliminary design of the simulation program must be complete. *InitialModel* spreadsheet nodes are the basis for creating the visualization model. To build the visualization worksheets, click on the Add-Ins option on the Home menu ribbon shown in Figure 14.1 to obtain the DEEDS drop-down menu shown in Figure 14.2. Clicking on DEEDS in Figure 14.2 shows the Program Manager, Start Simulation, and Start Visualization options in Figure 14.3. Click on Program Manager in the DEEDS drop-down menu to obtain *Program Manager*, shown in Figure 14.4. The two buttons in the lower right corner of the *Initial Model* tab in *Program Manager* are used to create the *VisualData* and *VisualPallet* worksheets. The Build Visual Data option is selected before the Build Icons option. Then click on Build Visual Data and Build Icons respectively. The user will be reminded when these options are not chosen in the correct sequence.

A *VisualData* spreadsheet similar to the one in Figure 14.5 enumerates, by node classification, all source, queue, facility, and delay nodes defined in the *Initial* spreadsheet of the model workbook. These nodes represent physical resources of the system. Node definitions in the *VisualData*

Figure 14.1 DEEDS program menu.

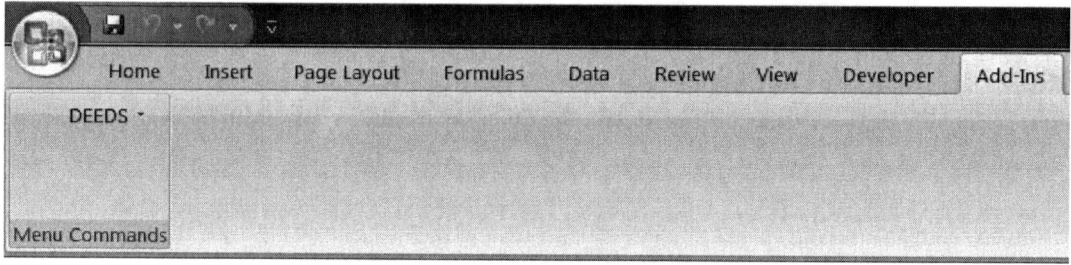

Figure 14.2 DEEDS drop-down menu.

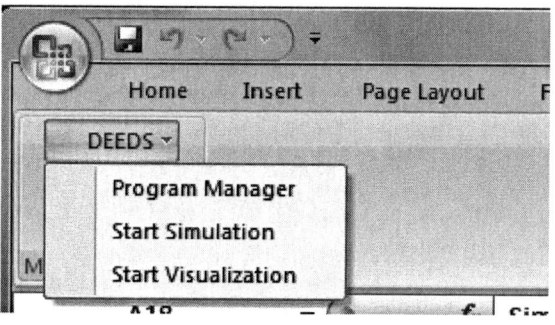

Figure 14.3 DEEDS options.

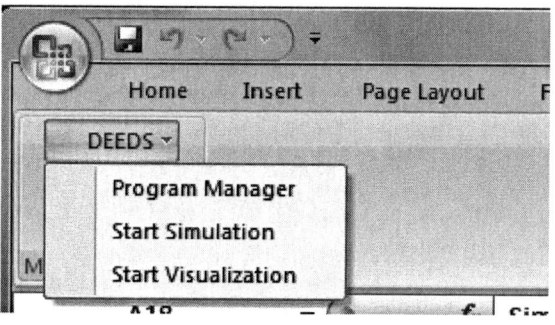

Figure 14.4 DEEDS *ProgramManager—Initial Model* tab.

spreadsheet that are not to be included in the visualization model should be deleted before selecting the Build Icons option.

Each of the remaining nodes is assigned a normal, stress, and strain level for that resource. During visualization program execution, the node color is a shade of green when current node transaction counts are from zero to normal level. The node color transitions from a shade of green to red, depending on the stress level and strain level specifications for that resource.

The Animation Clock (Seconds) cell in Figure 14.5 is the screen refresh rate for the visualization program. In this example the screen is refreshed at 0.25 second intervals. Animation Clock Time Units is the granularity of the simulation clock. The value of 5 indicates that after five units of simulation clock time the program is ready to update the screen. If Animation Clock Time Units is too large, changes in the state of the system between views will be dramatic. In the sense that the time to execute 10 simulation clock time units is longer than the time necessary for 5 time units, the delay time for refreshing the screen is cumulative. The specification of these two parameters must be tuned for a given simulation program.

Figure 14.6 is an example of a *VisualPallet* worksheet for the simulation program. This worksheet is created from resource information in the *VisualData* worksheet and each visual node icon represents a node in the *VisualData* worksheet. Numbers under the queue, facility, and delay

		Sources		Queues			
(Seconds) 0.25					Normal Level	Stress Level	Strain Level
Animation Clock Time Units		Name	Name				
	5	Terminal1	Terminal1		3	9	18
		Terminal2	Terminal2		3	9	18
		Return	Counter		6	18	36
			DropOff		1	2	3
			CheckedIn		3	9	18
			IdleVan		1	3	5
			Van1		3	9	18
			Van2		3	9	18
			Van3		3	9	18
			Van4		3	9	18
			Van5		3	9	18

Figure 14.5 DEEDS *VisualData* worksheet.

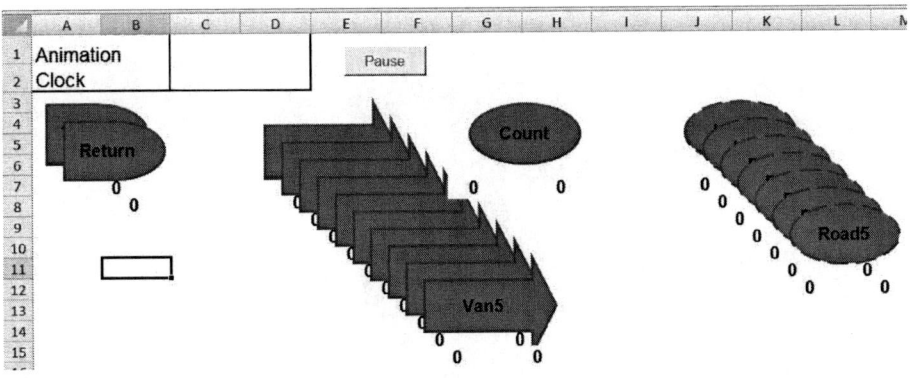

Figure 14.6 DEEDS *VisualPallet* worksheet.

nodes represent the number of transactions entering and residing in the node respectively during the simulation. In addition to changes in node colors after each visualization clock update, adjustments are also made to these node transaction counts.

In Figure 14.7 icons on the *VisualPallet* worksheet have been configured to represent the program logic. Any pictures, clip art, and shapes available in Excel may be used to enhance the logical representation of the model. Figure 14.8 is another example of a *VisualPallet* worksheet.

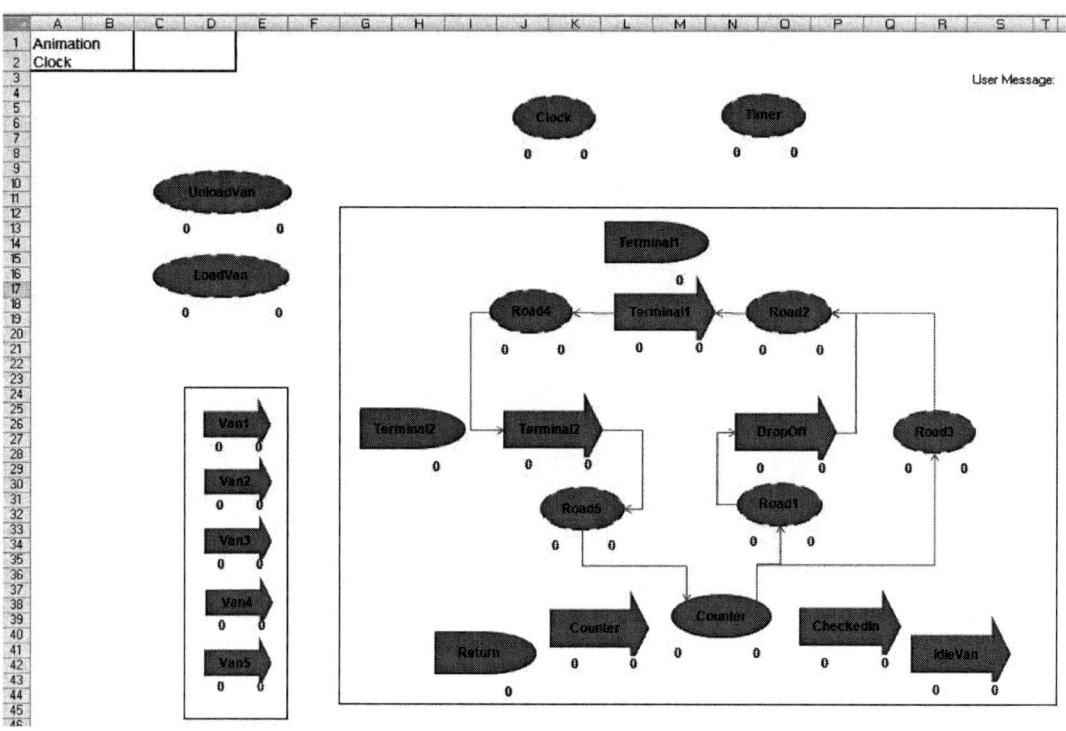

Figure 14.7   Sample configured *VisualPallet*.

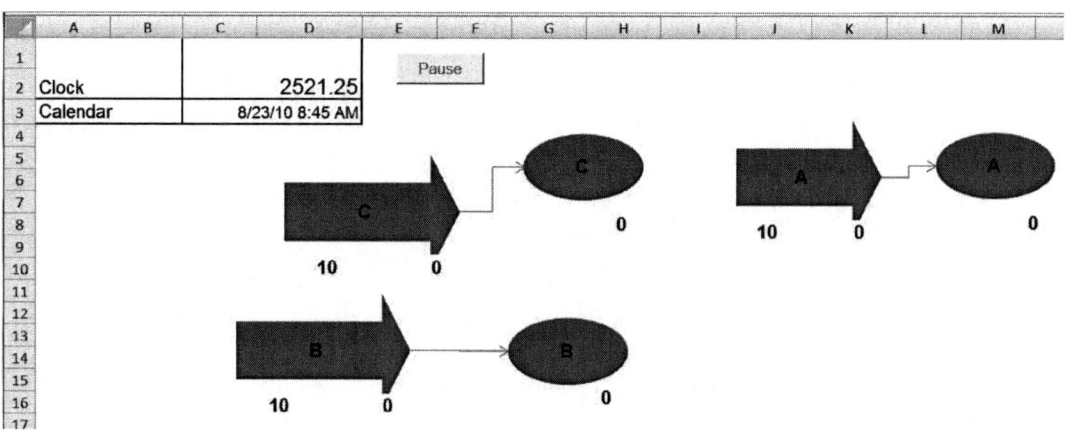

Figure 14.8   Sample configured *VisualPallet* with user-defined features.

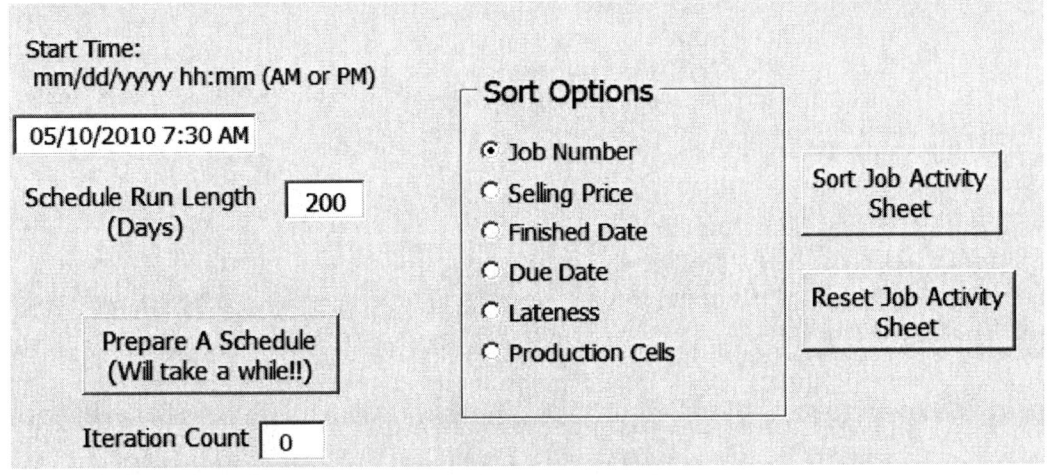

Figure 14.9 User form for visualization program.

	A	B	C	D	E	F	G
1	**Schedule***	**Begin1**	**End1**	**Begin2**	**End2**	**Begin3**	**End3**
2	**Sunday**						
3	**Monday**	7:30	10:30	10:45	12:00	13:00	17:00
4	**Tuesday**	7:30	10:30	10:45	12:00	13:00	17:00
5	**Wednesdy**	7:30	10:30	10:45	12:00	13:00	17:00
6	**Thursday**	7:30	10:30	10:45	12:00	13:00	17:00
7	**Friday**	7:30	10:30	10:45	12:00	13:00	17:00
8	**Saturday**						
9							
10	*** 24 hr. clock begins at midnight 0:00**						
11	**13:45 is 1:45 pm**						

Figure 14.10 Visualization program production schedule.

This program represents a job shop with cells A, B, and C. A job is routed through each cell for processing. The sequence of cells visited is dependent on the job type. The Excel workbook for this problem is included in Chapter 14 of the companion CD.

Notice the calendar on the *VisualPallet*. With the referenced beginning time for the simulation model in Figure 14.9 and the production schedule information in Figure 14.10, the visualization program is able to map simulation time to the actual production time and date displayed in the calendar. By viewing the visualization program the user will know precisely when a condition occurs in the model. Visualization parameters are defined in the data definition section of the simulation program along with other user-defined variables. Definitions in Figure 14.11 are an example of user-defined parameters for the model. *StartTime*, defined in Figure 14.11, is used to determine the actual time and date in the Figure 14.5 *Visualization Pallet*. The sub *InitialVisualization* in Figure 14.12 resides in the *EventsHandler* module of the simulation program. Before the visualization simulation begins, the user-defined visualization parameters are initialized in sub *InitialVisualization*.

```
Public _
  TmpCellName As String, _
  TmpCellNumber As String, _
  TmpCellIndex As Integer, _
  NumberJobs  As Integer, _
  NumberCells As Integer, _
  LengthOfRun As Single, _
  StartTime As Date, _
  EndTime As Date, _
```

Figure 14.11   Data definition for simulation program.

```
Public Sub InitialVisualization()
  DEEDSManager.InitialVisualization Simulator
  Worksheets("VisualPallet").Activate
  OpenVisualizationManager

  ' user defined visualization output

End Sub
```

Figure 14.12   DEEDS sub *InitialVisualization*.

```
Public Sub UpdateVisualization()

  ' update visualization output

  WS.Range("C3").Value = VisualizationClock / 24 + StartTime
  DoEvents
End Sub
```

Figure 14.13   DEEDS sub *UpdateVisualization*.

Also in the *EventsHandler* module of the simulation program is sub *UpdateVisualization*, shown in Figure 14.13. Subroutine *UpdateVisualization* enables user-defined changes to the *VisualPallet* during execution of the visualization program. In addition to changes by the visualization program to node colors and node transaction counts after each visualization clock update, the code segment in Figure 14.13 updates the actual time and date in the Figure 14.8 *VisualPallet*.

14.3 Program Execution

After detailing *Normal*, *Stress*, and *Strain* values in the *VisualData* worksheet and after the *VisualPallet* worksheet has been configured to represent the system, click on the Add-Ins option on the Home menu ribbon to obtain the DEEDS drop-down menu in Figure 14.3. In a similar manner to initiating the simulation program, *Start Visualization* launches DEEDS visualization that begins by presenting the *Visualization Manager* form in Figure 14.14.

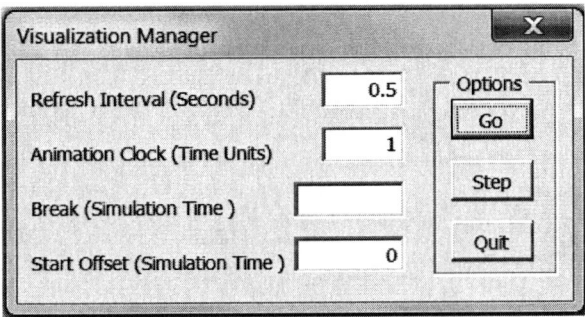

Figure 14.14 Visualization program manager.

Control of the visualization program toggles between the Pause button shown in Figure 14.8 and the *Visualization Manager*. During visualization, clicking the Pause option launches *Visualization Manager* and hides the Pause button. Similarly, the Go option in *Visualization Manager* shows the Pause button and hides *Visualization Manager*. After specifying parameters on the right side of the *Visualization Manager* click on the Go option.

Refresh Interval and Animation Clock in *Visualization Manager* enable the user to override these visualization parameters on the *VisualData* spreadsheet. With these options the user can speed up or slow down the screen refresh rate of the visualization program and focus on specific times that the system is operational.

Break, Start Offset, and Step enable the user to pause and start execution of the visualization program. Start Offset specifies when the visualization of the simulation begins. For example, an offset of 50 directs the visualization program to advance 50 time units before updates to worksheet *VisualizationPallet* occur. Of course Start Offset can be used only once when the visualization program is launched. Break directs the visualization program to interrupt the visualization presentation at the simulation clock time specified. A value of 100 directs the visualization program to suspend the program at simulation time 100 and return control to the *Visualization Manager*. The user is then able to reconfigure the *Visualization Manager* for the next phase of the visualization program.

Whenever *Visualization Manager* is visible, the user may override the automatic updates of the visualization program using the Step option to indicate when the next visualization update will occur. Of course, the user may terminate the program when *Visualization Manager* is visible by selecting the Quit option.

14.4 Summary

The visualization features of DEEDS enable the modeler to graphically represent simulation model activity as a function of time. Visualization is an invaluable development tool for isolating problems with respect to program logic and for identifying binding resource constraints and the operational environment.

The flexibility of the DEEDS *Visualization Manager* enables the user to strategically focus on visual details of resource utilization at specific points in time during the simulation at the expense of other portions of the simulation. Finally, this chapter demonstrates the features of the DEEDS visualization environment and suggests how visualization can be used to communicate information about the model.

Chapter 15

Modeling Special Effects

15.1 Introduction

In this chapter, special modeling techniques that demonstrate the capabilities of design environment for event-driven simulation (DEEDS) beyond the basic features covered in Chapter 11 are introduced. The objective is to provide a higher degree of diversification in the language implementation.

15.2 A Multiserver Facility to Represent Independent Facilities

The model uses a single multiple-server facility to represent any number of independent facilities with different mean service times. On spreadsheet *InitialModel*, are a source, *S1*, a queue, *Q1*, and a facility, *F1*. Corresponding to the number of sources and facilities, there are primary event subroutines for job arrivals and service completions. The queue is divided (dynamically) into subqueues. Similarly, each server in the facility represents an independent facility. Parallel servers in *F1* are values 2 and 4, corresponding to runs 1 and 2, respectively.

Figure 15.1 presents the primary event sub for arrivals from source *S1*. A transaction from *S1* has the following two attributes:

1. *Attribute (1)* is the subqueue index and associated subfacility.
2. *Attribute (2)* is the transaction service time.

The value assigned to *Attribute (1)* is determined from discrete distributions *Case1* and *Case2*, defined in spreadsheet *PDF*. For run 1 with two parallel servers, 80% of the customers go to facility 1 and 20% to facility 2. For run 2 with four parallel servers, the percentages among the four facilities are 30%, 40%, 20%, and 10%.

Because each subfacility has its own mean service time, the *MeanService* from spreadsheet *Tables* is used to provide the model with mean service time for each subfacility. The assignment statement

```
Sub S1SourceEvent(ByRef Arrival As Transaction)
   Dim _
      Index As Integer, _
      Found As Boolean
   Arrival.Assign JobType, Case1PDF.RV
   Arrival.Assign ServiceTime, Distribution.Exponential _
                   (MeanServiceTable.Value(Arrival.A(JobType)))
   Found = False
   For Index = 1 To F1Facility.NumberOfBusyServers
      If F1Facility.View (Index).A(JobType)= Arrival.A(JobType) _
      Then
         Found = True
         Exit For
      End If
   Next Index
   If Found _
   Then
      Q1Queue.JoinOnLowAttribute JobType, Arrival
   Else
      F1Facility.EngageAndService Arrival, 1, Arrival.A(ServiceTime)
   End If
End Sub
```

Figure 15.1   Multiserver facility—*S1SourceEvent*.

```
Arrival.Assign ServiceTime, Distribution.Exponential _
                (MeanServiceTable.F(Arrival.A(JobType)))
```

assigns attribute 2 of the arriving transaction a random variable from the exponential distribution using the value of attribute 1 as an index into the *MeanService* table. Notice that the difference between run 1 and 2 is simply changing the assignment statement for attribute 1 from *Case1PDF* to *Case2PDF*.

After the attributes are assigned a job type and service time, the sub *S1SourceEvent* determines whether the transaction enters its subqueue *Attribute (JobType)* or engages a server and begins service in its subfacility *Attribute (JobType)*. Transactions in the facility are examined using a *For* loop to determine whether there is a transaction being serviced that has an *Attribute (JobType)* value that matches *Attribute (JobType)* of the arriving transaction.

The segment *F1Facility.View (Index).A(JobType)* evaluates, right to left, the value of *Attribute (JobType)* of transaction *Index* in facility *F1*. Upon completion of the loop, if *Found* is *True*, the subfacility for the arriving transaction is busy and the transaction joins the queue, last in its respective subqueue, on the basis of the lowest value of *Attribute (JobType)*. Otherwise, the transaction engages its idle server and begins service using the service time in *Attribute (ServiceTime)*.

As defined in sub *Initial*, facility *F1* has two parallel servers for run 1 and four servers for run 2. Figure 15.2 presents sub *F1FacilityEvent* for *F1* service completions. After a service completion, the transaction disengages the facility. The queue is searched for a transaction that has the same *Attribute (JobType)* value as the transaction that just completed service. If such a transaction is found, it is detached from the queue and scheduled for a service completion using the service time in *Attribute (ServiceTime)*. Otherwise, the subfacility remains idle. In either case the transaction that completed service exits the system.

Inserting the program segment in Figure 15.3 before *DepartingCustomer* exits the system provides independent statistics for time in the system by job type. Section 15.9 shows how this type of model can be easily generalized using arrays of DEEDS classes.

```
Sub F1FacilityEvent(ByRef DepartingCustomer As Transaction)
  Dim _
    Index As Integer
    NextCustomer As Transaction
  F1Facility.Disengage DepartingCustomer
  For Index = 1 To Q1Queue.Length
    If Q1Queue.View(Index).A(JobType) = _
                DepartingCustomer.A(JobType) _
    Then
       Exit For
    End If
  Next Index
  If Index <= Q1Queue.Length _
  Then
     Set NextCustomer = Q1Queue.Detach(Index)
     F1Facility.EngageAndService _
                NextCustomer, 1, NextCustomer.A(ServiceTime)
  End If
  Set DepartingCustomer = Nothing ' exit system
End Sub
```

Figure 15.2   Multiserver facility—*F1FacilityEvent*.

```
Select DepartingCustomer.A(JobType)
  Case 1
    JobType1Statistic.Collect DepartingCustomer.SystemTime
  Case 2
    JobType2Statistic.Collect DepartingCustomer.SystemTime
  Case 3
    JobType3Statistic.Collect DepartingCustomer.SystemTime
  Else
    JobType4Statistic.Collect DepartingCustomer.SystemTime
End Select
```

Figure 15.3   Multiserver facility—collect statistics program segment.

15.3 Facility Preemption Operation

Messages arrive every *Exponential* (4) time units for transmission over a line. Transmission time is *Exponential* (3). Urgent messages arrive every *Exponential* (25) time units, in which case any message occupying the line is interrupted, processed for one time unit, and then placed at the front of the waiting messages queue to wait for retransmission. The transmission time for urgent messages is the same as for an ordinary message.

On spreadsheet *InitialModel*, the model includes two sources, one for urgent and another for regular calls; a single-server facility that represents the line; queues for urgent, interrupted, and regular calls; and a timer delay for conditioning interrupted messages before they are placed into the interrupt queue. Corresponding to the total number of sources, facilities, and delays, there are four primary event subs (arriving urgent calls, arriving regular calls, call completions, and a timer conditioning delay) used to model this system. Urgent messages arrive with priority level 1 and regular messages arrive with priority level 2.

In Figure 15.4 is sub *UrgentSourceEvent* for arriving urgent calls. When the message arrives, it checks line availability in *LineFacility*. If the line is idle, the message is immediately scheduled by

```
Sub UrgentSourceEvent (ByRef UrgentCall As Transaction)
  If LineFacility.NumberOfIdleServers > 0 _
  Then
    Call BeginService (UrgentCall)
  Else
    If LineFacility.View(1).Priority = 1 _
    Then
      UrgentQueue.Join UrgentCall
    Else
      InterruptDelay.Start LineFacility.Interrupt(1)
      Call BeginService (UrgentCall)
    End If
  End If
End Sub
```

Figure 15.4 Facility preemption—*UrgentSourceEvent*.

```
Sub RegularSourceEvent (ByRef RegularCall As Transaction)
  If LineFacility.NumberOfIdleServers > 0 _
  Then
    Call BeginService(RegularCall)
  Else
    RegularQueue.Join RegularCall
End Sub
```

Figure 15.5 Facility preemption—*RegularSourceEvent*.

sub *BeginService* shown in Figure 15.8. If the line is busy, a check is made to determine whether it is occupied by a regular message, which happens when the transaction residing in *LineFacility* has *Priority* = 2. If *LineFacility* is transmitting an urgent message, the arriving urgent call joins the urgent calls queue. Otherwise, the regular call is interrupted and placed in the interrupted calls queue. The urgent call is then scheduled by sub *BeginService*.

Figure 15.5 shows sub *RegularSourceEvent* that processes arriving regular calls. When the message arrives, the line availability in *LineFacility* is checked. If the line is free, the message is immediately scheduled by sub *BeginService*. Otherwise, the call joins the regular call queue. There is no need to check for call priority because regular calls are the lowest priority level.

The sub *LineEventFacility* for completed calls is in Figure 15.6. An array of destination queues, *NodeObject (1 to 3)*, is defined in sub *Initial*. The order of queues within *NodeObject* is urgent calls queue, interrupted calls queue, and regular calls queue. This order gives the highest priority to urgent messages, followed by those that were interrupted. The lowest priority is given to regular messages. Notice that interrupted messages are processed in the same order that they joined the interrupt queue. Upon a call completion, queues in *NodeObject* are examined. A call from the first nonempty queue is the next call to process. If all queues are empty, the line remains idle.

The sub *InterruptDelayEvent* in Figure 15.7 simply places the interrupted message into the interrupt queue after the 1 time unit delay. The sub *BeginService* in Figure 15.8 initiates service for a call. Sub *UrgentCallEvent*, sub *RegularCallEvent*, and sub *LineFacilityEvent* each schedules a call service. Rather than duplicating the program segment in each subroutine, *BeginService* handles those activities for all of these subs. The parameter passed to *BeginService* is simply which call to process.

```
Sub LineFacilityEvent(ByRef FinishedCall As Transaction)
  Dim _
    Index As Integer
  LineFacility.Disengage FinishedCall
  For Index = 1 To 3
    If NodeObject(Index).Length > 0 _
    Then
      BeginService NodeObject(Index).Depart
      Exit Sub
    End If
  Next Index
End Sub
```

Figure 15.6   Facility preemption—*LineEventFacility*.

```
Sub InterruptDelayEvent(ByRef InterruptedCall As Transaction)
  InterruptQueue.Join InterruptedCall
End Sub
```

Figure 15.7   Facility preemption—*InterruptDelayEvent*.

```
Sub BeginService(ByRef NextCall As Transaction)
  NextCall.Assign ServiceTime, Exponential(3)
  LineFacility.EngageAndService NextCall, 1,
NextCall.A(ServiceTime)
End Sub
```

Figure 15.8   Facility preemption—sub *BeginService*.

15.4 Limit on Waiting Time in Queues

DEEDS automatically allows for limiting the size of the queue through field specification in spreadsheet *InitialModel*. In some situations, however, it may be necessary to set a limit on the time a transaction will wait in the queue. Upon exceeding that limit, the transaction reneges and seeks service elsewhere. This model demonstrates how that result can be achieved.

On spreadsheet *InitialModel*, the model includes a source, *S1*, for arriving jobs; two queues, *Q1* and *Q2*, for waiting jobs; two facilities, *F1* and *F2*, for processing jobs; and a timer segment for jobs in *Q1*. Corresponding to the total number of sources, facilities, and delays, there are four primary event subroutines (arriving jobs from *S1*, service completions at *F1* and *F2*, and timer delay for jobs waiting in *Q1* to be processed in facility *F1*).

Figure 15.9 shows sub *S1SourceEvent* for arriving customers. For each arrival a copy of the customer transaction is routed to a three-minute delay. The parent transaction either enters facility *F1* or waits in queue *Q1*. Notice in Figures 15.10 and 15.11 that facilities *F1* and *F2* have identical features. The finished job exits the system and when the facility's respective queue has waiting transactions, the next job is scheduled for service.

```
Sub S1SourceEvent(ByRef Customer As Transaction)
  Dim _
    Image As Transaction
  Set Image = Customer.Copy
  TimerDelay.Start Image
  F1Facility.Destination Customer, 1, Q1Queue
End Sub
```

Figure 15.9   Waiting time limit—*S1SourceEvent*.

```
Sub F1FacilityEvent(ByRef FinishedCustomer As Transaction)
  F1Facility.Disengage FinishedCustomer
  If Q1Queue.Length > 0 _
  Then
    F1Facility.EngageAndService Q1Queue.Depart, 1
  End If
  Set FinishedCustomer = Nothing
End Sub
```

Figure 15.10   Waiting time limit—*F1FacilityEvent*.

```
Sub F2FacilityEvent(ByRef FinishedCustomer As Transaction)
  F2Facility.Disengage FinishedCustomer
  If Q2Queue.Length > 0 _
  Then
    F2Facility.EngageAndService Q2Queue.Depart, 1
  End If
  Set FinishedCustomer = Nothing
End Sub
```

Figure 15.11   Waiting time limit—*F2FacilityEvent*.

```
Sub TimerDelayEvent(ByRef QueueTimer As Transaction)
  Dim _
    Index As Integer
  For Index = 1 To Q1Queue.Length
    If QueueTimer.TransID = Q1Queue.View(Index).TransID _
    Then
      F2Facility.NextFacility Q1Queue.Detach(Index), 1, Q2Queue
      Exit For
    End If
  Next Index
  Set QueueTimer = Nothing
End Sub
```

Figure 15.12   Waiting time limit—*TimerDelayEvent*.

In Figure 15.12, sub *TimerDelayEvent*, the copy of the transaction finishes the timer delay and attempts to locate the original transaction in *Q1* whose *TransID* matches that of the transaction copy. The parent transaction either has already entered *F1* or continues to reside in *Q1*. If the original is in *Q1*, it is removed and rerouted for service in facility *F2*.

15.5 Time-Dependent Intercreation Times at a Source

Transactions are created from a source with rates that are dependent on the time of the day. Specifically, suppose that throughout the day, customers arrive at a one-window drive-in bank according to the distributions shown in Table 15.1.

The model includes a source, *S1*, for arriving jobs; a queue, *Q1*, for waiting jobs; a facility, *F1*, for processing jobs; and a timer segment to change interarrival times. Corresponding to the number of sources, facilities, and delays, there are three primary event subroutines. In sub *Initial*, arrival rates and time intervals are from spreadsheet *UserInitialed*.

The associated model statements for each subroutine are in Figures 15.13 through 15.16. As shown in Figure 15.13, for sub *S1SourceEvent*, arriving jobs either join the queue or immediately begin service. For sub *F1FacilityEvent* in Figure 15.14, the server becomes idle after a service completion when there are no waiting customers; otherwise, a customer departs the queue and is scheduled for service.

In Figure 15.15, sub *TimerDelayEvent*, *Interval* is a time of day index. *Interval* = 1 corresponds to the 8:00 a.m. to 10:00 a.m. time slot. The timer transaction finishes the delay, updates *Interval*, and resets the customer interarrival rate, depending on the current value of *Interval*.

The interarrival time at *S1* is *Exponential* (0.4) for the first two hours (10:00 a.m.) until the timer transaction finishes its delay. At that time, the interarrival time is set to 0.3, yielding an interarrival time of *Exponential* (0.3). After three more hours, the timer transaction again finishes its delay; at that time interarrival time assumes a value of 0.2. Again, after three additional hours, *S1Source.Suspend* will deactivate *S1* and it will remain deactivated for the next 15 hours (until 8:00 a.m. the next day). The timer

Table 15.1 Time-Dependent Creation Interarrival Time

Time Period	Interarrival Time
08:00 a.m. to 10:00 a.m.	*Exponential* (24) minutes
10:01 a.m. to 01:00 p.m.	*Exponential* (18) minutes
01:01 p.m. to 04:30 p.m.	*Exponential* (12) minutes
04:31 p.m. to 07:59 a.m.	No arrivals, bank is closed

```
Sub S1SourceEvent(ByRef Arrival As Transaction)
    F1Facility.NextFacility Arrival, 1, Q1Queue
End Sub
```

Figure 15.13 Time-dependent creation—*S1SourceEvent*.

```
Sub F1FacilityEvent(ByRef CompletedJob As Transaction)
    F1Facility.Disengage CompletedJob
    Set CompletedJob = Nothing
    If Q1Queue.Length > 0 _
    Then
        F1Facility.EngageAndService Q1Queue.Depart, 1
    End If
End Sub
```

Figure 15.14 Time-dependent creation—*F1FacilityEvent*.

```
Sub TimerDelayEvent(ByRef Timer As Transaction)
  Interval = Interval + 1
  If Interval > 4 Then
    Interval = 1
  End If
  Select Case Interval
    Case 1
      S1Source.Restart
      S1Source.NewDistribution "Exponential", IAT(1) / 60
    Case 2
      S1Source.NewDistribution "Exponential", IAT(2) / 60
    Case 3
      S1Source.NewDistribution "Exponential", IAT(3) / 60
    Case 4
      S1Source.Suspend
    Case Else
      SimulatorMessage "Undefined time interval"
      End
  End Select
  TimerDelay.Start Timer, LapsedTime(Interval)
End Sub
```

Figure 15.15   Time-dependent creation—*TimerDelayEvent*.

```
' User initialed input
Dim _
  Index As Integer, _
  DelayTrans As Transaction
With Worksheets("UserInitialed").Range("C1")
  For Index = 1 To 4
    LapsedTime(Index) = .Offset(Index, 0).Value
    IAT(Index) = .Offset(Index, 1).Value
  Next Index)
End With
Interval = 1
TimerDelay.Start NewTransaction ' initial timer
End Sub
```

Figure 15.16   Time-dependent creation—user-defined program segment.

will return with *Interval* = 1 to *Restart* arriving transactions from *S1* and repeat the process of changing interarrival times. The *Case* structure can be extended to cover any number of breaking points during the day. Figure 15.16 shows the program segment from sub *Initial* that initializes the timer segment.

15.6 Network Logic Change Using Queue Nodes

Solid waste is hauled to a processing facility by trucks that arrive randomly every *Exponential* (20) minutes. The facility has two compactors that are used to reduce the solid waste into ballets for transfer by train to a dry-fuel installation.

The compactors operate in successive shifts of eight hours each, with the idle shift used for maintenance. The capacity of the first compactor is one truckload per ballet and the second is two loads per ballet. The compacting time of the first compactor is *Exponential* (12) minutes and the second is *Exponential* (15) minutes. For this model, Figures 15.17 through 15.19 demonstrate the use of queue nodes to represent the two shifts with their specific load and compacting time requirements.

The model includes a source, *S1*, for arriving trucks; two queues, *Q1* and *Q2*, for jobs waiting for compactors 1 and 2, respectively; two facilities, *F1* and *F2*, for compactors 1 and 2, respectively; and an eight-hour timer segment. Corresponding to the total number of sources, facilities, and delays, there are four primary event subroutines. A user-defined variable, *Shift*, is initialized to 1 in sub *Initial* so that compactor 1 is on the first shift and compactor 2 is on the alternate shift.

Figure 15.17 shows sub *TruckSourceEvent*, the primary event procedure for arriving trucks. Depending on the shift, an arriving truck is routed to compactor 1 (*Comp1Facility*) or compactor 2 (*Comp2Facility*). Also depending on the status of the compactor, the load of waste may be serviced immediately or queued for service. Upon arrival at compactor 1, if the compactor is idle, the load is serviced immediately. Otherwise, the load is queued to wait for processing.

The decision to process loads arriving to compactor 2 is slightly more involved. Specifically, there are three unique situations:

1. The server may be busy.
2. The server may be idle and its queue is empty.
3. The server may be idle because only one load is in the queue.

```
Sub TrucksSourceEvent(ByRef Arrival As Transaction)
   If Shift = 1 _
   Then
      Comp1Facility.Destination Arrival, 1, Q1Queue
   Else
      If Q2Queue.Length > 0 And _
         Comp2Facility.NumberOfIdleServers > 0 _
      Then
         Q2Queue.Remove 1, 1
         Comp2Facility.EngageAndService Arrival, 1
      Else
         Q2Queue.Join Arrival
      End If
   End If
End Sub
```

Figure 15.17   Network logic change—*TruckSourceEvent*.

```
Sub Comp1FacilityEvent(ByRef CompletedJob As Transaction)
   Comp1Facility.Disengage CompletedJob
   Set CompletedJob = Nothing
   If Q1Queue.Length > 0 _
   Then
      Comp1Facility.EngageAndService Q1Queue.Depart, 1
   End If
End Sub
Sub Comp2FacilityEvent(ByRef CompletedJob As Transaction)
   Comp2Facility.Disengage CompletedJob
   Set CompletedJob = Nothing
   If Q2Queue.Length > 1 _
   Then
      Comp2Facility.EngageAndService Q2Queue.Depart, 1
      Q2Queue.Remove 1, 1
   End If
End Sub
```

Figure 15.18   Network logic change— *Comp1FacilityEvent*.

```
Sub TimerDelayEvent(ByRef Timer As Transaction)
  If Shift = 1 _
  Then
    Shift = 2
    While Q1Queue.Length > 0
      Q2Queue.Join Q1Queue.Depart
    Wend
    If Q2Queue.Length > 1 And _
           Comp2Facility.NumberOfIdleServers > 0 _
    Then
      Comp2Facility.EngageAndService Q2Queue.Depart, 1
      Q2Queue.Remove 1, 1
    End If
  Else
    Shift = 1
    While Q2Queue.Length > 0
      Q1Queue.Join Q2Queue.Depart
    Wend
    If Q1Queue.Length > 0 And _
           Comp1Facility.NumberOfIdleServers > 0 _
    Then
      Comp1Facility.EngageAndService Q1Queue.Depart, 1
    End If
  End If
End If
```

Figure 15.19   Network logic change—*TimerDelayEvent*.

As shown in the Figure 15.17 model segment, only in the third case will service begin as a result of the arriving load. Notice that as the transaction representing a load is scheduled for processing, another load transaction is removed from the queue and the system. In the other instances an arriving load simply joins the queue.

Primary event procedures for compactors 1 and 2 are shown in Figure 15.18. The job completion logic is similar to that for load arrivals. Only one load is needed to begin service by compactor 1. In contrast, compactor 2 must have two loads in the queue. For compactor 2, the first queued transaction is engaged for service and the second exits the system.

Figure 15.19 shows the sub *TimerDelaySegment*, the primary event procedure for the shift timer segment. After eight hours of operation, the transaction finishes the timer delay for shift 1 and moves all transactions from *Q1* to *Q2*. Then, trucks will begin sending loads to *Q2*. After another timer delay of eight hours the shift timer resets the stage for the *Q1* compactor 1 sequence.

An additional requirement for this subroutine that follows the movement of transactions from *Q2* to *Q1* (or *Q1* to *Q2*) is, when possible, to initiate service in the compactor coming on line. Initiating service after moving transactions between queues ensures that queued transactions receive service prior to any subsequently arriving transactions.

15.7 Controlled Blockage of a Facility

Facility blockage occurs whenever a facility is in front of another (busy) facility or a full (finite capacity) queue. In this case, the facility may become unblocked when the status of the successor node changes. For example, consider the case of three machines in series that produce a product. A single robot automatically moves material between machines. When a machine completes a job, it must wait for the robot to perform certain tasks before starting a new job. Since the robot may be busy elsewhere when the job is completed, the job completion and robot availability may not coincide. When this occurs, the job occupies the machine until the robot is available. In this section, how to model blockage of a facility is demonstrated.

A simplified case is a single machine that, upon completing its load, must request a crane to remove the finished job. At the instant the crane begins picking up the load, the machine can begin processing a waiting job. Jobs arrive at the machine every *Exponential* (25) minutes. *Exponential* (20) minutes are required to process a job on the machine. The time for the crane to arrive at the machine and pick up the load is *UniformContinuous* (4, 6) minutes. The performance measure is the total time a job is in the system.

One approach to modeling the above environment is to "inflate" the job processing time by the time for the crane to arrive at the workstation. This ensures that the machine is not able to begin a new job until the pickup occurs. The disadvantage of such a modeling strategy is that it biases the machine statistics. In the following approach the job processing is complete but the workstation is occupied while waiting on a pickup.

From spreadsheet *InitialModel*, the model includes a source, *SS*, for arriving jobs, a queue, *QQ*, for waiting jobs, a facility, *FF*, for the workstation, and a timer for the crane delay. Corresponding to the number of sources, facilities, and delays on the spreadsheet there are three primary event subroutines.

Figure 15.20 shows sub *SSSourceEvent*, the primary event procedure for arriving jobs. Upon arrival, a service time is assigned to *Attribute (1)* of the job. The job either engages the server and begins service or joins the queue of waiting jobs.

Sub *MMFacilityEvent* in Figure 15.21 is for the service completion. *TimerDelay* represents the lapsed time before the crane arrives. Notice that the job remains in the facility while waiting on the crane. Sub *TimerDelayEvent* in Figure 15.22 controls the crane. When the crane arrives (finishes its delay), the time in the system statistic for the job is collected. The job is then immediately removed from the facility. If there are jobs in the queue, the next job is scheduled for service and another crane request is issued. Otherwise, the server remains idle.

```
Sub SSSourceEvent(ByRef Arrival As Transaction)
   Arrival.Assign 1, Distribution.Exponential(25)
   If MMFacility.NumberOfIdleServers > 0 _
   Then
      MMFacility.EngageAndService Arrival, 1, Arrival.A(ServiceTime)
   Else
      QQQueue.Join Arrival
   End If
End Sub
```

Figure 15.20   Controlled blockage—*SSSourceEvent*.

```
Sub MMFacilityEvent(ByRef CompletedJob As Transaction)
   TimerDelay.Start CompletedJob, UniformContinuous(4, 6)
End Sub
```

Figure 15.21   Controlled blockage—*MMFacilityEvent*.

```
Sub TimerDelayEvent(ByRef Crane As Transaction)
  Dim _
    NextJob As Transaction
  CycleTimeStatistic.Collect MMFacility.View(1).SystemTime
  MMFacility.Disengage MMFacility.View(1)
  If QQQueue.Length > 0 _
  Then
    Set NextJob = QQQueue.Depart
    MMFacility.EngageAndService NextJob, 1, NextJob.A(ServiceTime)
    Call CraneRequest(NextJob)
  End If
End Sub
```

Figure 15.22   Controlled blockage—*TimerDelayEvent*.

15.8 Assemble and Match Sets with Common Queues

A factory manufactures two types of chairs. The first type requires two cushions; the second type is assembled with only one cushion. Frames for the chairs as well as the cushions are manufactured on three separate production lines. Production times per unit of the frames for chair 1 and chair 2 are exponentially distributed with means 40 and 30 minutes, respectively. Production time per cushion is *UniformContinuous* (10, 15) minutes. Figures 15.23 through 15.25 present the associated model procedures.

The model includes sources for arriving frames and cushions, queues for each frame and cushions, and facilities to assemble each type of chair. Corresponding to the total number of sources and facilities, there are five primary event subs. There are also three additional subroutines used to simplify the program logic and reduce duplicate coding effort.

Figure 15.23 has primary event procedures for arriving frames and cushions. The model shows that the frames for chair 1 and chair 2 are stored in queues *Frame1* and *Frame2*, respectively. Manufactured cushions are kept in a common cushion queue. Each time a frame arrives, it joins the queue and then the model attempts to begin a chair assembly. In a similar manner, each time a cushion arrives, it joins the cushion queue and then the model attempts to begin a chair assembly. Notice that because *StartChair1Assembly* precedes *StartChair2Assembly*, the model gives preference to assembling chair 1. When there is a shortage of cushions, the potential exists for "cushion starvation" of chair 2.

Figure 15.24 presents primary event procedures for assembled chairs. In each procedure the chairs are disengaged from the server and, assuming a frame and the appropriate number of cushions are available (two cushions for frame 2), another assembly is scheduled. Figure 15.25 shows procedures that initiate the assembly operation. As a reminder, these are not primary event procedures. However, they are used by primary event procedures when a new transaction enters a frame or cushion queue or when a transaction completes service (assembly) in facility *Chair1* or *Chair2*. To initiate an assembly, there must be at least the minimum number of units needed to assemble a chair and the assembly facility must be idle. In both procedures, when the above conditions are true, the appropriate number of cushion transactions is removed from the cushion queue and the transaction in the frame queue now represents the combination of frame and cushions necessary for an assembly.

```
Sub Frame1SourceEvent(ByRef NewFrame1 As Transaction)
    Frame1Queue.Join NewFrame1
    Call StartChair1Assembly
End Sub
Sub Frame2SourceEvent(ByRef NewFrame2 As Transaction)
    Frame2Queue.Join NewFrame2
    Call StartChair2Assembly
End Sub
Sub CushionSourceEvent(ByRef NewCushion As Transaction)
    CushionQueue.Join NewCushion
    Call StartChair1Assembly
    Call StartChair2Assembly
End Sub
```

Figure 15.23   Assemble set—*Frame1SourceEvent, Frame2SourceEvent,* and *CushionSourceEvent.*

```
Sub Chair1FacilityEvent(ByRef AssembledChair As Transaction)
    Chair1Facility.Disengage AssembledChair
    Set AssembledChair = Nothing
    Call StartChair1Assembly
End Sub
Sub Chair2FacilityEvent(ByRef AssembledChair As Transaction)
    Chair2Facility.Disengage AssembledChair
    Set AssembledChair = Nothing
    Call StartChair2Assembly
End Sub
```

Figure 15.24   Assemble set—*Chair1FacilityEvent* and *Chair2SourceEvent.*

```
Sub StartChair1Assembly()
    If Frame1Queue.Length > 1 And CushionQueue.Length > 2 And _
       Chair1Facility.NumberOfIdleServers > 0 _
    Then
        CushionQueue.Remove 1, 2
        Chair1Facility.EngageAndService Frame1Queue.Depart, 1
    End If
End Sub

Sub StartChair2Assembly()
    If Frame2Queue.Length > 1 And CushionQueue.Length > 1 And _
       Chair2Facility.NumberOfIdleServers > 0 _
    Then
        CushionQueue.Remove 1, 1
        Chair2Facility.EngageAndService Frame2Queue.Depart, 1
    End If
End Sub
```

Figure 15.25   Assemble set—sub *StartChair1Assembly* and *StartChair2Assembly.*

15.9 Network Models

The Jackson network In Figure 15.26 represents capacitated stations with *M/M/C* queue stations. Details of the *M/M/C* representation of basic queuing systems are described in Chapter 2. The network consists of four stations connected by arcs that indicate the allowable flow direction and probabilistic routing information. Jobs enter at station 1, are processed, and then routed probabilistically

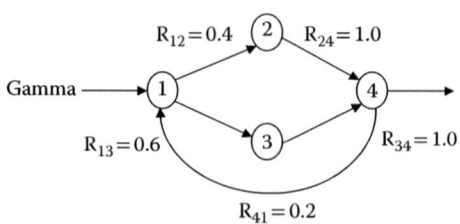

Figure 15.26 Jackson network of queueing nodes.

to station 2 or 3, and so on until exiting at station 4. At each station is a queue for jobs waiting to be processed at that station. The feedback loop from station 4 to 1 could represent failing inspection and returning for rework. In each station there are one or more servers, each capable of performing the operations. In a simulation model the *M/M/C* requirement can be relaxed to allow any probability distribution for arrival rate to the system and processing times in each node. Performance metrics for a Jackson network include the normal basic queue statistics for each node, average cycle time for jobs, and average work in progress (cumulative queue counts) for the network.

It is straightforward to generalize a network of queues and facilities because any DEEDS-defined class may be represented as a single dimension array of class instances. In spreadsheet *InitialModel*, a source describes jobs arriving at the network. There must also be a queue and facility for each node in the network. The facility description includes the number of servers and service time distribution. The following replaces the node declaration and initialization statements generated by *ProgramManager*:

```
Public _
    F(1 To 20) As Facility, _
    Q(1 To 20) As Queue
```

This defines an array of 20 facilities and queues. The following program segment is from sub *Initial*. It also replaces a portion of the sub *Initial* program produced by *ProgramManager*. The array of queues and facilities is in the same order as listed on spreadsheet *Initial*.

```
For Index = 1 To InFacilities.Count
    Set Q(Index) = Instance("Q" & Index, InQueues)
    Set F(Index) = Instance("F" & Index, InFacilities)
    Set PDFNode(Index) = Instance("Node" & Index, InPDFs)
Next Index
```

On the basis of the above definitions, there are only two primary event subroutines, one for arrivals and the other for service completions. As shown in the following, the case structure in sub *Parser* is modified to route all service completions to the same procedure:

```
Case "F1Facility", "F2Facility", _
     "F3Facility", "F4Facility", _
     "F5Facility"
        ServiceCompletionEvent CurrentEvent
```

Figure 15.27 has the arrival procedure. Upon arrival, *Attribute (1)* is assigned the number of the current node. The arriving transaction either is placed into service or joins the queue for node

```
Sub S1SourceEvent(ByRef Arrival As Transaction)
    Arrival.Assign NodeID, 1
    If F(1).NumberOfIdleServers > 0 _
    Then
        F(1).Engageandservice Arrival, 1
    Else
        Q(1).Join Arrival
    End If
End Sub
```

Figure 15.27   **Jackson network—*S1SourceEvent*.**

```
Sub ServiceCompletionEvent(Job As Transaction)
    Dim _
        LastNode As Integer, _
        Index As Integer, _
        NextNode As Integer
    LastNode = Job.A(NodeID)
    F(LastNode).disengage Job
    If (Q(LastNode).Length > 0) _
    Then
        F(LastNode).Engageandservice Q(LastNode).Depart, 1
    End If
    NextNode = PDFNode(LastNode).rv
    If NextNode <> 0 _
    Then
        Job.Assign NodeID, NextNode   ' send to next node
        F(NextNode).NextFacility Job, 1, Q(NextNode)
    Else
        CycleTimeStatistic.Collect Job.SystemTime
    End If
End Sub
```

Figure 15.28   **Jackson network—*ServiceCompletionEvent*.**

F(1), the first node. Figure 15.28 shows the subroutine for service completion events. *LastNode* is assigned the value of *Attribute (1)*, the node index for the node that completed service on the transaction. The completed job disengages from the *LastNode* facility. If there is a transaction in the queue for the disengaged facility, it departs the queue, engages the facility, and is scheduled for service completion.

To search for the next destination, *NextNode*, the *PDFNode (LastNode)* returns a node index, *NextNode*. If *NextNode* is equal to zero the transaction exits the network. Otherwise, the transaction is scheduled for a service completion at *NextNode* or joins the *NextNode* queue.

15.10 Sampling without Replacement

Consider the situation in which customers are served in a 10-story building by a single elevator. Customers may originate from any floor with equal probabilities. Their destination floor can be any of the remaining floors, also with equal probabilities. The situation calls for sampling from 10 floors to determine the "origin" floor and then from the remaining 9 floors to determine the "destination" floor. This case is equivalent to sampling without replacement. The model segment in Figure 15.29 demonstrates how the sampling process can be simulated.

```
Sub Main()
  Dim _
    Index As Integer, _
    OriginalValue(1 To 10) As Single, _
    SampledValue(1 To 10) As Single
  For Index = 1 To 10
    OriginalValue(Index) = Index
  Next Index
  SampleWoReplacement OriginalValue, SampledValue
End Sub

Sub SampleWoReplacement(ByRef OriginalValue() As Single, _
                        ByRef SampledValue() As Single)
  Dim _
    Index As Integer, _
    Jndex As Integer, _
    VectorLength As Integer

  VectorLength = UBound(OriginalValue())
  For Index = VectorLength To 1 Step -1
    Jndex = CInt(1 + Rnd() * (Index - 1))
    SampledValue(VectorLength + 1 - Index) = OriginalValue(Jndex)
    OriginalValue(Jndex) = OriginalValue(Index)
  Next Index
End Sub
```

Figure 15.29 Program segment—sampling without replacement.

Sub *Main* passes two arrays, the *OriginalValue* and *SampledValue* to subroutine *SampleWoReplacement*. When control is returned to *Main*, the *SampledValue* array is ordered by randomly selected values in *OriginalValue*. The first two values in *SampledValue* are an "origin" and "destination" pair for the elevator problem. However, the *SampleWoReplacement* procedure is extended to obtain M objects without replacement from N objects, where $M < N$.

Problems

15.1. Jobs arrive from a source every *Exponential* (2) minutes in two distinct types. Type I jobs constitute 60% of the arrivals. The remaining 40% are of Type II. Each job type has four priority classes that occur with probabilities 0.3, 0.2, 0.4, and 0.1 for classes 1, 2, 3, and 4, respectively. Class 4 represents the highest priority for processing. All jobs are processed on a single machine. The processing time is *UniformContinuous* (1, 1.5) minutes for Type I jobs and *Exponential* (1.8) for Type II jobs. When the machine becomes idle, it will choose a job from Type I or Type II depending on which one has the highest sum of class priorities. If both sums are equal, the job with higher priority is selected. If the tie persists, the job type is chosen randomly. Compute the total time each type job is in the system.

15.2. Two different types of jobs arrive at a shop. Type I jobs arrive every *Exponential* (20) minutes, and their processing time is *UniformContinuous* (15, 25) minutes. For Type II jobs, the interarrival time is *Exponential* (25) minutes, and their processing time is *Triangle* (15, 20, 25) minutes. The shop has three machines, each with its own operator. The third machine is used to relieve work buildup at the first two machines. Whenever the third machine completes a job, its next job is selected from the queue with the larger number of waiting jobs. A tie is broken randomly with equal probabilities. If an arriving job finds its preassigned machine busy, the job is assigned to the third machine if it is idle. Otherwise, the job stays in the preassigned queue. Study the impact of the third machine on the operation of the shop by computing the average time each job type is in the shop with and without the third machine

15.3. Modify the model in Section 15.2 to compute the utilization of each subfacility as well as the average length of each subqueue using arrays of time-based variables.

15.4. Revise the model in Section 15.2 using regular nodes and compare the results. Keep in mind that the two methods should produce the same numerical results when using the same random number sequence.

15.5. Revise the model in Section 15.2 using arrays of queues and facilities. Compare the results with those of Problem 15.4.

15.6. Revise the model in Section 15.2 to reflect the following conditions. Suppose that each server utilizes a resource and a maximum of six resources are available. The number of resources requested by each server is one, two, or three units with probabilities 0.3, 0.5, and 0.2, respectively. The same number is returned after the service is completed.

15.7. Revise the model in Section 15.5 assuming that interarrival times are as listed in Table 15.2.

15.8. Items arrive on a conveyor belt at a constant rate. The interarrival time is exponentially distributed with mean value equal to 0.24 times the inverse of the speed of the conveyor. Model several conveyor speeds, with each speed associated with a different run. Show how the arrival process can be simulated with conveyor speeds entered in the model via the *UserInitialed* spreadsheet.

15.9. Model the situation where jobs arrive at a facility every *Exponential* (1) minute. Each arrival is tagged with its estimated processing time. Historical data show that these processing times equal *Exponential* (8) minutes. Every two hours the operator will scan the queue of waiting jobs and choose the one with the longest processing time as the next job to process. The remaining jobs are serviced in the order of arrival for the next two hours.

15.10. Consider the situation where a product is assembled from two parts, A and B. The production time for A is *Exponential* (5) minutes and that for B is *Exponential* (4) minutes. Assembly of the two parts into a final product requires *Exponential* (4.5) minutes. The assembled product is inspected in approximately *Exponential* (1) minute. About 10% of inspected units are usually discarded. Develop a model that computes the following:
   a. The average in-process inventory for parts A and B
   b. The average in-process inventory for assembled products, including those being inspected

Table 15.2 Interarrival Times for Problem 15.7

Time Period	Interarrival Time (Minutes)
08:00 a.m. to 10:00 a.m.	*UniformContinuous* (20, 28)
10:01 a.m. to 01:00 p.m.	*Exponential* (18)
01:01 p.m. to 04:30 p.m.	*Triangular* (7, 9, 15)
04:31 p.m. to 07:59 a.m.	No arrivals

15.11. Three serial machines are used to process a product. Raw materials arrive at the shop every *Exponential* (12) minutes. *Exponential* (3) minutes are needed to process a job on each machine. A single robot performs all materials handling operations with priority given to loading the first machine. All other requests are processed on a strict first-in, first-out (FIFO) basis. No jobs are allowed to wait between machines. Thus, when a job is completed, it cannot be removed from the machine unless the next machine is free and the robot is not occupied elsewhere. *Exponential* (0.7) minutes are needed for the robot to load/unload a machine. However, the time needed for the arm of the robot to reach the requesting machine depends on the locations of the machines. The distances from the receiving area to machine 1, machine 1 to machine 2, machine 2 to machine 3, and machine 3 to receiving area are 20, 30, 30, and 20 inches, respectively. The robot moves at a constant speed of eight inches per second. Develop a model to predict the effect of robot availability on the productivity of the machines. (*Hint*: Use an array of facilities.)

Reference

1. Taha, H. A., *Operations Research: An Introduction.* Upper Saddle River, NJ: Prentice-Hall, 2007.

Chapter 16

Advanced Routing Techniques

16.1 Introduction

The class procedures described in Chapter 11 collect statistics, change system parameters, and cause transactions residing in queues, facilities, and delays, to resume motion in the network by affecting changes to transaction lists. The simulation program logic determines which changes should be made and when to affect the changes. Advanced modeling techniques for transaction routing are presented in this chapter.

16.2 Routing Transactions

Spawning and routing transactions in most transaction flow simulation languages are somewhat unwieldy because of the complex syntax to achieve synchronization. In contrast, for situations with complex flow logic, design environment for event-driven simulation (DEEDS) offers powerful and flexible procedures to assist with controlling and directing transaction flow in the network. And, as shown in Chapters 9, 10, and 11, DEEDS has no special syntax beyond Visual Basic for Applications (VBA).

Specifically, DEEDS supports the following seven routing mechanisms:

1. Always
2. Conditional
3. Select
4. Probabilistic
5. Dependent
6. Exclusive
7. Last choice

When a transaction exits a node, an event handler procedure determines the destination and then routes the transaction to that destination. The simple case is analogous to a passenger departing a flight and asking the gate attendant for directions. *A, C,* and *S* are used to model such decisions.

The next node destination must be a facility, queue, or delay. Otherwise, the destination must be to exit the system. A more complex modeling situation is a project manager simultaneously dispatching a number of assistants on project activities. There may be multiple instances of *Conditional, Always, Probabilistic*, and *Select (CAPS)* routes within the same procedure to achieve such a result.

16.2.1 Always Routing

Always routing means branching to a destination, without consideration (the destination will accept the node). An infinite capacity queue node and the delay node (which also has infinite capacity) are the only two situations that characterize an uncontested *Always* destination. Actually, a queue never refuses entry, but DEEDS destroys transactions that attempt to join a full queue. In all other situations, there must be an alternate destination node when a transaction is refused entry by a *CAPS* destination node. Otherwise, the abandoned transaction is reclaimed by VBA during garbage collection.

16.2.2 Conditional Routing

The *If–Then–Else–End If* structure is used to evaluate the conditions for a branch and specify an alternate destination when conditions are not satisfied for that branch. In a sense, the alternate destination may be viewed as a *Last Choice*. For example, a transaction, *Customer*, is routed to service in facility *F1* if a server is available. Otherwise, the transaction is routed to queue *Q1*. The program segment is shown in Figure 16.1. Because the situation described is, by far, the most common, DEEDS has a *Facility* class procedure *NextFacility* that achieves the same result. The format is

```
F1Facility.NextFacility Customer, 1, Q1Queue
```

In both instances, 1 indicates the number of servers requested by *Customer*.

16.2.3 Select Routing

Select routing is also a form of conditional routing in which a destination node is selected on the basis of policy or comparative conditions at each destination being considered. The select routing must have a last resort destination for when the selected destination refuses to accept the transaction.

The node selection criteria are used to choose a next node from among the set of candidate nodes. As summarized in Table 16.1, the node selection criteria may be independent of node characteristics; however, the criteria may apply to the current state or "recent history" of the node. *NodeObjects* separated by a comma is an indefinite list of *NodeObjects* to be considered by the function. The syntax for these functions is

Set *ObjectVariable* = Find.*FunctionName arglist*

```
If F1Facility.NumberOfIdleServers > 0 _
Then
    F1Facility.EngageAndService Customer, 1
Else
  Q1Queue.Join Customer
End If
```

Figure 16.1 Program segment—simple conditional routing.

Table 16.1 Destination Functions for Select Routing

Function	Returns
A. Node independent	
PreferredOrder (NodeObjects separated by comma)	*NodeObject*
Random (NodeObjects separated by comma)	*NodeObject*
Rotation (LastNodeIndex, NodeObjects separated by comma)	*NodeObject*
B. Current state of node	
MaxQueueLength (QueueObjects separated by comma)	*QueueObject*
MinQueueLength (QueueObjects separated by comma)	*QueueObject*
MaxNumberOfBusyServers (FacilityObjects separated by comma)	*FacilityObject*
MinNumberOfBusyServers (FacilityObjects separated by comma)	*FacilityObject*
MaxNumberOfIdleServers (FacilityObjects separated by comma)	*FacilityObject*
MinNumberOfIdleServers (FacilityObjects separated by comma)	*FacilityObject*
TotalLength (NodeObjects separated by comma)	Integer
C. "Recent history" of node	
MaxAverageLq (QueueObjects separated by comma)	*QueueObject*
MinAverageLq (QueueObjects separated by comma)	*QueueObject*
MaxAverageWq (QueueObjects separated by comma)	*QueueObject*
MinAverageWq (QueueObjects separated by comma)	*QueueObject*
MaxAverageBusyServers (FacilityObjects separated by comma)	*FacilityObject*
MinAverageBusyServers (FacilityObjects separated by comma)	*FacilityObject*

where, depending on the function, *ObjectVariable* is either VBA Object data type or a DEEDS reference variable for an instance of class *Queue, Facility*, or *Delay*. Of course, VBA data type Object may be used for all of these functions. *FunctionName* is a function from Table 16.1. These functions are in DEEDS module *Find*.

The function returns an object variable set to the first node in the list that satisfies the condition, unless none of the nodes will accept a transaction. In that case, the function returns a node object variable set to *Nothing*. Assume for this example that the number in the queue for *Q1, Q2*, and *Q3* is 3, 2, and 3, respectively. The function call

```
Set QueueObject = Find.MaxQueueLength(Q1Queue, Q2Queue,
Q3Queue)
```

sets *QueueObject* to *Q1Queue* because it is the first queue with length of three or *Nothing* if queue *Q1* has a capacity of three (i.e., *Q1* is full and will not accept a transaction).

16.2.3.1 Node Independent

A node-independent function receives a list of nodes that are considered in the order listed. The function returns an object variable set to the first destination that will accept the transaction. The program segment must explicitly indicate as a *Last Choice* what happens to the transaction when it is not accepted by any node. For example, consider the code given in Figure 16.2 for routing transaction *Customer*. Function *NodeType* returns a String indicating the type of node for instances of class *Queue, Facility,* and *Delay*. If *F1, F2,* or *Q1* will accept the customer, the *Last Choice* option may be omitted.

Notice that

```
NodeObject.Join Customer
```

and

```
NodeObject.EngageAndService Customer, 1
```

are examples of the flexibility derived from DEEDS. *NodeObject* is an indirect reference to the selected destination. In the above examples, *NodeObject* must reference a queue and facility, respectively. The difference between the previous example and the example in Figure 16.3 is that the value assigned to attribute 1 is dependent on the destination.

In function *Rotation*, a destination node that cannot accept the transaction must wait for its next turn consideration in the rotation. The first parameter of *Rotation* is an index to the last destination that accepted a transaction. *LastNodeIndex* is initialized to the number of nodes in the list. The procedure that calls *Rotation* must maintain the value of *LastNodeIndex*. Recall that to maintain variable values in procedures, the variable must be declared as *Private* in the *EventsManager* module or the procedure must be declared as *Static*.

16.2.3.2 Current State of Node

A current state measurement is based on the status of the node when the function is called. With the exception of *TotalLength*, these functions return an object variable set to the first network node in the parameter list that satisfies the condition or *Nothing* when the selected destination node will

```
Set NodeObject = _
    Find.Preferred(F1Facility, F2Facility, Q1Queue)
If Not(NodeObject is Nothing) _
Then
   Select Case NodeObject.NodeType
     Case "Facility"
      NodeObject.EngageAndService Customer, 1
     Case "Queue"
       NodeObject.Join Customer
     Case Else

   End Select
Else
   ' Last Choice destination
End If
```

Figure 16.2 Program segment—node-independent routing.

```
Set NodeObject = Find.Preferred(F1, F2, Q1)
If Not(NodeObject is Nothing) _
Then
  Select Case NodeObject.NodeType
    Case "Facility"
      If Facility.Name = "F1" _
      Then
        Customer.Assign (1) = 1
        NodeObject.EngageAndService Customer, 1
      Else
        Customer.Assign (1) = 2
        NodeObject.EngageAndService Customer, 1
      End IF
    Case "Queue"
      NodeObject.Join Customer
    Case Else

  End Select
Else
    ' Last Choice destination
End If
```

Figure 16.3   Program segment—node-independent routing and attribute assignment.

```
Set QueueObject = _
    Find.MinQueueLength(Q1Queue, Q2Queue)
If Not(QueueObject is nothing) _
Then
    QueueObject.Join Customer
Else
    D1Delay.Start Customer 'Last Choice destination
End If
```

Figure 16.4   Program segment—state of node routing.

```
If Customer.A(3) > 100 _
Then
  Set QueueObject = Find.MinQueueLength(Q1, Q2)
  If Not(QueueObject is Nothing) _
  Then
    QueueObject.Join Customer
  Else
    D1.Start Customer 'Last Choice destination
  End If
Else
    D1.Start Customer 'Last Choice destination
End If
```

Figure 16.5   Program segment—state of node routing with conditions.

not accept a transaction. *TotalLength* returns the sum of the transaction counts for nodes in the parameter list. In Figure 16.4, the transaction is routed to the shortest queue or to *D1* when the shortest queue is full and cannot accept the customer transaction.

The example in Figure 16.5 combines *Conditional* and *Select* routing. When the value of attribute 3 exceeds 100, choose between *Q1* and *Q2*. When the value of attribute 3 is less than or equal to 100, the transaction is routed to delay *D1*.

```
Set QueueObject = -
    Find.MinAverageLq(Q1Queue, Q2Queue, Q3Queue)
If Not(QueueObject is Nothing) _
Then
    QueueObject.Join Customer
Else
    'Last Choice destination
End If
```

Figure 16.6 Program segment—recent history routing.

16.2.3.3 "Recent History" of Node

Recent history data reflects activity in the node since the current observation of the simulation began. Recall that for a facility the server is considered busy when a transaction simply occupies a server (i.e., not necessarily on the primary events list). Similar to previous examples, unless at least one of these queues will accept the transaction, a *Last Choice* destination must be included. The concept is demonstrated in Figure 16.6. *Last Choice* must be included because the queue with the shortest queue length may be unable to accept the transaction.

16.2.4 Probabilistic Routing

The VBA structure *Select* is used to implement probabilistic routing. In the example of probabilistic routing in Figure 16.7 notice that *RandomValue* is assigned a value before the *Select* statement.

It is important that *RandomValue* be assigned prior to the *Select*; otherwise, each *Case* will invoke another call to function *Random* and a different random number will be used for each comparison. Notice also that the *Last Choice* destination was omitted because the assumption is that if the transaction is not scheduled for service, it will join the corresponding queue for that facility.

16.2.5 Dependent Routing

The concept of *dependent routing* increases the flexibility for synchronizing the flow of transactions. Dependent routing requires that the destination may be taken only when at least one previous route *CAPS* destination has been taken. The following VBA Collections are defined and initialized for dependent routing:

```
'user defined

Private _
    DestCAPS As Collection, _
    DestD As Collection
```

```
' User initialized

Set DestCAPS = New Collection
Set DestD = New Collection
```

```
RandomValue = Distribution.Random(1)
Select Case RandomDestination
  Case Is < 0.2
    FacilityF1.NextFacility Customer, 1, Q1Queue
  Case Is < 0.5
    FacilityF2.NextFacility Customer, 1, Q2Queue
  Case Is < 0.75
    FacilityF3.NextFacility Customer, 1, Q3Queue
  Case Else
    FacilityF4.NextFacility Customer, 1, Q4Queue
End Select
```

Figure 16.7 Program segment—probabilistic routing.

```
Sub RouteDependent(ByRef Parent As Transaction)
  Dim _
    TransactionList As Collection, _
    index As Integer
  Set TransactionList = New Collection
  If F1Facility.NumberOfIdleServers > 0 _
  Then
    DestCAPS.Add Item:=F1Facility
  End If
  If F2Facility.NumberOfIdleServers > 0 _
  Then
    DestCAPS.Add Item:=F2Facility
  End If
    ' F3 and F4 are Dependent destinations
    ' Verify acceptance by Dependent Destinations
  If F3Facility.NumberOfIdleServers > 0 _
  Then
    DestD.Add Item:=F3Facility
  End If
  If F4Facility.NumberOfIdleServers > 0 _
  Then
    DestD.Add Item:=F4Facility
  End If

  If DestCAPS.Count = 0 _
  Then
    'Parent to last resort
    Exit Sub
  End If
  For index = 1 To DestCAPS.Count + DestD.Count- 1
    TransactionList.Add Item:=Parent.Copy
  Next index

  TransactionList.Add Parent
  Route DestCAPS, TransactionList
  Route DestD, TransactionList
End Sub
```

Figure 16.8 Program segment—dependent routing.

As shown in Figure 16.8, the transaction being processed by sub *RouteDependent* is passed as a parameter, *Parent*. Sub *RouteDependent* accumulates a list of destination nodes that will accept a transaction in *DestCAPS* and a similar list of destination nodes in *DestD*. For each *DestCAPS* and *DestD* node, copies (offspring) of *Parent* are accumulated on *TransactionList*. The last transaction

```
Sub Route(ByRef NodeList As Collection, _
          ByRef TransactionList As Collection)
   Do While (NodeList.Count > 0)
      Select Case NodeList(1).NodeType
         Case "Facility"
            NodeList(1).EngageAndService TransactionList(1), 1
         Case "Queue"
            NodeList(1).Join TransactionList(1)
         Case "Delay"
            NodeList(1).Start TransactionList(1)
         Case Else
      End Select
      NodeList.Remove 1
      TransactionList.Remove 1
   Loop
End Sub
```

Figure 16.9   Program segment—CAPS routing.

on *TransactionList* is *Parent*. Sub *Route* then routes transactions on *TransactionList* to the list of nodes that have agreed to accept a transaction. When *DestCAPS* count is equal to zero, *Parent* is routed to the *Last Choice* destination and no transactions are routed to *DestD* nodes. Sub *Route* in Figure 16.9 routes transactions to the list of nodes in *DestCAPS* and *DestD* respectively.

The modeler copies sub *RouteDependent* and sub *Route* to *EventsManager* and edits the verification portion of sub *RouteDependent* to reflect the *CAPS* and *DestD* routes being considered.

16.2.6 Exclusive Routing

Exclusive routing is similar in concept and implementation to dependent routing. However, it differs, in that sub *RouteExclusive* routes to *CAPS* destination nodes only when at least one exclusive route is available. The following VBA Collections are defined and initialized for exclusive routing:

```
'user defined

Private _
   DestCAPS As Collection, _
   DestE As Collection
```

```
' User initialized

Set DestCAPS = New Collection
Set DestE = New Collection
```

Sub *RouteExclusive* in Figure 16.10 demonstrates exclusive routing. As previously indicated, sub *Route* in Figure 16.9 routes transactions to nodes in *DestCAPS*, *DestD*, and *DestE*.

```
Sub RouteExclusive(ByRef Parent As Transaction)
  Dim _
    TransactionList As Collection, _
    index As Integer
  Set TransactionList = New Collection
  If F1Facility.NumberOfIdleServers > 0 _
  Then
    DestCAPS.Add Item:=F1Facility
  End If
  If F2Facility.NumberOfIdleServers > 0 _
  Then
    DestCAPS.Add Item:=F2Facility
  End If
    ' F3 and F4 are Dependent destinations
    ' Verify acceptance by Dependent Destinations
  If F3Facility.NumberOfIdleServers > 0 _
  Then
    DestE.Add Item:=F3Facility
  End If
  If F4Facility.NumberOfIdleServers > 0 _
  Then
    DestE.Add Item:=F4Facility
  End If

  If DestCAPS.Count = 0 Or _
    DestE.Count = 0 _
  Then
    'Parent to last resort
    Exit Sub
  End If
  For Index = 1 to DestCAPS.Count + DestE.Count - 1
    TransactionList.Add Parent.Copy
  Next Index
  TransactionList.Add Parent
  Route DestCAPS, TransactionList
  Route DestE, TransactionList
End Sub
```

Figure 16.10 Program segment—identify exclusive routes.

16.2.7 Last Choice

Last Choice routing has only one destination and is attempted only when no other routes are traversed. By definition, if the *Exclusive* branch cannot be taken, the only available alternative route is the *Last Choice*.

From the previous discussion, *Dependent* and *Exclusive* routes depend on the outcome of *Conditional, Always, Probabilistic*, and *Select* routes. The *Last Choice* route depends on what happens with all other routes. Therefore, the *Last Choice* branch is considered last. The order of *Dependent* and *Exclusive* does not matter. The order of *Always, Select*, and *Conditional* routes is also unimportant.

16.3 Synchronized Queues

Another important synchronization feature is the ability to match or assemble transactions between and within queues. *Assemble* combines two or more transactions into a *single* transaction with the effect of removing transactions from the model. Transactions to be assembled may be from one or more queues. For a *Match*, matching transactions are removed from queues and routed to their next respective destinations. In either case, the procedure must ensure that the transaction (transactions) that results from an *Assemble (Match)* operation receives proper disposition.

16.3.1 Match

The simple case is when there is at least one transaction in queue$_1$, ... , queue$_m$ and all destinations will accept a transaction, then each transaction will be routed to its destination node. This matching situation is essentially synchronizing transaction departures. The selected transactions from the respective queues are from the front of each queue.

A DEEDS function determines when a match exists and an *If–Then–Else–End If* structure determines which destination nodes will accept a transaction. Each of the matched transactions may be routed to an independent destination. However, there may be situations where an associated transaction simply leaves the system. In the following example, when *Q1, Q2,* and *Q3* have one or more transactions and *F1, F2,* and *F3* will accept a transaction, a transaction from each queue is routed to *F1, F2,* and *F3*, respectively. In Figure 16.11, an *If–Then–Else–End If* structure determines whether there is a match and whether the destination nodes will accept a transaction.

In more complex matching conditions, a match may require that one or more attributes have equal values for all transactions leaving the queues. For this case, module *Find* has a *Match* function with the following format:

```
Find.Match(MatchList(), AttributeList(), _
     QueueObjects separated by comma) returns Boolean
```

The function returns *True* if there is at least one transaction in each queue with matching attributes specified in the *AttributeList* array. Also, when the function returns *True*, the *MatchList* array has an index into each queue to the matching transaction. In the function call, the first position in *AttributeList* may have the value "All." In that case, every attribute must match for transactions that satisfy the match condition. In Figure 16.12, transactions from queues *Q1* and *Q2* must have the same values for attributes 0 and 1 before proceeding to facilities *F1* and *F2*, respectively. In this example, an *If–Then–Else–End If* structure determines whether the destination nodes accept a transaction, and the *Match* function determines whether a match exists.

When using matching by attributes, DEEDS chooses the first *ordered* transaction in each queue that satisfies the matching conditions. However, be aware that the order in which the queues are scanned may lead to different results. To illustrate this point, consider the set of entries for *Q1* and *Q2* given in Table 16.2.

If *Q1* and *Q2* are matched on attributes 2 and 3, the matched entries will be the first in *Q1* (0, 11, 22) and the third in *Q2* (3, 11, 22). However, if we match *Q2* and *Q1* on attributes 2 and 3, the resulting matched entries will be second for *Q1* (1, 33, 44) and first for *Q2* (5, 33, 44). The variation stems from the fact that the first queue in the list is the "base" to which the other queue's entries are compared.

```
If (Q1Queue.length > 0 And F1Facility.NumberOfIdleServers > 0) And _
   (Q2Queue.length > 0 And F2Facility.NumberOfIdleServers > 0) And _
   (Q3Queue.length > 0 And F3Facility.NumberOfIdleServers > 0) _
Then
   F1Facility.EngageAndService Q1Queue.Depart,1
   F2Facility.EngageAndService Q2Queue.Depart,1
   F3Facility.EngageAndService Q3Queue.Depart,1
End If
```

Figure 16.11 Program segment—match queues.

```
If F1Facility.NumberOfIdleServers > 0 And _
   F2Facility.NumberOfIdleServers > 0 _
Then
   AttributeList (0) = 2
   AttributeList (1) = 3
   MatchFound = _
      Find.Match (MatchList(), AttributeList(), Q1Queue, Q2Queue)
   If MatchFound = True _
   Then
      F1Facility.EngageAndService Q1Queue.Detach(MatchList(0)), 1
      F2Facility.EngageAndService Q2Queue.Detach(MatchList(1)), 1
   Else
      Simulator.Message "No matches found."
   End If
```

Figure 16.12   Program segment—match with conditions.

Table 16.2   Queued Transactions with Attribute Values

Q1 Entry	A(1)	A(2)	A(3)	Q2 Entry	A(1)	A(2)	A(3)
1	0	11	22	1	5	33	44
2	1	33	44	2	0	10	13
3	4	44	55	3	3	11	22

16.3.2 Assemble

In the simple case, an *Assemble* is based on the availability of transactions in a queue(s). For example, assume a queue accumulates identical components one at a time and three components are combined to form a subassembly. Whenever three components are available, a subassembly is produced. The example in Figure 16.13 models that environment. The first transaction represents the assembly and the next two are removed from the system. In Figure 16.14, the last transaction represents the subassembly. Figure 16.15 represents three different components in separate queues *Q1*, *Q2*, and *Q3*. The first transaction will represent the subassembly. This model segment must be tested each time a transaction joins its respective queue.

Module *Find* also has an *Assemble* function with format

```
Find.Assemble(AssembleType As String, _
     TransactionObjects separated by comma) _
          As Transaction
```

AssembleType is a string with permissible values shown in Table 16.3. *TransactionObjects* represents an indefinite list of transactions to be assembled and the function returns an object reference variable for the combined transaction.

Figure 16.16 shows examples of *Assemble*. Omitted in these examples is a program segment that ensures that all transactions are available before *Assemble* is called. In each case, the transactions being removed from their respective queue are listed as parameters for *Assemble*. Assuming there are no other references to these transactions, memory for each of them will be reclaimed during the VBA garbage collection. Therefore, it is incumbent upon the modeler to ensure that the

```
Q1Queue.Join Component  ' queue then check
If Q1Queue.length >= 3 _
Then
   Set SubAssembly = Q1Queue.Depart
   Q1Queue.Remove 1, 2   ' next two queued transactions
End If
```

Figure 16.13 Program segment—first transaction assemble from same queue.

```
Q1Queue.Join Material
If Q1Queue.length >= 3 _
 Q1Queue.Remove 1, 2
 Set Material = Q1.Depart
End If
```

Figure 16.14 Program segment—last transaction assemble from same queue.

```
If Q1Queue.length > 0 And Q2Queue.length > 0 And _
   Q3Queue.length > 0_
Then
   Set SubAssembly = Q1Queue.Depart
   Q2Queue.Remove 1, 1
   Q3Queue.Remove 1, 1
End If
```

Figure 16.15 Program segment—transaction assemble from different queues.

Table 16.3 Complex Assembly Operations

Assemble Types
Sum: Sum of respective attributes
Product: Product of respective attributes
High/Index: Transaction with highest value of attribute (index)
Low/Index: Transaction with lowest value of attribute (index)

```
Set SubAssembly = _
        Find.Assemble ("Sum", Q1Queue.Depart, Q1Queue.Depart)
Set SubAssembly = _
        Find.Assemble ("Product", Q1Queue.Depart, Q1Queue.Depart)
Set SubAssembly = _
        Find.Assemble ("High/1", Q1Queue.Depart, Q2Queue.Depart, _
        Q3Queue.Depart)
Set SubAssembly = _
        Find.Assemble ("Low/2", Q1Queue.Depart, Q2Queue.Depart, _
                        Q3Queue.Depart)
```

Figure 16.16 Program segment—complex assemble examples.

transactions that are the basis for the *Assemble* operation have an appropriate disposition. In contrast to the above example, transactions in the *Assemble* operation do not necessarily reside in queues.

16.4 Summary

Primary event procedures route transactions between nodes, collect statistics, change system parameters, and affect changes to transaction lists. As shown in this chapter, DEEDS offers powerful and flexible mechanisms for controlling and directing transaction flow in the network and there is no special syntax beyond the VBA language requirements.

Problems

16.1. Specify the select condition that best describes each of the following situations:
   a. Players at a five-court racquetball facility are assigned to the first available court.
   b. Off-campus students may park their cars in any of several designated parking lots.
   c. The machine that has been idle the longest is selected to process the job.
   d. Drivers arriving at a multilane gas station will choose the shortest lane.
   e. Jobs taken from a waiting area will be assigned to one of the idle machining centers.
   f. A typing pool serving several departments will pay attention to the department with the highest work pile.
16.2. Specify the *Match* condition that describes each of the following:
   a. Disabled cars wait for one of two towing trucks.
   b. An online dating service matches couples for compatibility.
16.3. Specify the *Assemble* operation that describes each of the following:
   a. Trucks in a quarry operation are loaded by one of two mechanical loaders.
   b. Frames and cushions of chairs are manufactured in two different departments in a furniture factory.
16.4. Suppose that a model includes queues *Q1* and *Q2* and facility *F1*. The contents of each are given in Table 16.4. Determine the transactions in the three lists after each of the following (independent) cases:

a.
```
LocationIndex = Q2Queue.Location(2, "=", 5)
If LocationIndex > 0 _
Then
    Q1Queue.Join Q2Queue.Detach(LocationIndex)
End If
```

b.
```
LocationIndex = Q2Queue.Location(2, "=", 4)
If LocationIndex > 0 _
Then
    Q2Queue.JoinAfter 1, Q2Queue.Detach(LocationIndex)
End If
```

c.
```
LocationIndex = Q2Queue.Location(2, "=", 4)
If LocationIndex > 0 _
Then
    Q2Queue.JoinAfter 1, Q2Queue.Detach(LocationIndex)
End If
```

d.
```
Set TempTrans = F1Facility.View(1).Copy
For Index = 1 to Q1Queue.Length
  Found = True
  For Jndex = 1 to NumberAttributes
   If Q1Queue.View(Index).A(Jndex) < TempTrans.A(Jndex) _
   Then
      Found = False
      Exit For
   End If
  Next Jndex
  If Found = True
    Set TempTrans = F1Facility.Interrupt(1)
    Exit For
  End If
Next Index
  ' Found = True => facility interrupted
```

Table 16.4 Node Descriptions for Problem 16.4

Node	A(1)	A(2)	A(3)
Q1-1	1	2	4
2	5	4	6
3	3	7	2
4	4	8	-3
Q2-1	1	5	9
2	4	4	10
3	3	5	2
4	2	7	1
5	5	8	4
F1-1	2	4	3
2	5	1	6

Table 16.5 Node Descriptions for Problem 16.5

Entry	A(1)	A(2)	A(3)	A(4)
1	1	5	7	-1
2	4	3	9	5
3	6	8	8	7
4	3	9	1	9
5	5	7	9	0

16.5. With the contents of *Q1* given in Table 16.5, determine the *Index* of the transaction with the largest attributes among all entries in the file.

Table 16.6   Node Descriptions for Problem 16.6

Entry	A(1)	A(2)	A(3)	A(4)
1	2	4	3	2
2	5	7	9	0
3	10	19	20	11
4	1	2	4	3

16.6. Assume that for Problem 16.5 there is another queue, *Q2*, with the entries given in Table 16.6. Determine the entry *Index* in *Q1* and *Jndex* in *Q2* that satisfy the following conditions:
  a. *Index* and *Jndex* have the same *Attribute (1)* through *Attribute (4)* values.
  b. *Attribute (1)* and *Attribute (2)* of *Index* exceed *Attribute (1)* and *Attribute (2)* of *Jndex*.

APPLICATIONS

Chapter 17

Simulation Project Management

17.1 Introduction

Because a simulation model is a program to predict performance characteristics of a new system or the effects of changes to an existing system, the general impression is that simulation modeling is a programming activity; however, for a typical simulation project only about a third of the time is devoted to model development. The fact is that simulation modeling also requires project management skills. Therefore, this chapter presents an overview of simulation project activities with comments on how the structure of design environment for event-driven simulation (DEEDS) facilitates many of these activities. Before considering project management, a conceptual view of a simulation model is presented. The conceptual view is the basis of the models presented in the remaining chapters. Both of these perspectives are critical to the success of a simulation project. Chapter 1 introduced many of these topics at a conceptual level.

17.2 System Specification

Figure 17.1 is a conceptual representation of a system. The combination of workload and system specification determines the overall performance of the system as well as performance characteristics of the subsystems. The global performance metric, P, reflects the purpose of the system. In most discrete event simulation applications, P is defined as cycle time, an observation-based statistical variable. However, for more complicated expressions of P, subs *Initial* and *ShutDown* in *EventsManager* enable the modeler to define other complex metrics for P. Common subsystems such as manpower and equipment have unique performance characteristics. As presented in Section 17.3, subsystem performance characteristics are often reflected by information on queues, facilities, and delays.

The workload specification characterizes the loading on the system. The mix of customers and respective arrival rates are examples of workloads. The system also has production rules for scheduling and processing work. For example, jobs may be scheduled on a first come, first served basis or by shortest processing time.

A general representation of the performance, system, and workload relationship is

$$P = f(W_t, S_t)$$

The W_t and S_t notations suggest that the workload and system may be constant or expressed as a function of time. Discrete event simulation is used to model such systems because a closed-form mathematical model would be extremely difficult, if not impossible, to develop.

From a management perspective, the purpose of modeling is to analyze various strategies that affect P. An interest in modeling may be the result of anticipated changes in W_t and a desire to maintain performance level P. Another may be that even though no changes in W_t are anticipated, the objective is to affect an improvement in P. In either context, the interest is in a system, S_t, that achieves the desired level of performance P.

A system may have conflicting performance metrics. For a given product mix in a flexible production system, one facility layout may reduce the total distance production units are moved in the system. However, cycle time for low-volume products may increase when using that layout. In a similar manner, a configuration that reduces movement distances may increase processing time because of traffic congestion. An example of a dynamic system configuration is the cross dock problem presented in Chapter 24, where the number of tow motors on a loading dock is dependent on the time of day. The justification for varying the number of tow motors is that freight arrival rates are dependent on time of day.

Chapter 11 presented user-defined and common probability distribution functions used to model a workload. Another workload specification technique is to script the workload from historical data. For scripted workloads, actual arrival times define the workload. Interarrival times are embedded in the data. The user simply creates a script spreadsheet, and the script is read by the simulation program and scheduled as primary events. Scripted data from the cross dock problem in Chapter 24 is shown in Figure 17.2. Notice that the actual day and time of freight arrival are

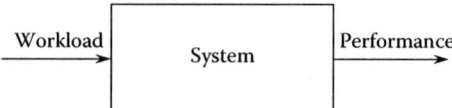

Figure 17.1   Basic system characterization.

Day	Time	TrailorID	Dest	Units	Dest	Units
1	7:05 AM	534183	BHM1	0	KNX1	0
1	12:11 PM	511735	BHM1	0	KNX1	0
1	1:11 PM	400021	BHM1	0	KNX1	2
1	4:22 PM	534180	BHM1	0	KNX1	0
1	4:47 PM	Z171	BHM1	17	KNX1	10
1	4:58 PM	534783	BHM1	0	KNX1	0
1	4:58 PM	535765	BHM1	0	KNX1	0
1	5:14 PM	534283	BHM1	4	KNX1	24
1	5:49 PM	483284	BHM1	3	KNX1	3
1	6:33 PM	534816	BHM1	12	KNX1	3
1	6:38 PM	534072	BHM1	1	KNX1	4
1	6:46 PM	535262	BHM1	5	KNX1	3
1	6:52 PM	533642	BHM1	9	KNX1	2
1	7:24 PM	2821	BHM1	6	KNX1	17
2	6:19 AM	536047-2	BHM1	0	KNX1	0

Figure 17.2   Scripted workload specification.

used to describe the workload. Program requirements for implementing a scripted workload are presented and discussed in Chapter 24. The major advantage of a script is that the need to describe a dynamic workload using a statistical probability distribution function is eliminated.

Section 17.3 uses the conceptual system model just described and terminology from optimization theory for characterizing data requirements for the model. Examples of performance, system, and workload specifications, along with data descriptions for selected problems from Chapters 18 through 24, are also presented at the end of the section.

17.3 Simulation Constants, Decision Variables, and Constraints

In a sense, P is an objective function to be maximized (minimized), and model parameters are a constant or decision variable associated with W_i and S_i. Constraints are expressed as a function of W_i and S_i. Values for decision variables are determined by the model, but, for a feasible solution, the set of decision variables must be within an acceptable operating range (satisfy constraints). As in an optimization model, the objective function, P, is dependent on the values of decision variables. Whether a parameter is modeled as a constant or decision variable depends on the strategy being considered. Tables 17.1 and 17.2 present system components and workload characterizations.

The system summarized in Table 17.1 is from a traditional assembly line problem described in Section 18.2. The performance measure is defined as *AverageCycleTime*, and the workload is a sequence of tasks needed to produce a unit of production. Each task has a random processing time

Table 17.1 System Characterization of an Assembly Line

Model	Line Balancing with Workstation Failures
P	AverageCycleTime
S	Workstations on the line Task allocation scheme System maintenance procedures Workstation performance characteristics
W	Number of tasks Distribution of task completion times
Decision variables	Tasks assigned to stations Lapsed time between maintenance
Constants	Number of stations Number of tasks and task distributions Workstation performance characteristics System maintenance characteristics
Constraints	Buffer capacity All tasks assigned

Table 17.2 System Characterization of an Inspection Program

Model	Assembly Inspection
P	Average cost per assembly Component specifications
S	Assembly tolerances Sensitivity and specificity levels Inspection procedures Inspection costs Rework costs Scrap costs
W	Component production parameters
Decision variables	Component production parameters Sensitivity and specificity levels
Constants	Inspection costs Rework costs Scrap costs Assembly tolerance
Constraints	Not applicable

described by a triangular distribution. Subsystems for the production system include a sequence of workstations and maintenance procedures. Workstations are identical; therefore, distributions for failure and repair times are the same for each workstation. The workstation performance measure is percentage utilization. Because there are finite-length work-in-process buffers between workstations, it is important that the line be balanced (i.e., each workstation should have a similar utilization).

The nature of the product dictates the shutdown procedures necessary to perform maintenance and the disposition of a unit of production when a workstation fails. For example, the maintenance shutdown procedure may allow each workstation to finish its unit of production before maintenance begins. Alternatives for the disposition of a unit of production when a workstation failure occurs include scrapping the part, completing the remaining work after the workstation is repaired, and completely reworking the part.

For this model, decision variables are which tasks are assigned to each workstation and time between scheduled maintenance. All other system components are assumed to be fixed by management or technological requirements of the product or system components. Constraints are the buffer capacity of work-in-process and the fact that all tasks must be completed to produce a unit of product. The buffer capacity limits the number of production units between workstations.

The system represented in Table 17.2 is an inspection system, presented in Section 23.2. A process produces an assembly from two wheels and an axle. There is variability in producing the individual components and components that are not produced to specifications will be rejected or reworked. Any component rejection causes the entire assembly to be rejected. There is also a tolerance level for fitting components that meet specifications. Errors are inherent to the inspection program. Components that meet specification may be classified as failing the inspection (sensitivity), and components that do not meet specifications may pass inspection (selectivity).

For this system, the performance measure is defined as *average cost per assembly*. The workload consists of two wheels and an axle, with their associated variability in actual diameter. Subsystems include the inspection procedures for component parts and the assembly operation. System constants are the assembly tolerance and component cost of inspection, rework, and scrap.

For this model, the decision variables are the production parameters, levels of sensitivity and specificity, and variability in the production process. The nature of the product produced dictates the details of rework and scrapping component parts. From a management perspective, the assumptions are that variability in component production can be reduced and improvements can be made in the ability to distinguish good and bad components. The simulation model enables the analyst to measure effects of these changes on the cost of production.

The implementation of the concepts described in Figures 17.1 through 17.3 is critical to the success of a simulation project. Characterizing the model in such a manner is also the basis for a user interface design for presenting data to the model and simulation results to the user. How data is presented has a dramatic effect on the overall effort needed to analyze the system. In Section 17.4, design considerations for presenting data are discussed.

17.4 Data Specifications

Well-designed input/output specifications enable someone using the program to know what information is needed by the model and to perform the analysis with minimal effort. Another advantage of using the system description in Tables 17.1 and 17.2 and optimization terminology from Section 17.3 is that it facilitates the design for data entry and program output. As part of the model design, types and sources of data as well as how the model will access the data must be carefully considered.

In addition to embedded data in the program, a DEEDS simulation model has the following default external data sources:

1. Spreadsheet *InitialModel* defines and describes source, queue, facility, and delay nodes as well as user-defined statistical variables.
2. Spreadsheets *PDFs* and *Tables* define and describe probability distribution functions and tables created by the user.

All instances of DEEDS classes (sources, queues, facilities, and delays) must be defined in the *InitialModel* spreadsheet. However, instance properties may be initialized in sub *Initial* before the simulation begins. For example, a facility and its corresponding service time for each server may be defined in *InitialModel* and then sub *Initial* initializes the number of servers with data from the *UserInitialed* spreadsheet. Spreadsheet *UserInitialed* and sub *Initial*, in *EventsManager*, facilitate the design of a flexible model by providing an easily accessible user interface to the model. It may be helpful to review Section 10.6.1, where sub *Initial* is introduced.

The spectrum of alternatives for providing all other data to the model varies from embedding data in the program to enabling the model to retrieve data from spreadsheets. The design trade-off is to include all W_t and S_t parameters in sub *UserInitialed* and the incremental effort required to modify embedded program data between simulation runs. The spreadsheet can become cluttered with an excessive amount of data, and an inexperienced user can be overwhelmed by all the data entry options.

To reduce the problem, constant parameters can be embedded in the program and variable parameters are initialized in spreadsheet *UserInitialed*. Sub *Initial* is then programmed to retrieve variable parameters from the spreadsheets. Numerous examples of sub *Initial* are presented in Chapters 18 through 25.

The modeler can easily create additional data entry spreadsheets and organize them on the basis of sources and types of data. Data in these spreadsheets may also be retrieved using sub *Initial* in *EventsManager*. As a practical matter, the limit to such an approach is the creativity of the modeler.

Figures 17.3 and 17.4 show data presentation approaches for the inspection problem described in Table 17.2. As described in Section 17.3, an axle and two wheels are produced and immediately inspected. The components are then assembled subject to a final inspection. The performance metric P is the average cost of the inspection program. There is also a sensitivity and selectivity specification for the inspection process. Notice that the input and output data are grouped by activity. The simulation program obtains simulation data from spreadsheet *UserDefined* and presents results to spreadsheet *UserOutput*.

Now that the system to be modeled and corresponding user interface have been specified, attention turns to the project management activities important to a simulation project. The project management principles presented in Section 17.5 are a blueprint to ensure that members of a simulation project agree with and support the objective of the simulation project.

Upper Specification Limit	1.0020	1.0050	0.0050
Lower Specification Limit	0.9980	1.0010	0.0010
Inspection Cost	$0.10	$0.05	$0.25
Rework Cost	$0.25	$0.30	$2.00
Scrap Cost	$0.75	$0.50	$4.00
Sensitivity (decimal)	0.8		
Specificity (decimal)	0.8		
Run Length	10000		

Click here to enter new or edit data

When data is correct, click here to generate Average Cost Summary

Tarrah Wilkerson and Austin Pinkstaff

Figure 17.3 Inspection program spreadsheet input.

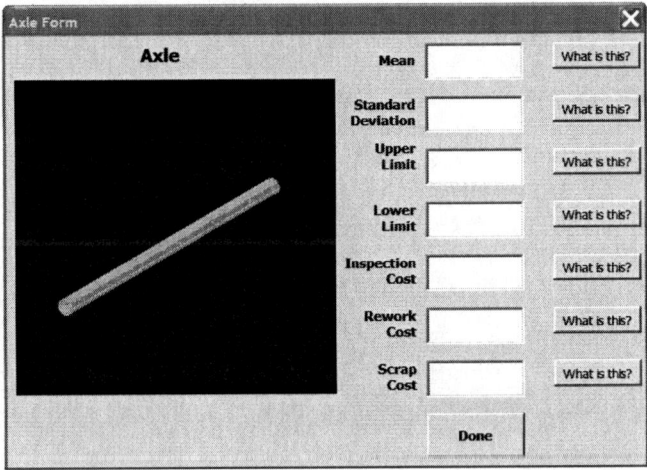

Figure 17.4   Inspection program spreadsheet output.

17.5 Project Management

Figure 17.5 shows the software engineering design process adapted for a typical simulation project. Only a brief description of each step is presented. However, there are several references [1–3] that provide extensive coverage on this topic.

The duration of a particular step depends on the system being modeled. For example, there may be more data collection and analysis for an existing system and more research for a new one. Also, because of differences in project goals and in data availability, models for new systems may be less detailed than for existing systems. In reality, the process is continuous rather than the discrete steps shown. The dashed lines indicate the cascaded nature of the process. For example, during the detailed design, additional data or reformulating project objectives may be necessary.

Although some revisiting of previous steps is unavoidable, an excessive number of iterations will lengthen the project and increase project costs. The situation that has the potential for jeopardizing a project is when backtracking involves more than one previous step. In concurrent engineering terminology it may be possible to compress the project length by initiating activities that are not necessarily sequential in nature, on the basis of anticipated model requirements.

17.5.1 Problem Definition

Critical to the success of a simulation project is a well-defined statement of project objectives. Managers, engineers, and operational personnel, as well as those responsible for the actual model development and statistical analysis, should participate in problem definition. Some of these individuals may not understand the nature of simulation or the time and effort required to perform a simulation study.

The issues to be addressed by the model dictate the level of model detail. A simulation model may be unable to address a broad range of objectives. Therefore, as described in Chapter 1, one

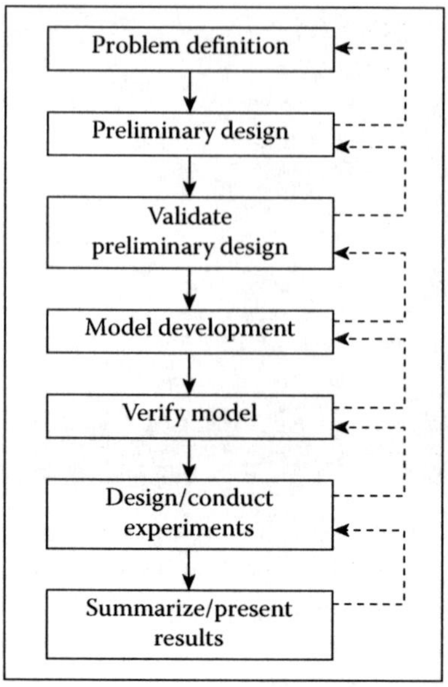

Figure 17.5 Simulation project activities.

model may be needed to assess subsystem details and another to evaluate the system. Before much work is done on the preliminary design, the following issues must be resolved:

- How will the model be used (e.g., one-time decision versus periodic decision support)?
- Who is the end user of the model (affects the user interface for the model)?
- What system configurations are to be modeled?
- What are the measures of performance (parameters are used to compare alternatives)?
- What are the problems with the current system?

With the above information, the project manager can plan the project requirements for the number of people, amount of time, and cost required for each phase of the project. However, the project duration is often longer than expected simply because system details are not well documented.

17.5.2 Preliminary Design

This step begins with a detailed description of the components of the system and their interrelationships. Sources of data may include operators, engineers, managers, and vendors. Data collection may be complicated by the lack of formal system specifications and operating procedures. Any data collected on the current system can be used to validate the final version of the model. As discussed in Chapter 4, sources of system randomness such as machine operating and repair times are represented by an appropriate probability distribution or distribution function derived from a histogram.

Model assumptions must be documented. Assumptions are generally the result of a lack of information on system specifications or the desire to reduce the effort to model complex relationships. Because appropriate simplifying assumptions will not compromise the validity of the model, it is not essential to have a one-to-one mapping between elements of the system and the model. Communications between members of the project ensure that assumptions are appropriate and level-of-detail issues are resolved. As indicated in Figure 17.5, because of the nature of the assumptions, it may be necessary to reformulate the problem statement.

The details of a preliminary model design evolve as the above information is accumulated. The first pass of the preliminary design is pseudo code, which is a combination of a natural language and a programming language. As the preliminary design details of the model become firm, they are converted into the simulation language.

17.5.3 Validate Design

The results of this step are an agreement on components of the preliminary design by team members and a reconciliation of the problem definition and preliminary design. Discrepancies are resolved by modifying the preliminary design or a restatement of the problem statement. A structured walkthrough by the project team is often the approach used to validate the preliminary design.

17.5.4 Model Development

With concurrent engineering, several portions of the model have already been developed. The pseudo code from the preliminary design is converted into the programming language. The remainder of the design is iteratively programmed using pseudo code and the programming language. Of course, the result of this step is a simulation model that produces reasonable results. However, individual components of the model may also be validated. Software design and development concepts ensure that the model is completed in a timely fashion and relatively free of logic errors. Probably the most important concept is the modular program design.

As discussed in Section 17.5.2, it is relatively easy to combine constructs of Visual Basic (VB) with a natural language to develop pseudo code. By its very nature, DEEDS is an object-oriented simulation language based on the concept of a modular design. DEEDS also uses the definition of a primary event as the basis for modular program design. In essence, each primary event is a major program segment that may be partitioned to increase clarity and improve maintainability of the program. Clarity is important for conducting a structured walkthrough of the program.

17.5.5 Model Verification

Verification is the reconciliation of the performance measures of the actual system performance data collected during the preliminary design with the results produced by the model. The simulation results from the test runs are reviewed for discrepancies, especially those caused by model assumptions. Discrepancies attributable to assumptions cause backtracking activities in the project.

The dilemma is that there is no definitive approach to validate a proposed system model. The general conclusion is that when the two sets of measures compare favorably, the model is considered valid. The model is then reconfigured to represent the proposed system. Of course, similarities between the current and proposed system increase the confidence in the model to predict performance measures of the proposed system.

Also, sensitivity analysis is used to determine which model parameters (e.g., number of servers, a probability distribution, or the level subsystem detail) have the most effect on the system performance measures. The effects of changes in those parameters should receive careful consideration in the verification process because they may likely cause major variations between performance measures of the proposed system and the simulation model.

17.5.6 Design/Conduct Experiments

Simulation experiments are an extension to sensitivity analysis in model verification. As discussed in Section 17.3, system performance metrics are characterized as a function of W_t and S_r. In certain simulation projects the focus is on steady-state performance metrics. Typically, simulations begin with the system idle. Therefore, as discussed in Chapter 13, a simulation must be run for a transient period before simulation results can be used to estimate steady-state performance metrics.

The interaction of the system configuration and workload specification (based on the workload probability distributions) affect the value and variability of the system performance measures. As a result, these performance measures are only an estimate of the actual performance metrics. It may be helpful to review details of obtaining statistically precise estimates (variance and mean equal of the performance measure) for these parameters in Chapter 13.

Of course, the preference would be to find an "optimum" system configuration. However, because system relationships are derived from domain knowledge of the system and whatever information that can be gleaned from conducting the simulation experiments, finding an optimum solution is problematic. Chapter 25 demonstrates the use of experimental design techniques and genetic algorithms to search for an "improved" system configuration for a given workload.

17.5.7 Summarize/Present the Results

The deliverables from the simulation project are numeric estimates of performance metrics, including confidence intervals, for each system configuration of interest. These estimates are used to determine the configuration that best satisfies the requirements enumerated in the problem definition. In addition to numeric estimates of performance metrics, it is often helpful to use graphical summaries (histograms, pie and bar charts, time plots) of simulation results to provide further insights into system response. Because the DEEDS platform is Excel/VBA, all of Excel's data presentation functions are readily available to prepare graphical presentations.

Good documentation is an essential part of the summary because a model may be used for more than one application. The summary should also include an assumptions document, documentation of the program, and a report summarizing the results and conclusions of the study.

Animation is useful for communicating the essence of a simulation model. However, an animation is not a substitute for a statistical analysis of simulation results, and a "correct" animation is no assurance of a verified model. For projects that have an animation component, DEEDS can be used for rapid prototyping of the model so that simulation experiments can begin before the animation is complete. For complex models, many of the VBA program segments from DEEDS can also be used in the animation software. The DEEDS program can also be used for model documentation.

17.6 Summary

As suggested in this chapter, simulation is much more than programming. The essential elements of a simulation project are modeling skills, knowledge of statistical analysis, project management skills, and domain knowledge of the system to be modeled. For a successful simulation project, the project manager must ensure that a sufficient level of competency in each of the areas is represented on the project. Also, the concepts presented in Sections 17.2 through 17.4 may be used as a checklist for steps in the project management process.

References

1. Law Averill, M., *Simulation Modeling and Analysis*, 4th ed. New York: McGraw-Hill Book Company, 2007.
2. Pressman, R. S., *Software Engineering: A Practitioner's Approach*, 6th ed. New York: McGraw-Hill Book Company, 2005.
3. Smith, K. A., *Teamwork and Project Management*, 2nd ed. New York: McGraw-Hill Book Company, 2004.

Chapter 18

Facilities Layout Models

18.1 Introduction

The primary objective of facilities layout is to design a physical arrangement that can economically produce the required quantity and quality of output. The general layout for product flow follows a pattern set by the type of production anticipated. Basic layouts are as follows:

Product layout: A line of facilities and auxiliary services that produces a progressively refined product. A sequence of operations that reduces material handling and inventories usually lowers production cost per unit and is easier to control and supervise. The pace of the line is determined by the slowest operation. Such a layout characterizes mass or continuous production.

Process layout: A grouping of machines and services according to common functions and distinct operations such as welding, grinding, polishing, and shipping. A function arrangement is characteristic of job and batch production. It allows more flexibility and reduces investment in equipment but increases handling, space requirements, production time, and the need for more supervision and planning.

The examples in this chapter are typical production facilities layout problems for a small manufacturing environment. However, the models for these systems can be easily scaled to a larger manufacturing environment or any other type of production system.

18.2 Line Balancing

This problem is from a senior design project by Mike Cunningham, a graduate of the Tennessee Tech Industrial Engineering Program. Recall that there is a brief description of this problem in Section 17.3. The metric for this simulation is *AverageCycleTime*.

Operator tasks from Table 18.1 and raw materials from Table 18.2 are assigned to a sequence of workstations. Typical task and material assignments are shown in Tables 18.3 and 18.4, respectively. Operator tasks and material replacement times are modeled by the triangular distribution.

Table 18.1   Triangular Distributions for Assembly Tasks

Task	Minimum	Middle	Maximum
1	4.87	5.87	6.87
2	4.00	5.09	6.00
3	6.30	7.68	9.10
4	6.52	7.56	8.76
5	5.12	7.09	8.30
6	2.00	2.91	4.00
7	5.89	6.89	8.02
8	7.76	8.87	10.23
9	15.68	18.24	21.42
10	2.00	2.82	4.00
11	5.89	6.89	8.02
12	3.00	4.00	5.00
13	15.68	18.24	21.42
14	2.00	2.82	3.52
15	5.89	6.89	8.02
16	7.45	10.04	12.25
17	15.00	18.00	21.00
18	2.00	2.85	3.52
19	5.89	6.89	8.02
20	2.00	3.00	4.00
21	16.57	19.07	22.02
22	2.50	3.50	4.50
23	7.67	9.11	11.12
24	12.07	13.44	15.00
25	5.52	7.64	9.52
26	2.00	3.50	5.00
27	4.86	6.45	8.05
28	2.89	3.89	4.89
29	2.34	3.83	5.43
30	1.00	1.44	2.00
31	7.31	8.49	9.62
32	3.00	3.41	4.21

Table 18.2   Material Summary for an Assembly

Number	Initial Level	Reorder Quantity	Reorder Level	Minimum	Middle	Maximum	Number/ Unit
1	22	27	1	20	23.71	26	1
2	50	108	2	40	44.61	48	1
3	2500	5000	2	45	49.00	54	1
4	300	6857	3	72	77.00	82	1
5	300	350	2	57	61.76	65	1
6	15	25	1	6	6.93	8	1
7	30	48	1	20	23.30	26	1
8	50	96	1	2	4.38	6	1
9	75	100	2	54	58.53	63	1
10	100	192	1	14	17.10	20	1
11	125	250	2	63	67.43	70	1
12	200	850	3	85	90.10	95	1
13	5	25	1	10	14.72	18	1
14	8	24	2	47	52.73	57	1
15	300	500	10	54	59.52	64	5
16	900	4000	15	76	81.39	86	5
17	1000	500	10	54	59.52	64	5
18	300	4000	15	76	81.39	86	5
19	2	3	1	33	37.23	41	1
20	100	144	2	45	50.59	55	1
21	300	504	18	77	82.26	87	6
22	600	3000	18	100	108.14	116	6
23	300	504	18	78	82.26	85	6
24	300	504	24	100	108.14	116	6
25	500	1000	2	46	52.80	57	1
26	750	1000	4	102	115.73	130	1
27	850	1000	4	102	115.73	130	1
28	2500	5000	4	90	99.14	110	1
29	2700	5420	3	71	77.06	83	1

(Continued)

Table 18.2   Material Summary for an Assembly (*Continued*)

Number	Initial Level	Reorder Quantity	Reorder Level	Minimum	Middle	Maximum	Number/ Unit
30	2	3	1	11	12.98	15	1
31	2	3	1	2	4.43	6	1
32	2	6	1	2	4.38	6	1
33	30	48	2	50	54.61	58	1
34	950	1950	2	40	43.79	46	1
35	1000	4500	4	98	104.24	110	1
36	2600	5142	4	98	104.24	110	1
37	3500	7500	2	40	43.79	46	1
38	69	168	10	230	271.81	300	1

Table 18.3   Task Allocation by Workstation

Workstation	Start Index	End Index
1	1	6
2	7	10
3	11	14
4	15	18
5	19	22
6	23	26
7	27	32

Table 18.4   Material Allocation by Workstation

Workstation	Start Index	End Index
1	1	13
2	14	16
3	17	18
4	19	22
5	23	24
6	25	27
7	28	38

Table 18.2 also shows order size, replenishment quantity, and data on delay time to replenish each raw material, as well as the number of raw materials used to produce a unit of product.

When a raw material level reaches its reorder point, the operator issues a request for material replenishment. The request is queued and these requests are serviced on a first come, first served basis by a line server who has responsibility for replenishing raw materials for each workstation. When a workstation operator must stop work because of an insufficient level of that resource, and that material request is in the queue, the operator cancels the request and assumes responsibility for replenishing that material. A unit of production is initiated in the first workstation.

To account for statistical variations in workstation cycle times, each downstream workstation has a work-in-process buffer capacity of three production units. When an operator completes a unit of production and the downstream buffer has three units, the operator becomes blocked with the completed unit occupying the workstation. With the exception of the first operator, there must be a unit of production available in the upstream buffer or the operator remains idle. The first operator continues to initiate a unit of production until it becomes blocked by its downstream buffer.

Task data is read from spreadsheet *TaskData* into an array *TaskTime* (*1 To StationCount, 1 To 3*). The columns represent the minimum, middle, and maximum time for each task. Material data is read from spreadsheet *MaterialData* into an array *RawMaterial* (*1 To StationCount, 1 To 7*). The Visual Basic for Applications (VBA) constants to reference the array *RawMaterial* are

> *1 = CurrentLevel*
> *2 = ReOrderQuantity*
> *3 = ReOrderLevel*
> *4 = Minimum*
> *5 = Average*
> *6 = Maximum*
> *7 = NumberUsed*

The data in Tables 18.3 and 18.4 are read from spreadsheet *WorkStationData* into arrays *WSTask(1 To StationCount, 1 To 2)* and *WSMaterial(1 To StationCount, 1 To 2)*. The columns represent the first and last task (raw material), respectively. The workstations and their associated queue are defined as arrays *WSQ(1 To StationCount)* and *WS(1 To StationCount)*. Of course, the first workstation queue is not used because the first workstation introduces jobs into the system.

The workstation states are as follows:

- Blocked when the operator has finished a job and the downstream queue is full
- Idle (workstation empty) when the operator has no waiting jobs or is waiting on the line server to restock needed materials
- Busy when the operator is working on a job or busy restocking material

To simplify program logic Boolean arrays, *MaterialOrdered(1 To MaterialCount)* and *WSBlocked(1 To StationCount)* are defined. When *MaterialOrdered()* is *True*, the operator has placed an order for material replacement. Otherwise, *MaterialOrdered()* is *False*. When *WSBlocked()* is *True*, the workstation has finished a job but its succeeding workstation's buffer is full. Otherwise, *WSBlocked()* is *False*.

Table 18.5   Assembly Line—Primary Events

Event	Description
JobCompletionEvent	Workstation job completions
LineServerFacilityEvent	Line server replenished a part

Transactions represent units of production or an operator when the operator is restocking material. Transaction attributes are:

1 = WSID
2 = ActivityType
3 = MaterialID
4 = ActivityTime

Table 18.5 describes each primary event in the model. As implied in the table, subroutine *Parser* routes all completed workstation jobs to sub *JobCompletionEvent*.

In addition to the primary events subroutines, the following subroutines are included in the model:

■ *ScheduleLineServer* initiates material replenishment by the line server.
■ *UpStreamCheck* initiates service on jobs at upstream workstations blocked by their respective downstream queue.
■ *CreateNewJob* initiates service for a new job at the first workstation.
■ *StartNextJob* at the current workstation removes a job from its queue and begins service.

The model also has the following functions:

■ *PartsAvailable*, depending on the situation, allocates parts to a job, initiates a request for the line server to replenish material, or directs the operator to replenish material.
■ *ServiceTime* determines the time required to replenish material at a workstation.

The remainder of this section provides an overview of model procedures and a brief description of program output.

JobCompletionEvent: In Figure 18.1, after identifying which workstation is affected, the subroutine determines whether the event is a job completion or the operator restocking raw material. When restocking has occurred, the workstation is disengaged, *MaterialOrdered()* is changed to *False*, and the current level of that item is incremented by *ReOrderQuantity*. After the job is completed at the last workstation, the cycle time for the job is tabulated and it exits the system. For other workstations, the workstation disengages the job and moves it to the next workstation or next workstation queue. Of course, the workstation becomes blocked, with the job in the workstation, when the downstream queue is full. After disposing of the completed job, the workstation is ready to begin work on another job. For the first workstation, a new job is initiated upon completion of a job. For other workstations, a job must be from an upstream workstation.

```
'SECONDARY EVENTS FOR EACH PRIMARY EVENT
Sub JobCompletionEvent(ByRef Job As Transaction)
  Dim _
    CurrentWS As Integer, _
    Index As Integer
  CurrentWS = Job.A(WSID)
  If Job.A(ActivityType) = "Restock" _
  Then
    WS(CurrentWS).Disengage Job
    Index = Job.A(MaterialID)
    MaterialOrdered(Index) = False
    RawMaterial(Index, CurrentLevel) = _
      RawMaterial(Index, CurrentLevel) + RawMaterial(Index,ReOrderQuantity)
    Set Job = Nothing
  Else
    If CurrentWS = StationCount _
    Then
      WS(CurrentWS).Disengage Job
      CycleTimeStatistic.Collect Job.SystemTime
      Set Job = Nothing
    Else
      If WSQ(CurrentWS + 1).Full _
      Then
        WSStatus(CurrentWS) = "Downstream Block"
      Else
        WS(CurrentWS).Disengage Job
        Job.Assign WSID, CurrentWS + 1
        WSQ(CurrentWS + 1).Join Job
        StartNextJob (CurrentW S + 1)
      End If
    End If
  End If
  If CurrentWS = 1 _
  Then
    Call CreateNewJob
  Else
   StartNextJob (CurrentWS)
  End If
End Sub
```

Figure 18.1   Assembly line—*JobCompletionEvent*.

LineServerFacilityEvent: In Figure 18.2, when restocking has occurred, *LineServer* is disengaged, *MaterialOrdered()* is changed to *False*, and the current level of that item is incremented by *ReOrderQuantity*. *LineServer* then attempts to initiate a job for the workstation that was waiting on the material just restocked. Finally, the *LineServer*, by calling *ScheduleLineServer*, checks for other material replenishment requests.

ScheduleLineServer: Also in Figure 18.2, when *LSQ*, the line server queue, has requests and *LineServer* is idle, *LineServer* removes the request from the queue, determines the replenishment time, and initiates the replenishment process.

UpStreamCheck: As shown in Figure 18.3, upon removal of a blocking condition, *UpStreamCheck* begins by removing previous workstation blocking conditions and continues until either the first workstation is reached or a workstation is not blocked. For each workstation, the

```
Sub LineServerFacilityEvent(ByRef Job As Transaction)
  Dim _
    CurrentWS As Integer, _
    Index As Integer
  LSFacility.Disengage Job
  Index = Job.A(MaterialID)
  MaterialOrdered(Index) = False
  RawMaterial(Index, CurrentLevel) = _
      RawMaterial(Index, CurrentLevel) + RawMaterial(Index, ReOrderQuantity)
  CurrentWS = Job.A(WSID)
  Set Job = Nothing
  If CurrentWS = 1 _
  Then
    Call CreateNewJob
  Else
    Call StartNextJob(CurrentWS)
  End If
  Call ScheduleLineServer
End Sub
Sub ScheduleLineServer()
  Dim _
    Job As Transaction
  If LSQ.Length > 0 And LSFacility.NumberOfIdleServers > 0 _
  Then
    Set Job = LSQ.Depart
    LSFacility.EngageAndService Job, 1, Job.A(ActivityTime)
  End If
End Sub
```

Figure 18.2   Assembly line—*LineServerFacilityEvent* and *ScheduleLineServer*.

```
Sub UpStreamCheck(ByVal CurrentWS As Integer)
  Dim _
    Job As Transaction, _
    Index As Integer
  Do While (CurrentWS > 0)
    If WSStatus(CurrentWS) = "Downstream Block" _
    Then
      WSStatus(CurrentWS) = "Idle"
      Set Job = WS(CurrentWS).View(1)
      WS(CurrentWS).Disengage Job
      Job.Assign WSID, CurrentWS + 1
      WS(CurrentWS + 1).NextFacility Job, 1, WSQ(CurrentWS + 1)
      If WSQ(CurrentWS).Length > 0 And PartsAvailable(CurrentWS) _
      Then
        WSStatus(CurrentWS) = "Busy"
        WS(CurrentWS).EngageAndService WSQ(CurrentWS).Depart, 1, _
                                   ServiceTime(CurrentWS)
      End If
      If CurrentWS = 1 _
      Then
        Call CreateNewJob
      Else
        CurrentWS = CurrentWS -1
      End If
    Else
      Exit Do ' no additional blocking conditions
    End If
  Loop
End Sub
```

Figure 18.3   Assembly line—sub *UpstreamCheck*.

blocked transaction is moved to the downstream queue or workstation because the downstream station may have been blocked with an empty upstream buffer. As mentioned, for the first workstation, a new job is created when the blocking condition is removed. For all other workstations, a job is pulled from the upstream queue or the workstation becomes idle.

CreateNewJob: In Figure 18.4, when the first workstation is idle and parts are available, work begins on a new job using service time from Figure 18.7.

StartNextJob: In Figure 18.5, whenever the workstation server is idle, its queue has a job, and parts are available for a job, work begins on the job removed from the queue. Because a job has been removed from the queue, beginning at the previous workstation, the system is checked for upstream workstations that may have had a blocking condition removed.

PartsAvailable: As shown in Figure 18.6, when parts are available, *CurrentLevel* for each part is decremented by *NumberUsed*. When *CurrentLevel* is below *ReOrderLevel*, a replenishment request is issued to *LineServer*. The request includes information on which material is requested, which workstation issued the request, and the time needed to service the request. The function returns the value *True* to indicate that materials have been allocated to the next job. When an insufficient number of parts are available at a workstation, *LSQ*, the service

```
Sub CreateNewJob()
  Dim _
    Job As Transaction
  Set Job = NewTransaction
  Job.Assign WSID, 1
  If WS(1).NumberOfIdleServers > 0 _
  Then
    If PartsAvailable(1) _
    Then
      WSStatus(1) = "Busy"
      Job.Assign ActivityType, "Assemble"
      WS(1).EngageAndService Job, 1, ServiceTime(1)
    End If
  End If
End Sub
```

Figure 18.4   Assembly line—sub *CreateNewJob*.

```
Sub StartNextJob(ByVal CurrentWS As Integer)
  Dim _
    Job As Transaction
  If WS(CurrentWS).NumberOfIdleServers > 0 _
  Then
    If WSQ(CurrentWS).Length > 0 And PartsAvailable(CurrentWS) _
    Then
      Set Job = WSQ(CurrentWS).Depart
      Job.Assign WSID, CurrentWS
      Job.Assign ActivityType, "Assemble"
      WS(CurrentWS).EngageAndService Job, 1, ServiceTime(CurrentWS)
      Call UpStreamCheck(CurrentWS - 1)
    End If
  End If
End Sub
```

Figure 18.5   Assembly line—sub *StartNextJob*.

```
Function PartsAvailable(ByVal CurrentWS As Integer) As Boolean
  Dim _
    Index As Integer, _
    Jndex As Integer, _
    Job As Transaction
  PartsAvailable = True
  For Index = WSMaterial(CurrentWS, First) To WSMaterial(CurrentWS, Last)
    If RawMaterial(Index, CurrentLevel) < RawMaterial(Index, NumberUsed) _
    Then
      PartsAvailable = False
      WSStatus(CurrentWS) = "Material Block"
      For Jndex = 1 To LSQ.Length
        Set Job = LSQ.View(Jndex)
        If Job.A(MaterialID) = Index _
        Then
          Set Job = LSQ.Detach(Jndex)
          Job.Assign ActivityType, "Restock"
          WS(CurrentWS).EngageAndService Job, 1, Job.A(ActivityTime)
          Exit Function ' only 1 item
        End If
      Next Jndex
    End If
  Next Index
  If PartsAvailable _
  Then
    For Index = WSMaterial(CurrentWS, First) To WSMaterial(CurrentWS, Last)
      RawMaterial(Index, CurrentLevel) = RawMaterial(Index, CurrentLevel) -_
                                         RawMaterial(Index, NumberUsed)

      If RawMaterial(Index, CurrentLevel) <= _
                                    RawMaterial(Index, ReOrderLevel) And _
        Not MaterialOrdered(Index) _
      Then
        MaterialOrdered(Index) = True
        Set Job = NewTransaction
        Job.Assign WSID, CurrentWS
        Job.Assign MaterialID, Index
        Job.Assign ActivityTime, Triangle(RawMaterial(Index, Minimum), _
                                   RawMaterial(Index, Maximum), _
                                   RawMaterial(Index, Average))

        LSQ.Join Job
        Call ScheduleLineServer
      End If
    Next Index
  End If
End Function
```

Figure 18.6 Assembly line—function *PartsAvailable*.

request queue, is searched for the workstation request that can be serviced by the operator. If a request is not in *LSQ*, the workstation becomes idle, waiting on material to begin the next job. In either case, the function returns the value *False*.

The output of the model is shown in Figures 18.8 and 18.9. *AverageCycleTime* is 579.59. Figure 18.9 presents workstation statistics for workstations with utilization exceeding 90%. Of course, additional analysis is needed but because four of seven workstations have a higher utilization, reallocating tasks and materials may improve *AverageCycleTime*.

```
Function ServiceTime(ByVal CurrentWS As Integer) As Single
  Dim _
    Index As Integer
  ServiceTime = 0
  For Index = WSTask(CurrentWS, First) To WSTask(CurrentWS, Last)
    ServiceTime = ServiceTime + _
      Distribution.Triangle(TaskTime(Index, 1), TaskTime(Index, 3),
TaskTime(Index, 2))
  Next Index
End Function
```

Figure 18.7   Assembly line—function *ServiceTime*.

Statistic	(O) Cycle Time	
Number In	488	
Min Value	233.270	
Mix Value	781.412	
Average	579.589	
StdDev	98.072	
LCL		
UCL		
Observation	Mean	S
0		
1	579.589	98.072

Figure 18.8   Assembly line—statistical variables output.

Facility	F1		F2		F3		F4	
Number In	530		500		495		517	
Number Out	529		499		494		516	
Residing	1		1		1		1	
Interrupted	0		0		0		0	
Destroyed	0		0		0		0	
Minimum	0		0		0		0	
Maximum	1		1		1		1	
	AvgBusy Servers		AvgBusy Servers		AvgBusy Servers		AvgBusy Servers	
Average	0.982		0.995		0.981		0.969	
StdDev	0.134		0.072		0.136		0.174	
LCL								
UCL								
Observation	Avg	StdDev	Avg	StdDev	Avg	StdDev	Avg	StdDev
0								
1	0.982	0.134	0.995	0.072	0.981	0.136	0.969	0.174

Figure 18.9   Assembly line—facilities output.

18.3 Flexible Manufacturing Environment

This process layout example is adapted from [1]. It is the use of automated guided vehicles (AGVs) for transporting materials within a flexible manufacturing system (FMS) environment. Figure 18.10 is an example of the system being modeled. The plant is assumed to have eight machine centers and eight "tree-type" track segments. The model is general in the sense that it applies to any track tree as well as any number of centers and AGVs. As a result, the only changes are to the initial data in the model.

The three machine centers and the receiving and loading areas are encoded from 1 to 5. Locations 1 and 5 are reserved for receiving and shipping, respectively. To increase the flexibility of the model, a location map is used to map machine centers to actual locations in the plant. In Table 18.6, manufacturing cells are mapped to actual plant locations.

Jobs arrive at a receiving area every *Exponential* (3) hours for processing in three machining centers. Experience has shown that there are four job sequences. Table 18.7 summarizes the sequences for each job type along with the probability of their occurrence.

The mean processing times at machine centers 1, 2, and 3 are estimated as 2, 1.5, and 3 hours, respectively. After processing is complete, jobs are transferred to a loading terminal for shipping. The mode of transportation in the plant is a fleet of AGVs traveling on an eight-segment track. Each segment of the track is assigned a unique identification number, shown in Figure 18.10. These segment numbers are also used in the model to define an AGV route. Each segment has a track length that depends on the plant layout. Table 18.8 shows the service time (in hours) for an AGV to traverse each segment.

For simplicity, assume that the segments intersect at control points where vehicles may bypass one another. Thus, a newly arriving vehicle may not enter a segment occupied by another vehicle and instead must wait at a control point until the segment is available. This assumption makes it possible to ignore the effects of interference among vehicles. Otherwise,

Table 18.6 Process Sequence by Job Type

Manufacturing Cell	Plant Location
	2
3	3
4	4

Table 18.7 Process Sequence by Job Type

Job Type	Processing Sequence	Probability
1	1, 2, 3	.2
2	1, 3, 2	.3
3	2, 3, 1	.4
4	2, 1, 3	.1

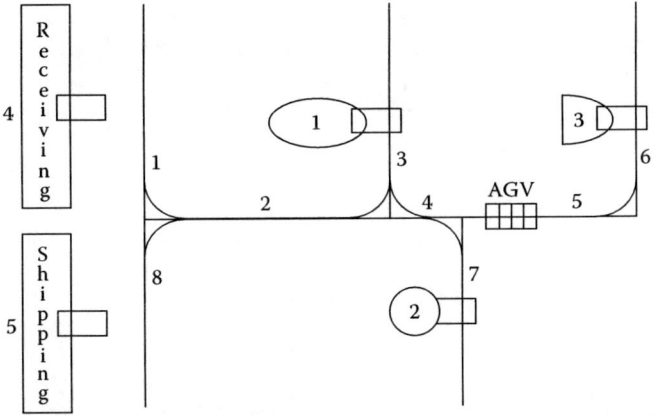

Figure 18.10 Flexible manufacturing layout.

Table 18.8   Traversal Time by AGV Line Segment

Segment	Time (hour)	Segment	Time (hour)
1	0.05	5	0.07
2	0.09	6	0.05
3	0.05	7	0.05
4	0.03	8	0.06

the resulting high level of detail may make it prohibitive to provide a concise explanation of the system at the macro level.

The processing sequence is read from the *UserInitial* spreadsheet into a two-dimensional array of integers, *JobSequence (1 To 4, 1 To 3)*. The distribution of jobs is in *JobTypePDF*. Each center is modeled as a *DropOffCenterQueue()*, *CenterFacility()*, and *PickUpCenterQueue()* sequence. For completeness, the receiving area and loading terminal are included as centers. However, in this model neither of these centers has a service time. Also, jobs arriving at the model immediately join the *PickUpCenterQueue* and completed jobs exit the system.

A *SegmentQueue()* and *SegmentFacility()* are used to model each segment of the track. AGVs wait in *SegmentQueue()* for access to a single server *SegmentFacility()*. An array *SegmentPath(1 To 19, 1 To 5)* summarizes the track segments connecting the five areas. For example, machine 1 is connected to machine 2 by the segments 3-4-7. This information is stored in row 1 of the array *SegmentPath()*. In general, each row of the *SegmentPath()* corresponds to a route in the plant. In this example, there are 19 such routes.

A table lookup in *SegmentMapTable* is used to cross-reference routes with areas they connect. A function is defined with entries $k = f(i + 10j)$, where i and j are numeric codes of the two connected areas and k is the row defining the route in the *SegmentPath()* array. This information is stored in table lookup function. For example, for $i = 1$ and $j = 2$ (that is, from center 1 to center 2) *SegmentMapTable (k = 12)* returns row 1 that maps to connecting track route 3-4-7 in *SegmentPath()*. Similarly, for $i = 4$ and $j = 3$, *SegmentMapTable (k = 43)* returns row 15 that maps to the route 1-2-4-5-6 in *SegmentPath()*. This notation completely defines AGV routes in the plant.

Transactions represent jobs and AGVs. Job attributes are

> 1 = *FirstStop*
> 2 = *SecondStop*
> 3 = *ThirdStop*
> 4 = *FourthStop*
> 5 = *FifthStop*
> 6 = *StopIndex* (1-5)
> 7 = *AGVID* (-1=> job not waiting for an AGV, 0=> job waiting for an AGV, > 0 => assigned AGV)

AGV attributes are

> 1 = *Origin*
> 2 = *Destination*
> 3 = *SegmentIndex*
> 4 = *SegmentID*
> 5 = *VehicleID*
> 6 = *Status*

Finally, there is a *JobQueue* for all active jobs in the system. It is used to order job requests for an AGV. Table 18.9 describes each primary event in the model. As indicated in the table, subroutine *Parser* routes all completed jobs to sub *GeneralCenterFacilityEvent* and both AGVs to *GeneralSegmentFacilityEvent*.

In addition to the primary events subroutines, the following subroutines are included in the model:

- *StartAGV*: Starts the AGV on the next track segment.
- *PickUpJob*: AGV picks up a job ready for transport to next center.
- *DropOffJob*: Delivers jobs to a processing center.
- *CheckWaitingJobs*: Searches for jobs waiting on an AGV.
- *FindAnAGV*: Locates an AGV for a job that has completed service.

There is also a function, *NextSegment*, that returns an identifier for the next segment to be traversed by the AGV. *PickUpJob*, *DropOffJob*, and *CheckWaitingJob* are separate procedures simply to reduce the length of sub *GeneralSegmentFacility*.

Table 18.9   Flexible Manufacturing Environment—Primary Events

Event	Description
JobSourceEvent	Arriving job
GeneralSegmentFacilityEvent	AGV crossed a segment
GeneralCenterFacilityEvent	Job completion

```
              ' EVENT HANDLERS
   Sub JobSourceEvent(ByRef Job As transaction)
      Dim _
         JobType As Integer, _
         ShopDestination As Integer, _
         ShopLocation As Integer, _
         Index As Integer
      JobType = JobTypePDF.RV
      Job.DimAttributes 7
      Job.Assign FirstStop, 4
            'Index for job route; first stop always receiving => 4
      For Index = 1 To 3
         ShopDestination = JobSequence(JobType, Index)
         Job.Assign Index + 1, LayoutMapTable.F(ShopDestination)
      Next Index
      Job.Assign FifthStop, 5
      JobQueue.Join Job
      Job.Assign AGVID, -1 ' waiting for service
      Job.Assign (StopIndex), 1
      CurrentStop = Job.A(StopIndex)
      CenterFacility(Job.A(CurrentStop)).NextFacility Job, 1, _
                             DropOffCenterQueue(Job.A(CurrentStop))
   End Sub
```

Figure 18.11   Flexible manufacturing environment—*JobSourceEvent*.

The remainder of this section provides an overview of model procedures and a brief description of program output.

JobSourceEvent: In Figure 18.11, *JobTypePDF* randomly assigns attribute *JobType*. The value of *JobType* corresponds to a row in *JobSequence*. Because the first and last stops are always 4 and 5, respectively, these centers are implied in array *JobSequence*. Attribute *StopIndex*, initialized to 1, is used to determine the next machine center to service the job. *LocationMapTable* maps the machine center to the center's location in the plant. All jobs join the queue of active jobs. An attribute *AGVID* value of -1 indicates that the job is not waiting for an AGV. Although the first stop is center 4, recall that center 4 has 0 service time. Therefore, there is never a job delay time in center 4. The net effect is that a job "immediately" joins the *PickUpCenterQueue* and, as discussed in *GeneralCenterFacilityEvent*, requests an AGV.

GeneralSegmentFacilityEvent: This segment represents jobs at the end of a track segment. Indirect addressing is used to determine which track segment was traversed. As shown in Figure 18.12, an AGV in *SegmentQueue()* may now start the delay to traverse the segment. An updated AGV *SegmentIndex*, along with attributes *Origin* and *Destination*, is used to identify the next *SegmentID*. A *SegmentID* of 0 indicates that an AGV has reached its final destination. If the AGV attribute *Status* is "Empty," the AGV has arrived to transport a job to its next destination; otherwise, it is there to drop off a job. After dropping off a job, sub *CheckWaitingJobs* searches for the next waiting job to access an AGV. If the AGV is at an intermediate destination, the AGV engages the next track segment *SegmentFacility(AGV.A (SegmentID))* or it must wait in *SegmentQueue(AGV.A(SegmentID))* because the track segment is occupied.

GeneralCenterFacilityEvent: A job completing service at a center is shown in Figure 18.13. The center facility is determined for the job that completed service. The job disengages the corresponding facility and when jobs are in the corresponding *DropOffCenterQueue()*, a job

```
Sub GeneralSegmentFacilityEvent(ByRef AGV As transaction)
  Dim _
    Job As transaction
  CurrentSegment = AGV.A(SegmentID)
  SegmentFacility(CurrentSegment).Disengage AGV
  If SegmentQueue(CurrentSegment).Length > 0 _
  Then ' AGV waiting on released segment
    SegmentFacility(CurrentSegment).EngageAndService _
                            SegmentQueue(CurrentSegment).Depart, 1
  End If
  AGV.Assign SegmentIndex, AGV.A(SegmentIndex) + 1
  AGV.Assign SegmentID, NextSegment(AGV.A(Origin), _
                             AGV.A(Destination), _
                             AGV.A(SegmentIndex))
  If AGV.A(SegmentID) = 0 _
  Then   ' AGV reached destination
    If AGV.A(Status) = "Empty" _
    Then
      PickUpJob AGV
    Else
      DropOffJob AGV
      CheckwaitingJobs AGV
    End If
    Exit Sub
  End If
  SegmentFacility(AGV.A(SegmentID)).NextFacility AGV, 1, _
                            SegmentQueue(AGV.A(SegmentID))
End Sub
```

Figure 18.12   Flexible manufacturing environment—*GeneralSegmentFacilityEvent*.

```
Sub GeneralCenterFacility(ByRef Job As transaction)
  Dim _
    Index As Integer
  CurrentStop = Job.A(StopIndex)
  CenterFacility(Job.A(CurrentStop)).Disengage Job
    'check job queue
  If DropOffCenterQueue(Job.A(CurrentStop)).Length > 0 _
  Then
    CenterFacility(Job.A(CurrentStop)).EngageAndService _
              DropOffCenterQueue(Job.A(CurrentStop)).Depart, 1
  End If
    ' process Job
  If Job.A(CurrentStop) = 5 _
  Then 'job is complete
    For Index = 1 To JobQueue.Length
      If Job.SerialNumber = JobQueue.View(Index).SerialNumber _
      Then
        JobQueue.Detach Index
        Set Job = Nothing
        Exit Sub
      End If
    Next Index
  End If
  Job.Assign AGVID, 0 ' job ready for pickup
  PickUpCenterQueue(Job.A(CurrentStop)).Join Job
  If FreeAGVQueue.Length > 0 _
  Then
    FindAnAGV Job
  End If
End Sub
```

Figure 18.13   Flexible manufacturing environment—*GeneralCenterFacilityEvent*.

departs the queue and is scheduled for a service completion. If the service was at center 5, the job is finished. It is detached from the active *JobQueue* and exits the system. Otherwise, the job joins the corresponding *PickUpCenterQueue()* with *AGVID* attribute equal to 0 to indicate that the job is waiting for an AGV. From the queue the job issues a request for an idle AGV, or the job waits for an AGV to find it when the AGV searches *JobQueue*.

StartAGV: In Figure 18.14, AGV attributes *Origin, Destination*, and *SegmentIndex* are sent to sub *NextSegment* in return for *SegmentID*. The AGV then engages *SegmentFacility()* or joins the corresponding *SegmentQueue()* when the segment is occupied.

PickUpJob: The procedure in Figure 18.15 loads the job on the AGV and begins the trip to the next job destination. Recall that *PickUpJob* is called from *GeneralSegmentFacilityEvent*. When the AGV has arrived at a center with attribute *Status* equal to "Empty," the AGV is at the center to transport a job to its next destination.

ThisStop (i.e., the current center) becomes the origin. *ThisStop* is based on the value of *StopIndex*. The Job departs *PickUpCenterQueue()* for *ThisStop*. The AGV attribute *Origin* is *ThisStop* and attribute *NextStop* is determined after incrementing *StopIndex* by 1. *NextStop* becomes the *Destination*. AGV attribute *SegmentIndex* is 1 to indicate a new origin–destination sequence

```
Sub StartAGV(ByRef AGV As transaction)
  'start AGV to destination
  AGV.Assign SegmentID, NextSegment(AGV.A(Origin), _
                                     AGV.A(Destination), _
                                     AGV.A(SegmentIndex))
  SegmentFacility(AGV.A(SegmentID)).NextFacility AGV, 1, _
                                SegmentQueue(AGV.A(SegmentID))
End Sub
```

Figure 18.14   Flexible manufacturing environment—sub *StartAGV*.

```
Sub PickUpJob(ByRef AGV As transaction)
  Dim _
    Index As Integer, _
    Job As transaction
  For Index = 1 To JobQueue.Length
    If JobQueue.View(Index).A(AGVID) = AGV.A(VehicleID) _
    Then
      Set Job = JobQueue.View(Index)
      Exit For
    End If
  Next Index
  CurrentStop = Job.A(StopIndex)
  PickUpCenterQueue(Job.A(CurrentStop)).Depart
  AGV.Assign Origin, Job.A(CurrentStop)
  AGV.Assign Status, "Loaded"
  Job.Assign StopIndex, Job.A(StopIndex) + 1
  NextStop = Job.A(StopIndex)
  AGV.Assign Destination, Job.A(NextStop)
  AGV.Assign SegmentIndex, 1
  StartAGV AGV
End Sub
```

Figure 18.15   Flexible manufacturing environment—sub *PickUpJob*.

```
Sub DropOffJob(ByRef AGV As transaction)
  Dim _
    Index As Integer, _
    Job As transaction
  For Index = 1 To JobQueue.Length
    If JobQueue.View(Index).A(AGVID) = AGV.A(VehicleID) _
    Then
      Set Job = JobQueue.View(Index)
      JobQueue.Join JobQueue.Detach(Index) ' reorder job queue
      Exit For
    End If
  Next Index
  AGV.Assign Status, "Empty"
  Job.Assign AGVID, -1 ' job again waiting service
  CenterFacility(AGV.A(Destination)).NextFacility Job, 1, _
          DropOffCenterQueue(AGV.A(Destination))
End Sub
```

Figure 18.16 Flexible manufacturing environment—sub *DropOffJob*.

and *Status* is now "Loaded." Sub *StartAGV* routes the AGV to the first track segment for the origin–destination sequence.

DropOffJob: As shown in Figure 18.16, the job is delivered to its next processing center. *DropOffJob* is also called from *GeneralSegmentFacilityEvent*. The AGV has arrived at a center with attribute *Status* equal to "Loaded." It is therefore at the center to deliver the job. A search of *JobQueue* is made for the job with attribute *AGVID* that matches the AGV. The job is detached from *JobQueue* and then immediately rejoins *JobQueue*. The effect is to move the most recently transported job to the end of the queue.

The AGV *Status* is changed to "Empty" and the job is either serviced by *CenterFacility()* or waits in the *DropOffCenterQueue()*. In either case, job *AGVID* becomes -1 to indicate that the job is not waiting for an AGV.

CheckWaitingJobs: The procedure in Figure 18.17 searches for a job in *JobQueue* waiting for an AGV. *CheckWaitingJobs* is also called from *GeneralSegmentFacilityEvent*. The AGV has arrived at a center with attribute *Status* equal to "Loaded." It is therefore at the center to deliver the job. After dropping off the job, the AGV will either find another job to transport or become idle. *JobQueue* is searched for a job with attribute *AGVID* equal to 0 (i.e., ready for transporting). If no job is found, the AGV is detached from *BusyAGVQueue* and joins *FreeAGVQueue*. The attribute *Status* is changed to "Empty" and attribute *Origin* is changed to *Destination* so that when the next request for an AGV is received the AGV *Origin* is its current location.

StopIndex references *CurrentStop* and after incrementing *StopIndex*, it references the job's next destination. A job waiting for an AGV must be at the AGV's current location or the AGV must travel to the center where the job resides, pick up the job, and transport it to another center. In either case, the AGV *Origin* (now starting here) is from the AGV *Destination*. The job *AGVID* becomes *VehicleID* and the AGV *SegmentIndex* is 1 to indicate a new origin–destination track sequence.

When the job and AGV are in the same center, the AGV *Destination* becomes the job's next center and the AGV *Status* is "Loaded." The job departs *PickUpCenterQueue()* and is destined for its next center. Recall that the transaction permanently resides in *JobQueue* so nothing about the job changes except its attribute values.

```
Sub CheckwaitingJobs(ByRef AGV As transaction)
  Dim _
    Index As Integer, _
    NextJob As transaction
Set NextJob = Nothing
  For Index = 1 To JobQueue.Length
    If JobQueue.View(Index).A(AGVID) = 0 _
    Then 'job waiting
       Set NextJob = JobQueue.View(Index)
       Exit For
    End If
  Next Index
  If NextJob Is Nothing _
  Then ' no waiting jobs
    For Index = 1 To 2
      If BusyAGVQueue.View(Index).A(VehicleID) = _
                                       AGV.A(VehicleID) _
      Then
         FreeAGVQueue.Join BusyAGVQueue.Detach(Index)
         AGV.Assign Status, "Empty"
         AGV.Assign Origin, AGV.A(Destination)
        Exit Sub
      End If
    Next Index
  End If
  NextJob.Assign AGVID, AGV.A(VehicleID)
  AGV.Assign SegmentIndex, 1
  CurrentStop = NextJob.A(StopIndex)
  If NextJob.A(CurrentStop) = AGV.A(Destination) _
  Then ' agv at job stop
    AGV.Assign Origin, NextJob.A(CurrentStop)
    PickUpCenterQueue(NextJob.A(CurrentStop)).Depart
    NextJob.Assign StopIndex, NextJob.A(StopIndex) + 1
    NextStop = NextJob.A(StopIndex)
    AGV.Assign Destination, NextJob.A(NextStop)
    AGV.Assign Status, "Loaded"
  Else ' AGV must travel to job stop
    AGV.Assign Origin, AGV.A(Destination)
    AGV.Assign Destination, NextJob.A(CurrentStop)
    AGV.Assign Status, "Empty"
  End If
  StartAGV AGV
End Sub
```

Figure 18.17 Flexible manufacturing environment—*CheckWaitingJobs*.

When the AGV and jobs reside in different centers, the AGV *Destination* is the center where the job resides. The AGV *Status* is "Empty" because it is in transit to pick up the job. In either case, subroutine *StartAGV* routes the AGV to the next track segment and the AGV is inserted into the primary events list (PEL).

FindAnAGV: In Figure 18.18 the procedure searches for an idle AGV in *FreeAGVQueue*. *FindAnAGV* is called from *GeneralCenterFacilityEvent* after a job has completed service and is ready for routing to its next service center. The job has checked *FreeAGVQueue* and therefore knows that there is an idle AGV in the system. The job is in *PickUpCenterQueue()* waiting for an AGV assignment.

Similar to sub *CheckWaitingJobs*, a job waiting for an AGV must be at the AGV's current location or the AGV must travel to the center where the job resides, pick up the job, and transport it to another center. As before, the AGV *Origin* (now starting here) is from the AGV

```
Sub FindAnAGV(ByRef Job As transaction)
  Dim _
    Index As Integer, _
    AGV As transaction
  Set AGV = Nothing
  For Index = 1 To FreeAGVQueue.Length
    If FreeAGVQueue.View(Index).A(Origin) = _
                                      Job.A(CurrentStop) _
    Then 'AGV at current location
      Set AGV = FreeAGVQueue.Detach(Index)
      Exit For
    End If
  Next Index
  CurrentStop = Job.A(StopIndex)
  If AGV Is Nothing _
  Then  ' AGV not local
    Index = 1
    Set AGV = FreeAGVQueue.Detach(Index)
    AGV.Assign Destination, Job.A(CurrentStop)
    AGV.Assign Status, "Empty"
  Else
    AGV.Assign Origin, Job.A(CurrentStop)
    Job.Assign StopIndex, Job.A(StopIndex) + 1
    NextStop = Job.A(StopIndex)
    AGV.Assign Destination, Job.A(NextStop)
    AGV.Assign Status, "Loaded"
    PickUpCenterQueue(Job.A(CurrentStop)).Depart
  End If
  BusyAGVQueue.Join AGV
  AGV.Assign (SegmentIndex), 1
  Job.Assign (AGVID), AGV.A(VehicleID)
  StartAGV AGV
End Sub
```

Figure 18.18   Flexible manufacturing environment—*FindAnAGV*.

Destination. The job *AGVID* becomes *VehicleID* and the AGV *SegmentIndex* is 1 to indicate a new origin–destination track sequence.

A successful search of *FreeAGVQueue* indicates that the AGV and job are in the same center. In that case, the AGV is detached from the *FreeAGVQueue* and assigned to the job. For an unsuccessful search (AGV and job not in the same center), the first AGV in *FreeAGVQueue* is assigned the job. In either case, the AGV has the current center as *Origin*.

When the job and AGV are at the same center, the AGV *Destination* is the job's next center and the AGV *Status* is "Loaded." The job departs *PickUpCenterQueue()* destined for its next center. Again, the transaction permanently resides in *JobQueue* so for that job nothing changes except attribute values of the job, and the AGV is placed on the PEL. When the AGV and job are in different centers, the AGV *Destination* is where the job is located. The AGV *Status* is "Empty" because the AGV travels to pick up the job. In either case, sub *StartAGV* routes the AGV to the next track segment. In both cases, the AGV joins the *BusyAGVQueue*, the AGV *SegmentIndex* is 1, and sub *StartAGV* routes the AGV to the next track segment.

NextSegment: Figure 18.19 is a function that receives *Origin* center, *Destination* center, and *SegmentIndex*, and returns the *NextSegment* from the array *SegmentPath*.

```
Function NextSegment(ByVal Origin As Integer, _
                     ByVal Destination As Integer, _
                     ByVal SegmentIndex As Integer)
   Dim _
      SegmentCode As Integer, _
      PathIndex As Integer
   SegmentCode = 10 * Origin + Destination
   PathIndex = SegmentMapTable.Value(SegmentCode)
   NextSegment = SegmentPath(PathIndex, SegmentIndex)
End Function
```

Figure 18.19   Flexible manufacturing environment—sub *NextSegment*.

Queue	BusyAGV		FreeAGV	
Number In	227		229	
Number Out	0		0	
Residing	0		2	
Destroyed	0		0	
Removed	0		0	
Detached	227		227	
Minimum	0		0	
Maximum	2		2	
	Lq	Wq	Lq	Wq
Average	0.416	0.365	1.584	1.377
StdDev	0.638	0.014	0.638	0.093

Figure 18.20   Flexible manufacturing environment—queues output.

Facility	Center1		Center2		Center3	
Number In	66		64		55	
Number Out	65		64		54	
Residing	1		0		1	
Interrupted	0		0		0	
Destroyed	0		0		0	
Minimum	0		0		0	
Maximum	1		1		1	
	AvgBusy Servers		AvgBusy Servers		AvgBusy Servers	
Average	0.502		0.531		0.708	
StdDev	0.500		0.499		0.455	

Figure 18.21   Flexible manufacturing environment—facilities output.

The output of the model is shown in Figures 18.20 and 18.21. Notice that the average number of free vehicles is 1.584 and the average number of busy vehicles is 0.416. This, in essence, indicates a utilization of 20.8%. Two of the three centers have average utilization of approximately 50% and although not shown, the network's average rail utilization is 4.96%. These results suggest that one AGV may be sufficient for the operation of this plant.

Reference

1. Cheng, T. C. E., Simulation of flexible manufacturing systems, *Simulation*, 1985, 45(6), 298–302.

Chapter 19

Material-Handling Models

19.1 Introduction

Materials handling deals with the movement of materials between work centers. Such movement may be manual or by machinery (e.g., overhead cranes, forklift trucks, robots, automated guided vehicles [AGVs], and conveyors). Simulation models are useful for identifying possible bottlenecks in a materials-handling system.

This chapter presents four materials-handling examples. The first is the use of a transporter to transfer piles of loads from a collection area to a repair facility. The second example is similar to the first except that the transporter is replaced by an overhead crane. The third and fourth examples model carrousel and belt conveyors in a manufacturing setup and a log mill operation, respectively.

19.2 Transporter Car

This example is adapted from [1] (pp. 596–598), where it is modeled using the materials-handling extension of SLAM II. A transporter is used to transfer rejected steel pipes from a collection pile to a repair facility. Rejected pipes arrive every *Exponential* (2.7) minutes as individual units with each weighing *UniformContinuous* (5, 8) tons. The maximum capacity of the transporter is 30 tons. Transporter loading begins when 20 tons or more becomes available. If the pile contains more than the capacity of the car, as many pipes as necessary, in their order of arrival, are loaded until a maximum of 30 tons is secured. The trip to the repair facility is *UniformContinuous* (2.13, 3.9) minutes. The return trip is only *Triangle* (1, 1.9, 2.3) minutes. The performance metrics for this model are the average weight of the pipes at the collection pile and average carload weight.

A transaction represents a bundle of defective pipes and the value of attribute 1 is the weight of the bundle. Arriving defective pipe is stored in a queue. The transporter is also represented by a transaction. The time to load and unload the pipe is reflected in delay times to and from the repair facility.

A user-defined Boolean variable, *TransporterIdle*, indicates the status of the transporter. Two integer variables, *PileWeight* and *LoadWeight*, represent the total weight of pipe in the pile and on the transporter, respectively. *QueueWeightStatistic* and *CarLoadWeightStatistic* provide the desired model output. *QueueWeightStatistic* is time based, whereas *CarLoadWeightStatistic* is an observation-based statistic.

Table 19.1 describes each of the primary events in the model. In addition to subroutines for each of the primary events, a *LoadTransporter* subroutine details events related to loading the transporter. The remainder of this section provides an overview of model procedures and a brief description of program output.

PipeSourceEvent: This subroutine represents arriving defective pipe. As shown in Figure 19.1, upon arrival, *PipeWeight* is determined and *PileWeight* is incremented by *PipeWeight*. Thereafter, the time-based *QueueWeightStatistic* is updated. Whenever the transporter is idle and *PileWeight* exceeds 20 tons, *LoadTransporter* initiates a trip to the repair area.

TransporterFacilityEvent: In Figure 19.2, there are no statements in this procedure. The facility is used simply to collect utilization information on the transporter. Transporter travel times are reflected in the delay events to and from the repair area.

TransporterDelayEvent: In Figure 19.3, the transporter finishes the trip to the repair area. *LoadWeight* is set to zero and the transporter starts the return trip to the pile.

ReturnDelayEvent: In Figure 19.4, the transporter finishes the return trip to the pile. For data collection purposes, the transporter disengages the facility to indicate that the transporter is idle. If *PileWeight* exceeds 20 tons, *LoadTransporter* again initiates a trip to the repair area.

Table 19.1   Transporter Car—Primary Events

Event	Description
PipeSource	Generates pipe arrivals
TransporterFacility	Not used
TransporterDelay	Finish unloading in the repair area
ReturnDelay	Return trip to the pile of pipe

```
Sub PipeSourceEvent(ByRef ArrivingPipe As Transaction)
  Dim _
      PipeWeight As Single
  Pipe Weight = UniformContinuous(5, 8)
  PileWeight = PileWeight + PipeWeight
  ArrivingPipe.Assign 1, PipeWeight
  PipePileQueue.Join ArrivingPipe
  QueueWeightStatistic.Collect PileWeight
  If TransporterFacility.NumberOfIdleServers > 0 And _
        PileWeight >= 20 _
  Then
      Call LoadTransporter
  End If
End Sub
```

Figure 19.1   Transporter car—*PipeSourceEvent*.

LoadTransporter: In Figure 19.5, bundles are moved to the transporter. The transporter is loaded while there are additional bundles in the pile and the transporter load is less than 30 tons. After the transporter is loaded, *TransporterFacility* is engaged. Statistical variables are collected and the trip to the repair area begins. Notice that *TransporterFacility* is engaged but the transporter movement is reflected in delay times.

```
Sub TransporterFacilityEvent(ByRef CurrentEvent As Transaction)
End Sub
```

Figure 19.2   Transporter car—*TransporterFacilityEvent*.

```
Sub TransportDelayEvent(ByRef Transporter As Transaction)
  LoadWeight = 0
  ReturnDelay.Start Transporter
End Sub
```

Figure 19.3   Transporter car—*TransporterDelayEvent*.

```
Sub ReturnDelayEvent(ByRef Transporter As Transaction)
  TransporterFacility.Disengage Transporter
  If PileWeight >= 20 _
  Then
    Call LoadTransporter
  End If
End Sub
```

Figure 19.4   Transporter car—*ReturnDelayEvent*.

```
Sub LoadTransporter()
  Dim _
    NextPipe As Transaction
  Do While PipePileQueue.Length > 0
    Set NextPipe = PipePileQueue.View(1)
    If LoadWeight + NextPipe.A(1) > 30 _
    Then
      Exit Do
    Else
      Set NextPipe = PipePileQueue.Depart
      PileWeight = PileWeight - NextPipe.A(1)
      LoadWeight = LoadWeight + NextPipe.A(1)
    End If
  Loop
  TransporterFacility.Engage Transporter, 1
  TransportDelay.Start Transporter
  CarLoadWeightStatistic.Collect LoadWeight
  QueueWeightStatistic.Collect PileWeight
End Sub
```

Figure 19.5   Transporter car—sub *LoadTransporter*.

In Figure 19.6, the average transporter load is 23.75 tons and the net average weight at the collection pile is 9.14 tons. From Figure 19.7, transporter utilization is approximately 50%. For a consistency check, from Figure 19.8, the average number of pipes in the queue is 1.42. The average *PipeWeight* is 6.5 tons for an estimated *PileWeight* of 9.23 tons as compared to 9.136 in Figure 19.6.

Facility	Transporter	
Number In	501	
Number Out	500	
Residing	1	
Interrupted	0	
Destroyed	0	
Minimum	0	
Maximum	1	
	AvgBusy	Servers
Average	0.507	
StdDev	0.500	

Figure 19.6 Transporter car—facility output.

Statistic	(T) QueueWeight	(O) CarLoadWeight
Number In	2331	501
Min Value	0.00	20.0016
Max Value	44.66	28.8173
Average	9.136	23.747
StdDev	7.204	2.404

Figure 19.7 Transporter car—statistics output.

Queue	PipePile	
Number In	1830	
Number Out	1828	
Residing	2	
Destroyed	0	
Removed	0	
Detached	0	
Minimum	0	
Maximum	7	
	Lq	Wq
Average	1.425	3.892
StdDev	1.126	0.098

Figure 19.8 Transporter car—queue output.

19.3 Overhead Crane

This example deals with the use of an overhead (bridge) crane for materials handling in a factory. An overhead crane on elevated runways provides movement of material in any one of three directions: longitudinal, transversal, and vertical. As a result, the crane can reach any spot on the plant floor. A bridge that spans the width of the bay where the crane operates affects longitudinal movement. A trolley mounted on the bridge provides the lateral movement of the crane. The vertical movement is provided by a hoist mechanism with cable attachment. Normal operation of the crane may allow for all three types of motion concurrently.

The situation in this example is an adaptation from [1] (pp. 614–616). It deals with an overhead crane used to transport coils from a bulk storage area to a load storage area where the coils are processed by two spindles. The coil widths are 36 and 48 inches. The load storage area has two piles. The first is large enough for either the 36- or 48-inch coils, whereas the second can accommodate only two 36-inch coils. The two piles are positioned at (135, 10) and (145, 10) feet from the center of the bulk storage area. Crane speeds are 195 feet per minute for the bridge and 200 feet per minute for the trolley.

Coils arrive at the bulk storage area every *Exponential* (1.7) minutes. There are equal probabilities that arriving coils will be either 36 or 48 inches wide. Approximately 0.7 minutes are needed to position a coil before it is picked. The picking and drop-off times are 0.1 and 0.15 minutes, respectively. The loading time of the coil on the spindle is 0.17 minutes, and the crane is not used. The processing time at the spindles depends on the size of the coil. The time to process a 36-inch coil is *UniformContinuous* (2.7, 3.3) minutes. The processing time for the 48-inch coil is (48/36)**UniformContinuous* (2.7, 3.3) minutes.

Transactions are used to represent the crane and arriving coils. The coil attributes are

1 = *CoilSize* (36 or 48 inches)
2 = *Destination* area (1 or 2)
3 = *TransportTime*

BulkQueue coils are ordered on arrival. *LoadQueue* is in front of the spindles. *SmallStorageCount* and *LargeStorageCount* are integer variables that represent the number of coils at the respective

Table 19.2   Overhead Crane—Primary Events

Event	Description
BulkCoilSource	Coil arrived
CraneFacilityEvent	Not programmed
SpindleFacilityEvent	Spindle service completion
PositionPickDelay	Pick and positioning of coil finished
TransportDropDelay	Crane transport and drop finished
LoadCoilDelay	Coil loaded onto spindle
ReturnDelay	Crane finished return to bulk storage

```
Sub BulkCoilSourceEvent(ByRef Arrival As transaction)
  If Randomnumber(1) < 0.5 _
  Then
    Arrival.Assign CoilSize, 36
  Else
    Arrival.Assign CoilSize, 48
  End If
  BulkQueue.Join Arrival
  Call ServiceBulkQueue
End Sub
```

Figure 19.9 Overhead crane—*BulkCoilSourceEvent*.

```
Sub CraneFacilityEvent(ByRef CurrentEvent As transaction)
End Sub
```

Figure 19.10   Overhead crane—*CraneFacilityEvent*.

```
Sub SpindleFacilityEvent(ByRef ProcessedCoil As transaction)
  Dim _
    NextCoil As transaction
  SpindleFacility.Disengage ProcessedCoil
  CycleTimeStatistic.Collect ProcessedCoil.SystemTime
  Set ProcessedCoil = Nothing
  If LoadQueue.Length > 0 _
  Then
      Set NextCoil = LoadQueue.Depart
      SpindleFacility.engage NextCoil, 1
      LoadCoilDelay.Start NextCoil
  End If
End Sub
```

Figure 19.11   Overhead crane—*SpindleFacilityEvent*.

spindle. A Boolean variable, *CraneIdle*, represents the status of the crane. *DistToStorageAreas* and *CraneSpeeds* are user-defined arrays used to determine the delay times for moving coils. Table 19.2 describes the primary events in the model. In addition to primary events subroutines, a *LoadTransporter* subroutine details events related to loading the transporter.

The remainder of this section provides an overview of model procedures and a brief description of program output.

BulkCoilSourceEvent: This procedure represents a coil arrival in the system. Upon arrival, as shown in Figure 19.9, attribute *CoilSize* is assigned the value 36 or 48 with equal probabilities using *RandomNumber(1)*. The arriving coil joins *BulkQueue*, and subroutine *ServiceBulkQueue* is called. Details of subroutine *ServiceBulkQueue* are explained below.

CraneFacilityEvent: This subroutine represents the service completion of the crane. In Figure 19.10, there are no secondary events for this primary event. It is defined only to collect crane utilization data.

```
Sub PositionPickDelayEvent(ByRef PickedCoil As Transaction)
   BulkQueue.Depart
   Call TravelTime(PickedCoil)
   TransportDropDelay.Start PickedCoil, _
                    PickedCoil.A(TransportTime) + DropDelay
End Sub
```

Figure 19.12 Overhead crane—*PositionPickDelayEvent*.

```
Sub TransportDropDelayEvent(ByRef DroppedCoil As transaction)
   Dim _
      Crane As Transaction
   CraneFacility.Disengage DroppedCoil
   If SpindleFacility.NumberOfIdleServers > 0 _
   Then
      SpindleFacility.Engage DroppedCoil, 1
      LoadCoilDelay.Start DroppedCoil
   Else
      LoadQueue.Join DroppedCoil
   End If
   Set Crane = DroppedCoil.Copy
   CraneFacility.Engage Crane, 1
   ReturnDelay.Start Crane, Crane.A(TransportTime)
End Sub
```

Figure 19.13 Overhead crane—*TransportDropDelayEvent*.

SpindleFacilityEvent: This procedure represents a spindle service completion. Upon completion, as shown in Figure 19.11 the coil disengages the spindle and collects cycle time for the coil. If there is a coil in *LoadQueue*, it departs the queue and engages the spindle for loading. The spindle is busy when engaged even though service does not begin until the spindle is loaded.

PositionPickDelayEvent: This subroutine indicates that the coil has been positioned in the bulk queue area and picked by the crane. As shown in Figure 19.12 it is now that the coil departs *BulkQueue*. The *TravelTime* subroutine places the coil transport time to the spindle area in attribute *TransportTime*. When the crane (coil) finishes the *TransportDrop* delay, the coil is ready for loading onto the spindle and the crane returns to the bulk queue area. Notice that the crane resides in *CraneFacility*; however, all time the crane spends on the PEL is controlled by delays. The effect is that the crane is busy from the time it begins the pick operation until it completes the return trip to the bulk coil area.

TransportDropDelayEvent: In Figure 19.13 the coil is now dropped in the spindle area. If the spindle is idle, then the coil engages the spindle to reflect that the spindle is no longer idle and the load coil delay is initiated. Otherwise, the coil joins *LoadQueue*. In either case the crane is now finished with that coil and returns to the bulk queue area. Notice that the crane remains busy even though it is finished with the dropped coil.

LoadCoilDelayEvent: In Figure 19.14 the coil is now loaded on the spindle. The number of coils in the appropriate storage area is decremented to indicate that a storage space is now available. Thereafter, spindle service begins. The service time is dependent on the size of the coil.

```
Sub LoadCoilDelayEvent(ByRef LoadedCoil As transaction)
  If LoadedCoil.A(Destination) = "SmallStorage" _
  Then
    SmallStorageCount = SmallStorageCount - 1
  Else 'Destination is LargeStorage
    LargeStorageCount = LargeStorageCount - 1
  End If
  If LoadedCoil.A(CoilSize) = 36 _
  Then
    SpindleFacility.Service LoadedCoil
  Else 'CoilSize = 48
    SpindleFacility.Service LoadedCoil, _
                (48 / 36 * Deeds.UniformContinuous(2.7, 3.3))
  End If
End Sub
```

Figure 19.14   Overhead crane—*LoadCoilDelayEvent*.

```
Sub ReturnDelayEvent(ByRef Crane As transaction)
  CraneFacility.Disengage Crane
  Call ServiceBulkQueue
End Sub
```

Figure 19.15   Overhead crane—*ReturnDelayEvent*.

ReturnDelayEvent: In Figure 19.15 the crane has returned to the bulk storage area. For statistical purposes, the crane disengages the facility and is now idle. If there are coils in the *BulkQueue*, the subroutine *ServiceBulkQueue* is called to determine whether another coil can now be transported to the spindle area. Recall that a coil is transported to the spindle area only when the staging area in front of the spindle has room for the coil.

ServiceBulkQueue: The program segment in Figure 19.16 determines when a coil may be transported to the spindle area. The coils are processed in the order that they arrive. If the crane is idle, the following two mutually exclusive scenarios are evaluated in the order presented:
1. If the small spindle storage area has an empty space and the next coil is a 36-inch coil, the coil destination is the small storage area and the occupied small coil storage area is incremented by 1. The crane status is changed to busy, the crane engages the crane facility, and the position coil delay is started.
2. If the large storage area has an empty space, the coil destination is the large storage area and the occupied large coil storage area is incremented by 1. The crane status is changed to busy, the crane engages the crane facility, and the position coil delay is started.

In either case, the crane is busy when it engages *CraneFacility* and the *PositionPick* delay begins. Notice that when the coil transaction engages *CraneFacility*, the transaction represents the crane and the coil. When the coil transaction is dropped in the spindle area, a copy of the coil represents the return trip to the bulk queue area.

TravelTime: The program segment in Figure 19.17 uses the coil destination to determine the delay time for moving the coil from the *BulkQueue* to the spindle storage area. The delay time is stored in attribute 3 of the coil.

In Figure 19.18, crane utilization is 0.99; therefore, output validation is fairly straightforward. The crane cycle time is 2.28 time units ((135/195)*2 + 0.7 + 0.1 + 0.1). At 100% utilization for

```
Sub ServiceBulkQueue()
  Dim _
    WaitingCoil As Transaction
  If CraneFacility.NumberOfIdleServers > 0 And _
    BulkQueue.Length > 0 _
  Then
    Set WaitingCoil = BulkQueue.View(1)
    If SmallStorageCount < 2 And _
       WaitingCoil.A(CoilSize) = 36 _
    Then
      WaitingCoil.Assign Destination, "SmallStorage"
      SmallStorageCount = SmallStorageCount + 1
      CraneFacility.Engage WaitingCoil, 1
      PositionPickDelay.Start WaitingCoil, _
                                    PositionDelay + PickDelay
      Exit Sub
    End If
    If LargeStorageCount < 2 _
    Then
      WaitingCoil.Assign Destination, "LargeStorage"
      LargeStorageCount = LargeStorageCount + 1
      CraneFacility.Engage WaitingCoil, 1
      PositionPickDelay.Start WaitingCoil, _
                                    PositionDelay + PickDelay
    End If
  End If
End Sub
```

Figure 19.16 Overhead crane—sub *ServiceBulkQueue*.

```
Sub TravelTime(ByRef Coil As transaction)
  Dim _
    BridgeTravelTime, _
    TrolleyTravelTime
  If Coil.A(Destination) = "SmallStorage" _
  Then
    BridgeTravelTime = _
                DistToStorageAreas(1, 1) / CraneSpeeds(1, 1)
    TrolleyTravelTime = _
                DistToStorageAreas(1, 2) / CraneSpeeds(1, 2)
  Else   ' large storage
    BridgeTravelTime = _
                DistToStorageAreas(2, 1) / CraneSpeeds(1, 1)
    TrolleyTravelTime = _
                DistToStorageAreas(2, 2) / CraneSpeeds(1, 2)
  End If
  Coil.Assign TransportTime, _
    WorksheetFunction.Max(BridgeTravelTime, TrolleyTravelTime)
End Sub
```

Figure 19.17 Overhead crane—sub *TravelTime*.

1000 time units, the crane could deliver approximately 439 coils as compared to the 411 shown in Figure 19.19. With an average arrival rate in the bulk coil area of 1.7 time units per arrival, an average of 588 will arrive in 1000 time units. In Figure 19.20, the actual number of arrivals is 610. Clearly, the crane is the bottleneck in this system, and as a result, the system is unstable (the crane cannot handle the workload).

Facility	Crane		Spindle	
Number In	825		412	
Number Out	824		411	
Residing	1		1	
Interrupted	0		0	
Destroyed	0		0	
Minimum	0		0	
Maximum	1		2	
	AvgBusy	Servers	AvgBusy	Servers
Average	0.999		1.504	
StdDev	0.028		0.503	

Figure 19.18   Overhead crane—facilities output.

Statistic	(O) CycleTime
Number In	411
Min Value	4.611
Max Value	342.466
Average	174.877
StdDev	97.950

Figure 19.19   Overhead crane—statistics output.

Queue	Bulk	
Number In	610	
Number Out	412	
Residing	198	
Destroyed	0	
Removed	0	
Detached	0	
Minimum	0	
Maximum	200	
	Lq	Wq
Average	102.546	167.833
StdDev	56.351	3.955

Figure 19.20   Overhead crane—queue output.

19.4 Carrousel Conveyor

This situation, adapted from [2], deals with a product feeding randomly from an assembly line into a carrousel conveyor. Each arriving item occupies a unit of space on the conveyor. Once all the conveyor space is occupied, feeding into the conveyor is automatically blocked, and items must wait in a preconveyor storage area until space is available. A number of workstations are positioned around the conveyor, and each station has an operator. An operator grabs an item for processing as the item passes the station. Items completed by the operator are gravity-fed to a shipping area, and the operator reaches for a new item to be processed. The circular conveyor design allows unprocessed items to recycle until they are eventually retrieved by one of the workstations. The metrics for this simulation model are workstation utilization and the time an item spends in the system before reaching the shipping area.

The model is designed to accommodate any number of workstations, as well as any conveyor capacity, by defining the number of conveyor and workstation servers in the *Initial* spreadsheet. An item entering the conveyor arrives at the first workstation in two minutes. The stations are spaced one minute apart, with the exception of the last station, which is five minutes away from the first station. (Notice that the times between the stations can be generalized by associating them to the capacity and the speed of the conveyor.) For simplicity, assume that once an item is removed from the conveyor, a waiting item can immediately be "squeezed" by gravity action onto the carrousel. As such, it is unnecessary for the item to wait for the space to arrive at the loading chute. Including this detail in the model is by no means a trivial task.

The number of servers in *ConveyorFacility* represents the capacity of the conveyor. Items arrive every *Exponential* (0.6) minutes and wait in *ConveyorQueue* when the conveyor is full. When allocated a conveyor space, the item engages a conveyor server. As such, *ConveyorFacility* may be regarded as an extension of queue *ConveyorQueue*.

The attributes for an item transaction are

1 = *NextStation*
2 = *NextStationDelay*

Workstations are also represented by a multiserver *WorkStationsFacility*. An array of Boolean variables, *WorkStationStatus()*, indicates the status of an individual workstation. Table 19.3 describes the primary events in the model.

The remainder of this section provides an overview of model procedures and a brief description of program output.

Table 19.3 Carrousel Conveyor—Primary Events

Event	Description
ItemSource	Item arrivals
ConveyorFacility	Conveyor status
WorkStationFacility	Operator picks
ConveyorDelay	Moved conveyor items

```
Sub ItemSourceEvent(ByRef Arrival As Transaction)
    ' server is a conveyor slot
    Arrival.Assign NextStation, 1
    Arrival.Assign NextStationDelay, 2
    If ConveyorFacility.NumberOfIdleServers > 0 _
    Then
        ConveyorFacility.Engage Arrival, 1 ' NOT on PEL
        ConveyorDelay.Start Arrival, Arrival.A(NextStationDelay)
    Else
        CarrouselQueue.Join Arrival
    End If
End Sub
```

Figure 19.21 Carrousel conveyor—*ItemSourceEvent*.

```
Sub ConveyorFacilityEvent(ByRef CurrentEvent As Transaction)
  ' ConveyorFacility transactions are never on the PEL
End Sub
```

Figure 19.22 Carrousel conveyor—*ConveyorFacilityEvent*.

```
Sub WorkStationsFacilityEvent(ByRef CompletedItem As Transaction)
  WorkStationStatus(CompletedItem.A(NextStation)) = Idle
  WorkStationsFacility.Disengage CompletedItem
  CycleTimeStatistic.Collect CompletedItem.SystemTime
End Sub
```

Figure 19.23 Carrousel conveyor—*WorkstationFacilityEvent*.

```
Sub ConveyorDelayEvent(ByRef ConveyorItem As Transaction)
If WorkStationStatus(ConveyorItem.A(NextStation)) = Idle _
  Then
    WorkStationStatus(ConveyorItem.A(NextStation)) = Busy
    ConveyorFacility.Disengage ConveyorItem
    WorkStationsFacility.EngageAndService ConveyorItem, 1
    Set ConveyorItem = Nothing
    If CarrouselQueue.Length > 0 _
    Then  ' next item on conveyor
      Set ConveyorItem = CarrouselQueue.Depart
      ConveyorFacility.Engage ConveyorItem, 1 ' NOT on PEL
      ConveyorDelay.Start ConveyorItem, _
                              ConveyorItem.A(NextStationDelay)
    End If
  Else
    If ConveyorItem.A(NextStation) = _
                        WorkStationsFacility.NumberOfservers _
    Then
      ConveyorItem.Assign NextStation, 1
      ConveyorItem.Assign NextStationDelay, 5
      ConveyorDelay.Start ConveyorItem, _
                              ConveyorItem.A(NextStationDelay)
    Else
      ConveyorItem.Assign NextStation, _
                              ConveyorItem.A(NextStation) + 1
      ConveyorItem.Assign NextStationDelay, 1
      ConveyorDelay.Start ConveyorItem, _
                              ConveyorItem.A(NextStationDelay)
    End If
  End If
End Sub
```

Figure 19.24 Carrousel conveyor—*ConveyorDelayEvent*.

ItemSourceEvent: This program segment models an arriving customer. In Figure 19.21, an arriving item has attribute *NextStation* equal to 1, indicating that the item will pass station 1 first. Attribute *NextStationDelay* represents the time to reach station 1. If there is an idle *ConveyorFacility* server, the arrival engages a server and starts the delay to station 1. Otherwise, the arrival joins *ConveyorQueue*.

ConveyorFacilityEvent: This procedure models the status of the conveyor. As shown in Figure 19.22, when all *ConveyorFacility* servers are engaged, the conveyor is full. Because

Queue	Carrousel	
Number In	526	
Number Out	526	
Residing	0	
Destroyed	0	
Removed	0	
Detached	0	
Minimum	0	
Maximum	23	
	Lq	Wq
Average	0.98	3.72
StdDev	3.30	0.13

Figure 19.25   Carrousel conveyor—queue output.

ConveyorFacility is engaged by an item but never serviced, there are no primary events associated with the *ConveyorFacility*. The effect is that aggregate conveyor utilization information is included in the simulation output.

WorkStationsFacilityEvent: This subroutine represents an operator pulling items from the conveyor. In Figure 19.23, when the operator processes an item, the corresponding *WorkStationStatus()* is changed to *Idle*. The item disengages a *WorkStation* facility and the *CycleTimeStatistic*, which measures the average time an item spends in the system, is collected.

ConveyorDelayEvent: This procedure models an item arriving at a workstation. The corresponding workstation is either busy or idle. In Figure 19.24, these scenarios are described as follows:

- Workstation is *Idle*: The *WorkStationStatus()* is changed to *Busy*, and the item disengages the *ConveyorFacility*. The item is removed from the conveyor and then engaged and serviced by a *WorkStationFacility* server. There is now an additional space on the conveyor. If an item is in *ConveyorQueue*, the item departs the queue, engages a *ConveyorFacility* server, and begins the delay to the first workstation.
- Workstation is *Busy*: If the item has reached the last workstation, it is again delayed on its way to the first workstation. Otherwise, the *NextStation* and *NextStationDelay* attributes are updated and the item is delayed (travels) to the next station.

The data presented is based on a conveyor capacity of 15 units and 3 workstations. Figure 19.25 indicates an average of 0.98 units wait for access to the conveyor. In Figure 19.26, the average number of busy servers of 9.03 is the same as a conveyor utilization of 0.60. An average number of busy workstations of 1.66 represents a workstation utilization of 0.55. From Figure 19.27, a total of 3291 units exited the conveyor and the average time on the conveyor was 7.09 time units.

19.5 Belt Conveyor—Plywood Mill Operation

This example is from [3]. The environment is a belt conveyor in a plywood mill. Trucks arrive at the mill every *Exponential* (120) minutes with loads of 15 logs. The standard log lengths are 8, 16, and 24 feet with the respective ratios of 0.2, 0.5, and 0.3. Log radius is *Normal* (1.3, 0.3) feet.

Facility	Conveyor		WorkStations	
Number In	3300		3293	
Number Out	3293		3291	
Residing	7		2	
Interrupted	0		0	
Destroyed	0		0	
Minimum	0		0	
Maximum	15		3	
	AvgBusy Servers		AvgBusy Servers	
Average	9.03		1.66	
StdDev	3.93		0.86	

Figure 19.26   Carrousel conveyor—facilities output.

Statistic	(O) CycleTime
Number In	3291
Min Value	2.00
Max Value	65.75
Average	7.09
StdDev	6.38

Figure 19.27   Carrousel conveyor—statistics output.

Arriving logs are stacked in a bin that feeds a 56-foot belt conveyor moving at a speed of 40 feet per minute. The belt transports the logs to a peeling machine that strips them into sheets for plywood production. The peeling machine can handle only eight-foot logs. Other logs (16 and 24 feet) must be cut into eight-foot sections before reaching the peeler. The cutting operation requires 0.2 minutes and occurs automatically at 32 feet from the start of the belt, whenever a log more than eight feet in length is detected by sensors along the side of the conveyor. The eight-foot logs are ready to be loaded on the peeler when their leading end reaches the end of the belt. It takes 0.2 minutes to complete the loading. The peeling time is a function of the log's volume. The peeler processes 20 cubic feet per minute. The belt automatically stops whenever empty. It also stops to allow the cutting of long logs into eight-foot sections or when a log reaches the end of the conveyor and waits to be loaded onto the peeler. At the instant a blockage is lifted, the belt automatically resumes motion. When a truck arrives, 15 logs are stacked in a bin. To move a log from the bin to the conveyor belt, the starting section of the belt must have sufficient space immediately in front of the bin to accommodate the log length.

In addition to transactions that manage belt movement and cutting logs, transactions in *BinQueue* represent logs waiting for loading onto the conveyor and transactions in *BeltQueue* are logs on the conveyor. The log attributes are as follows:

1 = *Length* (log)
2 = *Volume* of an eight-foot log (peeler accepts eight-foot sections)
3 = *EndPoint* (leading end of the log on the conveyor)

User-defined variable *LoaderFreeSpace* is the linear space in front of the bin and *BeltUsed* the basis for a time-based statistic on belt utilization. Boolean variables to control and assess the conveyor status are *CuttingLog*, *LoadingPeeler*, and *MovingBelt*. Table 19.4 describes primary events in the model.

There are three additional subroutines: *CheckFrontFreeSpace*, *CheckPeeler*, and *CheckCutLocation*. The remainder of this section provides an overview of model procedures and a brief description of program output.

TruckSourceEvent: This procedure models a truck arriving at *BinQueue*. In Figure 19.28, the truck loads 15 logs into *BinQueue*. For each log, attribute *Length* is assigned a value from *LogLengthPDF*. The value of the attribute *Volume* assumes that the log is eight feet in length. The volume calculation is done now to avoid duplicate calculations when the log is split further down the belt. After unloading the truck, subroutine *CheckFrontFreeSpace* loads a log onto the belt conveyor (*BeltQueue*) when sufficient *LoaderFreeSpace* is available. If the entire system is idle (this would occur when the belt was empty), the conveyor is started by *BeltDelay.Start*. When global variable *MovingBelt* equals *True* it indicates that the conveyor is in motion.

Table 19.4 Belt Conveyor—Primary Events

Event	Description
TruckSource	Truck arrived
PeelerFacility	Completion of peeler service
CutDelay	Finished cutting a log
LoadPeelerDelay	Finished loading peeler
BeltDelay	Finished advancing the conveyor

```
        ' EVENT HANDLERS
Sub TruckSourceEvent(ByRef Truck As transaction)
Dim _
    Log As transaction, _
    Index As Integer
For Index = 1 To 15  ' Truck drops logs into bin
    Set Log = Truck.Copy
    Log.Assign Length, LogLengthPDF.RV(1)
    Log.Assign Volume, _
        WorksheetFunction.Pi * 8 * DEEDS.Normal(1.3, 0.3) ^ 2
    Log.Assign EndPoint, 0
    BinQueue.Join Log
Next Index
Set Truck = Nothing
Call CheckFreeSpace
If Not LoadingPeeler And Not MovingBelt And Not CuttingLog _
Then  ' entire system is idle
    BeltDelay.Start NewTransaction
    MovingBelt = True
End If
End Sub
```

Figure 19.28 Belt conveyor—*TruckSourceEvent*.

PeelerFacilityEvent: This procedure represents the completion of a peeling operation. In Figure 19.29, the finished log disengages the peeler and departs the system. *LoadingPeeler* is *True* when the conveyor belt is blocked waiting for a log to be moved from the end of the conveyor. In that case, *LoadPeelerDelay.Start* begins loading the next log. However, a new transaction is used for the timer to ensure that the log does not actually depart the conveyor until *LoadPeelerDelay* is finished.

CutDelayEvent: This program segment represents the cut completion of a log longer than eight feet. In Figure 19.30 the cut timer transaction, *Cut*, that finished the delay is set to *Nothing*. The conveyor belt, *BeltQueue*, is searched for the log that has attribute *EndPoint* value of 40 (*SawLocation* + 8). A copy of the *OriginalLog* transaction is created and named *NewLog*. However, the *NewLog*'s attribute *Length* is reduced by eight, and for *OriginalLog* attribute *Length* is assigned the value of eight. *NewLog* is placed in *BeltQueue* after *OriginalLog*. Global variable *CuttingLog* is now *False*. The only other potential blockage is *LoadingPeeler*. If it is *False*, *BeltDelay* starts the conveyor with a new transaction and *MovingBelt* is changed to *True*.

```
Sub PeelerFacilityEvent(ByRef FinishedLog As transaction)
PeelerFacility.Disengage FinishedLog
  LogVolumeStatistic.Collect FinishedLog.A(Volume)
  Set FinishedLog = Nothing
  If LoadingPeeler _
  Then ' another log waiting
    LoadPeelerDelay.Start NewTransaction
  End If
End Sub
```

Figure 19.29 Belt conveyor—*PeelerFacilityEvent*.

```
Sub CutDelayEvent(ByRef Cut As transaction)
  Dim _
  OriginalLog As transaction, _
  NewLog As transaction, _
  Index As Integer
  Set Cut = Nothing
  For Index = 1 To BeltQueue.Length
    If BeltQueue.View(Index).A(EndPoint) = SawLocation + 8 _
    Then
      Set OriginalLog = BeltQueue.View(Index)
      Exit For  'found cut log
    End If
  Next Index
  Set NewLog = OriginalLog.Copy  ' process new log first
  NewLog.Assign Length, OriginalLog.A(Length) - 8
  NewLog.Assign EndPoint, SawLocation
  OriginalLog.Assign Length, 8
  BeltQueue.JoinAfter Index, NewLog 'new log behind original log
  CuttingLog = False
  If Not LoadingPeeler _
  Then
    BeltDelay.Start NewTransaction
    MovingBelt = True
  End If
End Sub
```

Figure 19.30 Belt conveyor—*CutDelayEvent*.

LoadPeelerDelay: This program segment represents a log moved from the conveyor to the peeler. In Figure 19.31, *EndLog* departs the conveyor belt and *BeltUsed* is decremented by eight. *BeltUsedStatistic*, *AverageConveyorLength* in use, is collected. *PeelerFacility* engages *EndLog* for a service time that is a function of log volume. If there are logs on the conveyor and no cutting operation is in progress, the *BeltDelay* timer is started and *MovingBelt* is again *True*.

BeltDelayEvent: This program segment represents a stopped conveyor. In Figure 19.32, *BeltDelay* is finished and *MovingBelt* is changed to *False*. Attribute *EndPoint* is updated for each log on the conveyor. As necessary, *LoaderFreeSpace* is updated to a maximum value of 24. If there are logs in *BinQueue*, subroutine *CheckFreeSpace* attempts to move a log from *BinQueue* to

```
Sub LoadPeelerDelayEvent(ByRef Loading As transaction)
  Dim _
     EndLog As transaction
  LoadingPeeler = False
  Set EndLog = BeltQueue.Depart
  BeltUsed = BeltUsed - 8
  BeltUsedStatistic.Collect BeltUsed
  PeelerFacility.EngageAndService _
                      EndLog, 1, EndLog.A(Volume) / 20
    'Start conveyor ?
  If BeltQueue.Length > 0 And Not CuttingLog _
  Then
     BeltDelay.Start NewTransaction
     MovingBelt = True
  End IfEnd
Sub
```

Figure 19.31   Belt conveyor—*LoadPeelerDelayEvent*.

```
Sub BeltDelayEvent(ByRef Conveyor As transaction)
Dim _
    Index As Integer
MovingBelt = False
  For Index = 1 To BeltQueue.Length 'Move logs on conveyor
   BeltQueue.View(Index).Assign EndPoint, _
       BeltQueue.View(Index).A(EndPoint) + 8
  Next Index
  If LoaderFreeSpace < 24 _
  Then 'Update space in front of bin
    LoaderFreeSpace = LoaderFreeSpace + 8
  End If
     'move bin log to conveyor ?
  If BinQueue.Length > 0 _
  Then
   CheckFreeSpace
  End If
  Call CheckPeeler
  Call CheckCutLocation
   'Start conveyor
  If Not LoadingPeeler And Not CuttingLog _
  Then
     BeltDelay.Start Conveyor
     MovingBelt = True
  End If
End Sub
```

Figure 19.32   Belt conveyor—*BeltDelayEvent*.

BeltQueue. Subroutine *CheckPeeler* determines whether a log's attribute *EndPoint* is at the end of the conveyor, and subroutine *CheckCutLocation* determines whether a log must be cut. *CheckPeeler* or *CheckCutLocation* may reset their respective global variable to indicate that the conveyor cannot be restarted. If neither of these occurs, the *BeltDelay* timer is started and the conveyor is in motion with *MovingBelt* equal to *True*.

The delay of 0.2 minutes is equal to an eight-foot forward movement of the conveyor with a speed of 40 feet per minute. The rationale for moving the conveyor in segments of eight feet is that this length represents the minimum "cell" length for this environment. For example, moving the belt in increments of one foot will needlessly increase the computational time because events only occur at eight-foot spacing.

CheckFreeSpace: Checks for sufficient space to move a log from *BinQueue* to *BeltQueue*. As shown in Figure 19.33, if the length of the next log in *BinQueue* is less than or equal to the *LoaderFreeSpace*, the log is moved to *BeltQueue*. *LoaderFreeSpace*, with an initial value of 24, represents the area directly in front of the bin.

Once a log seizes the starting location on the belt, log attribute *EndPoint* is the log length and, of course, *LoaderFreeSpace* is now 0. *BeltUsed* is incremented by the log length and the average belt used statistic is collected. Another log cannot be loaded until the present log has cleared sufficient space as it moves forward on the belt.

CheckPeeler: Checks for a log that may have reached the end of the conveyor belt and is ready to be loaded onto the peeler. As shown in Figure 19.34, if log attribute *EndPoint* for the log at the front of *BeltQueue* is equal to *BeltLength*, the log is ready for loading onto the peeler.

```
Sub CheckFreeSpace
Dim _
    NextLog As transaction
    'when space available, log goes onto conveyor
    Set NextLog = BinQueue.View(1)
    If NextLog.A(Length) <= LoaderFreeSpace _
    Then
        Set NextLog = BinQueue.Depart
        NextLog.Assign EndPoint, NextLog.A(Length)
        LoaderFreeSpace = 0
        BeltQueue.Join NextLog
        BeltUsed = BeltUsed + NextLog.A(Length)
        BeltUsedStatistic.Collect BeltUsed
    End If
End Sub
```

Figure 19.33 Belt conveyor—sub *CheckFreeSpace*.

```
Sub CheckPeeler()
    If BeltQueue.View(1).A(EndPoint) = EndOfBelt _
    Then  'Load log into peeler
        LoadingPeeler = True
        If PeelerFacility.NumberOfIdleServers > 0 _
        Then
            LoadPeelerDelay.Start NewTransaction
        End If
    End If
End Sub
```

Figure 19.34 Belt conveyor—sub *CheckPeeler*.

LoadingPeeler set to *True* indicates that the conveyor cannot be restarted until the log has been removed from the conveyor. However, the peeler may be busy processing another log. In that case, loading the peeler must wait until *PeelerFacility* is disengaged. If the peeler is idle, loading delay begins but the log remains on the conveyor until loading is finished. *LoadingPeeler* is also *True* until loading is finished.

CheckCutLocation: Checks for a log in *BeltQueue* that is ready to cut. As shown in Figure 19.35, when the length exceeds eight feet and is properly positioned for cutting, a *CutDelay* is started and *CuttingDelay* is *True* to indicate that the conveyor belt cannot be restarted. A log is ready to cut when its *Length* attribute is greater than eight and attribute *EndPoint* is equal to *SawLocation* + 8.

```
Sub CheckCutLocation()
Dim _
    Index As Integer
  For Index = 1 To BeltQueue.Length 'Check for log to be cut
    If BeltQueue.View(Index).A(Length) > 8 And _
       BeltQueue.View(Index).A(EndPoint) = SawLocation + 8 _
    Then
       CutDelay.Start NewTransaction
       CuttingLog = True
    End If
  Next Index
End Sub
```

Figure 19.35   Belt conveyor—sub *CheckCutLocation*.

Statistic	(T) BeltUsed
Number In	1016
Min Value	0.00
Max Value	56.00
Average	41.47
StdDev	18.87

Figure 19.36   Belt conveyor—statistics output.

Queue	Bin		Belt	
Number In	330		698	
Number Out	323		693	
Residing	7		5	
Destroyed	0		0	
Removed	0		0	
Detached	0		0	
Minimum	0		0	
Maximum	58		7	
	Lq	Wq	Lq	Wq
Average	14.27	86.24	3.95	11.29
StdDev	14.08	4.24	1.80	0.15

Figure 19.37   Belt conveyor—queues output.

Facility	Peeler	
Number In	693	
Number Out	692	
Residing	1	
Interrupted	0	
Destroyed	0	
Minimum	0	
Maximum	1	
	AvgBusy Servers	
Average	0.79	
StdDev	0.41	

Figure 19.38   Belt conveyor—facility output.

The model is run for 2400 minutes. Program output includes Figures 19.36 through 19.38. In Figure 19.36 the average length of belt used is 41.5 feet. As a validity check, the average belt used is approximately a multiple of eight feet.

References

1. Pritsker, A. A. B., *Introduction to Simulation and SLAM II*, 3rd ed. New York: Wiley, 1986.
2. Pritsker, A. and C. Sigal, *Management Decision Making*. Englewood Cliffs, NJ: Prentice-Hall, Inc., 1983, pp. 178–182.
3. Pegden, C., *Introduction to SIMAN*. State College, PA: Systems Modeling Corporation, 1985, pp. 145–413.

Chapter 20

Inventory Control Models

20.1 Introduction

This chapter presents simulation models for inventory problems. Inventory may be items in a warehouse, spare parts inventories, office supplies, or maintenance materials. The purpose of inventory control is to administer an economic balance between the size of an order, frequency of ordering, and customer service. When these functions are well managed, an appropriate level of customer service can be maintained without excessive inventory levels.

20.2 Discount Store Model

This example is adapted from [1], where it is modeled using the discrete event language SLAM II. A simplified version of the problem is given in [2] as a General Purpose Simulation System application of a library operation. This presentation closely follows Pritsker's formulation.

Customers arrive at a discount store every *Exponential* (2) minutes. The store has three clerks who receive a batch of customer orders and then proceed to an adjacent warehouse to fill them. Customers are served as they arrive. However, when the customer queue builds up, a clerk can service a maximum of six customer orders in any one trip. Travel time to and from the warehouse is *UniformContinuous* (0.5, 1.5) minutes each way. Once in the warehouse, the processing time depends on the number of customers being served. It is estimated that the acquisition time is *Normal* (μ, σ^2), where the mean μ equals the number of customers times three minutes per customer, and the standard deviation σ is estimated as $0.2 * \mu$. Upon return from the warehouse, the clerk delivers orders to respective customers in the same sequence in which the orders were received. The time for each customer to receive an order is *UniformContinuous* (1, 3) minutes. After all waiting customers have received their order, the clerk begins another trip to the warehouse.

The model represents clerk activities with a facility and a queue for waiting customers. Because the three clerks are identical, arrays of queues, $Q()$, and facilities, $F()$, model the system. The basic idea for the model is that arriving customers will join one of the three queues $Q(1)$, $Q(2)$, and $Q(3)$ when the clerk is idle (not on a trip to the warehouse or not delivering items to waiting customers).

The server *F()* is idle when a clerk is idle. If all three clerks are busy when a customer arrives, then that customer joins *StoreFrontQueue* and waits for a clerk.

In the model, transactions represent clerks and customers. The transaction attributes that define a clerk are

$1 = ID$ (1, 2, or 3)
$2 = Destination$

Table 20.1 describes the primary events in the model. In addition to subroutines for each primary event, *GetNextJob* details events related to dispatching a clerk. The remainder of this section provides an overview of model procedures and a brief description of program output.

CustomerSourceEvent: This program segment models an arriving customer. In Figure 20.1 the arrival joins *StoreFrontQueue* and uses *GetNextJob* to notify the system of a customer arrival.

ClerkFacilityEvent: This subroutine represents the clerk's completion of warehouse activities. In Figure 20.2, the clerk begins the trip back to "StoreFront." Notice that because the clerk did not disengage the facility, the clerk now resides in the facility and delay simultaneously. The effect of this approach is the clerk utilization includes delay time to the warehouse, service time in the warehouse, delay time back to the storefront, and cumulative delay times to distribute orders to customers. While in any one of these states, the server is "busy." The server disengages the facility after all customers have received an order.

Table 20.1   Discount Store—Primary Events

Event	Description
CustomerSource	Customer arrives
ClerkFacility	Clerk completes warehouse tasks
WarehouseDelay	Finishes delay to/from warehouse
OrderCompletionDelay	Finishes a customer service

```
Sub CustomerSourceEvent(ByRef Arrival As transaction)
   StoreFrontQueue.Join Arrival
   Call GetNextJob
End Sub
```

Figure 20.1   Discount store—*CustomerSourceEvent*.

```
Sub ClerkFacilityEvent(ByRef Clerk As transaction)
   Clerk.Assign Destination, "StoreFront"  ' orders filled
   WarehouseDelay.Start Clerk  ' return to store front
End Sub
```

Figure 20.2   Discount store—*ClerkFacilityEvent*.

WarehouseDelayEvent: This procedure represents a finished trip to or from the warehouse. In Figure 20.3, when attribute *Destination* equals "Warehouse," the clerk processes the orders. Because the clerk has previously engaged the facility, *Service* is used, rather than *EngageAndService*. The service time is a function of the number of customer orders (i.e., length of the respective *Q()*). If *Destination* is not "Warehouse," the clerk arrives at the store-front. The clerk starts the first of consecutive delays to process customer orders. The number of delays is equal to the number of customers in the clerk's respective queue, *Q()*.

OrderCompletionDelayEvent: This subroutine models an individual customer receiving an order. In Figure 20.4, each time the delay is finished, a customer has completed service. When the corresponding *Q()* is empty, the clerk disengages the facility to indicate an idle clerk who is available to service orders from more customers. Otherwise, the next customer in *Q()* is serviced by initiating another delay.

GetNextJob: This program segment allocates orders to clerks and dispatches the clerk. In Figure 20.5, if there are no jobs in *StoreFrontQueue*, the clerk remains idle. Otherwise, a maximum of six orders or the number of orders in *StoreFrontQueue* may be allocated to a clerk. The procedure then searches for a facility with an idle clerk. When an idle clerk is found, a clerk transaction is created and attributes *ID* and *Destination* are initialized. Orders are moved from *StoreFrontQueue* to the appropriate clerks *Q()*. The clerk engages the facility and starts the delay for the trip to the warehouse.

```
Sub WarehouseDelayEvent(ByRef Clerk As transaction)
  Dim _
    ProcessingTime As Single
  If Clerk.A(Destination) = "Warehouse" _
  Then
    ItemsPerTripStatistic.Collect Q(Clerk.A(ID)).Length
    ProcessingTime = _
     Normal(3 * Q(Clerk.A(ID)).Length, 0.6 * Q(Clerk.A(ID)).Length)
    F(Clerk.A(ID)).Service Clerk, ProcessingTime
  Else ' back to StoreFront
    OrderCompletionDelay.Start Q(Clerk.A(ID)).Depart
  End If
End Sub
```

Figure 20.3   Discount store—*WarehouseDelayEvent*.

```
Sub OrderCompletionDelayEvent(ByRef Order As transaction)
  CycleTimeStatistic.Collect Customer.SystemTime
  If Q(Customer.A(ID)).Length = 0 _
  Then   ' all orderes delivered
    F(Customer.A(ID)).Disengage F(Customer.A(ID)).View(1)
    Call GetNextJob
  Else  'customer leaves with order
    OrderCompletionDelay.Start Q(Customer.A(ID)).Depart
  End If
End Sub
```

Figure 20.4   Discount store—*OrderCompletionDelayEvent*.

```
Sub GetNextJob()
  Dim _
    Clerk As Transaction, _
    Customer As Transaction, _
    OrderCount As Integer, _
    Index As Integer, _
    Jndex As Integer
  If StoreFrontQueue.Length = 0 _
  Then
    Exit Sub ' no waiting customers
  End If
  If StoreFrontQueue.Length > 5 _
  Then   'no more that 6 customers
    OrderCount = 6
  Else
    OrderCount = StoreFrontQueue.Length
  End If
  For Index = 1 To 3
    If F(Index).NumberOfIdleServers > 0 _
    Then
      Set Clerk = NewTransaction
      Clerk.Assign Destination, "Warehouse"
      Clerk.Assign ID, Index
      For Jndex = 1 To OrderCount
        Set Customer = StoreFrontQueue.Depart
        Customer.Assign ID, Index
        Q(Index).Join Customer
      Next Jndex    'to wait for order
      F(Index).Engage Clerk, 1
      WarehouseDelay.Start Clerk
      Exit Sub
    End If
  Next Index
    '!!!! no servers were available
End Sub
```

Figure 20.5   Discount store—sub *GetNextJob*.

Queue	StoreFront		Clerk1		Clerk2		Clerk3	
Number in	980		334		320		321	
Number out	975		328		314		317	
Residing	5		6		6		4	
Destroyed	0		0		0		0	
Removed	0		0		0		0	
Detached	0		0		0		0	
Minimum	0		0		0		0	
Maximum	17		6		6		6	
	Lq	Wq	Lq	Wq	Lq	Wq	Lq	Wq
Average	1.85	3.77	2.27	13.58	2.34	14.60	2.16	13.41
StdDev	2.79	0.16	1.92	0.39	2.02	0.48	1.92	0.44

Figure 20.6   Discount store—queues output.

In Figure 20.6, each of the three clerks' queues has a little over an average of two customers. Figure 20.7 indicates that servers are busy over 90% of the time. In Figure 20.8 the average number of orders per trip is about 2.17 items. One would expect the average items per trip to equal the average of the lengths of *Clerk (1)*, *Clerk (2)*, and *Clerk (3)*. The discrepancy is because *ItemsPerTrip* is an observation-based statistic.

Facility	Clerk1		Clerk2		Clerk3	
Number in	154		146		149	
Number out	153		145		148	
Residing	1		1		1	
Interrupted	0		0		0	
Destroyed	0		0		0	
Minimum	0		0		0	
Maximum	1		1		1	
	AvgBusy servers		AvgBusy servers		AvgBusy servers	
Average	0.96		0.94		0.93	
StdDev	0.19		0.23		0.25	

Figure 20.7   Discount store—facilities output.

Statistic	(O) CycleTime		(O) ItemsPerTrip	
Number in	959		449	
Min value	4.69		1.00	
Max value	53.26		6.00	
Average	19.55		2.17	
StdDev	10.88		1.59	

Figure 20.8   Discount store—statistics output.

As a validity check, the simulation results should agree with intuition. Waiting time in a clerk queue should on average equal the time for the clerk to make a round trip to the warehouse, plus the average time to deliver orders to the customer. The delay includes the following components:

- Average item acquisition time $= 2.17 \times 3 = 6.52$ minutes
- Average round trip time to warehouse $= 1 + 1 = 2$ minutes
- Average delivery time of items $= 2.17 \times 2 = 4.34$ minutes

The sum of these times, 12.86 minutes, is consistent with the waiting times in clerk queues, which range from 13.41 to 14.60. Another check involves cycle time, which includes waiting time in the storefront queue and clerk queue. In Figure 20.6, the total time in the queues (approximately 3.77 + 14.0 = 17.77 minutes) is consistent with the independent cycle time calculation of 19.55 minutes.

20.3 Periodic Review Model

An inventory item is stocked according to the following specifications: Every six weeks the inventory position (amount on hand + on order + backlogged demand) is reviewed. When the position falls below the reorder level of 25 units, an order is placed to bring the net inventory level back to 95 units. Orders are typically delivered after a four-week delay.

Customers arrive every *Exponential* (0.4) weeks to purchase items. Because items are offered at a discount price, customers are usually willing to wait for stock replenishment. Orders are processed in the order received by the store. However, a later order whose size is small enough to be filled from available stock is given priority over earlier orders that must wait for stock replenishment. Because there is a four-week delay until an order is received, it is likely that additional customers will be in the wait queue before the order is received. Consequently, inventory replenishment may not be sufficient to fulfill the demands of all waiting customers. The objective of the simulation is to determine the following performance metrics of the system:

- The average safety stock level at the time of a replenishment order
- The average inventory level
- The average inventory position
- The average number of waiting customers

For this model, user-defined variables are *InventoryOnHand, InventoryPosition, InventoryOnOrder, InventoryBackLog, OrderAmount, ReorderLevel,* and *MaximumInventoryLevel.* Transactions represent customers and reorders. Attribute 1 of both types of transaction represents *OrderQuantity.* A FIFO queue *WaitQueue* enables customers to wait for inventory on hand replenishment. *InventoryOnHand, ReorderLevel,* and *MaximumInventoryLevel* are initialized from the spreadsheet with values of 95, 25, and 95, respectively. Statistical variables are *AvgInvOnHandStatistic, AvgInvPosStatistic, AvgSafetyStockStatistic,* and *AvgOrderSizeStatistic.* Table 20.2 describes the primary events in the model.

In addition to subroutines for each of the primary events, there are two other subroutines, *CheckCustomerQueue* and *UpdateInventoryPosition.* The remainder of this section provides an overview of model procedures and a brief description of program output.

> *CustomerSourceEvent*: This event handler models arriving customers. In Figure 20.9 the customer order size, *Poisson* (3), is stored in attribute *A(1).* If *InventoryOnHand* is sufficient, the customer begins service after reducing the *InventoryOnHand* by customer order level and collecting the *AvgInvOnHandStatistic.* Otherwise, a backorder is placed as the *InventoryBackLog* is incremented by *CustomerOrder* and the customer joins *WaitQueue.* The last step is to *UpdateInventoryPosition.* Notice that if any portion of the customer order can be filled, the inventory level is sufficient to service the customer.
>
> *ReviewSourceEvent*: This event handler models the inventory review and ordering process. In Figure 20.10, if *InventoryPosition* is less than *ReorderLevel,* an order of size *InventoryOnOrder* is scheduled for delivery. *AverOrderSizeStatistic* and *AvgSafetyStockStatistic* are collected. Again, the last step is to *UpdateInventoryPosition.*

Table 20.2 Periodic Inventory Review—Primary Events

Event	Description
CustomerSource	Generates customer arrivals
ReviewSource	Periodic review arrivals
FillOrderDelay	Fill order
ShippingDelay	Inventory replenishment

```
            ' EVENT HANDLERS
Sub CustomerSourceEvent(ByRef Arrival As Transaction)
   Arrival.Assign OrderQuantity, Poisson(3)
   If Arrival.A(OrderQuantity) <= InventoryOnHand _
   Then    'fill order
      InventoryOnHand = InventoryOnHand - Arrival.A(OrderQuantity)
      AvgInvOnHandStatistic.Collect InventoryOnHand
      FillOrdersDelay.Start Arrival
   Else 'place a backorder
      InventoryBackLog = InventoryBackLog + Arrival.A(OrderQuantity)
      WaitQueue.Join Arrival
   End If
   Call UpdateInventoryPosition
End Sub
```

Figure 20.9   Periodic inventory review—*CustomerSourceEvent.*

```
Sub ReviewSourceEvent(ByRef Reorder As Transaction)
   If InventoryPosition < ReorderLevel _
   Then 'place an order
      Reviewer.Assign OrderQuantity, _
               MaximumInventoryLevel - InventoryPosition
      InventoryOnOrder = InventoryOnOrder + Reviewer.A(OrderQuantity)
      AvgOrderSizeStatistic.Collect Reviewer.A(OrderQuantity)
      UpdateInventoryPosition
      AvgSafetyStockStatistic.Collect InventoryOnHand
      ShippingDelay.Start Reviewer
   End If
End Sub
```

Figure 20.10   Periodic inventory review—*ReviewSourceEvent.*

```
Sub FillOrdersDelayEvent(ByRef Customer As Transaction)
   ' simply a counter for filled orders
End Sub
```

Figure 20.11   Periodic inventory review—*FillOrdersDelayEvent.*

FillOrdersDelayEvent: This event handler models the completion of customer service. The delay time is 0; therefore, notice in Figure 20.11 that there is no lapse time in the delay mode.

ShippingDelayEvent: This event handler models the receipt of an inventory order. In Figure 20.12, the *ShippingDelay* is finished and *InventoryOnHand* is updated by the order size. *AvgInvOnHandStatistic* is collected and after an *UpdateInventoryPosition* is performed, *CheckCustomerQueue* searches for pending unfilled orders.

CheckCustomerQueue: This program segment searches *WaitQueue* for a waiting customer who can be serviced. In Figure 20.13, a FIFO search is made for a customer whose order size does not exceed *InventoryOnHand*. Such a customer is taken from the queue and serviced. *InventoryOnHand* and *InventoryBackLog* are reduced by the customer order. The *AvgInvOnHandStatistic* is collected and then *UpdateInventoryPosition* is performed.

```
Sub ShippingDelayEvent(ByRef ReceivedOrder As Transaction)
  InventoryOnHand = _
       InventoryOnHand + ReceivedOrder.A(OrderQuantity)
  InventoryOnOrder = _
       InventoryOnOrder - ReceivedOrder.A(OrderQuantity)
  UpdateInventoryPosition
  AvgInvOnHandStatistic.Collect InventoryOnHand
  CheckCustomerQueue
End Sub
```

Figure 20.12 Periodic inventory review—*ShippingDelayEvent*.

```
Sub CheckCustomerQueue()
Dim _
   Index As Integer, _
   WaitingCustomer As Transaction
 For Index = 1 To WaitQueue.Length
  If WaitQueue.View(Index).A(OrderQuantity) < InventoryOnHand _
  Then
    Set WaitingCustomer = WaitQueue.Detach(Index)
    FillOrdersDelay.Start WaitingCustomer
    InventoryBackLog = InventoryBackLog - WaitingCusto-
mer.A(OrderQuantity)
    InventoryOnHand = InventoryOnHand - WaitingCustomer.A(OrderQuantity)
    AvgInvOnHandStatistic.Collect InventoryOnHand
    UpdateInventoryPosition
    Exit For
  End If
Next Index
```

Figure 20.13 Periodic inventory review—sub *CheckCustomerQueue*.

```
Sub UpdateInventoryPosition()
  InventoryPosition = _
            InventoryOnHand + InventoryOnOrder - InventoryBackLog
  AvgInvPosStatistic.Collect InventoryPosition
End Sub
```

Figure 20.14 Periodic inventory review—sub *UpdateInventoryPosition*.

UpdateInventoryPosition: This model segment collects, as shown in Figure 20.14, *AvgInvPosStatistic* and maintains *InventoryPosition* using the following:

$$InventoryPosition = InventoryOnHand + InventoryOnOrder - InventoryBackLog$$

The output is shown in Figures 20.15 and 20.16.

Queue	Wait	
Number in	178	
Number out	0	
Residing	0	
Destroyed	0	
Removed	0	
Detatched	178	
Minimum	0	
Maximum	22	
	Lq	Wq
Average	2.00	2.68
StdDev	3.93	0.14

Figure 20.15 Periodic inventory review—queue output.

Statistic	(T) AvgInvOnHand	(T) AvgInvPos	(O) AvgSafetyStock	(O) AvgOrderSize	
Number in	620	816	18	18	
Min value	0.00	-45.00	0.00	75.00	
Max value	140.00	95.00	20.00	140.00	
Average	26.47	48.40	6.39	97.94	
StdDev	25.24	29.72	8.08	18.68	

Figure 20.16 Periodic inventory review—statistics output.

20.4 Continuous Review Model

In this model, the inventory level is reviewed on a continuous basis and whenever the level is below the reorder point, a new order is placed. The size of the order is the difference between a specified maximum level and the inventory on hand at the time the order is placed. There is a delay period of a week and a half between the placement and receipt of the order. There is no more than one outstanding order at a time.

The time interval between customer arrivals is *Exponential* (0.6), and the size of each order is *Poisson* (4). An order may be partially filled. However, whenever the stock is replenished, partially filled orders are completed before other waiting orders are serviced.

User-defined integer variables that describe inventory are *InventoryOnHand*, *InventoryBackLog*, *OrderAmount*, *ReorderLevel*, and *MaximumInventoryLevel*. *InventoryOrdered* is a Boolean variable that indicates an order status. User-defined integer variables for customers are *CustomerOrderSize*, *CustomerBackOrder*, and *PartialOrderCount*. Transactions represent customers in the system. The customer attributes are

1 = *OrderSize*
2 = *BackOrderSize*

Customer orders are filled in the order of arrival as long as the level of *InventoryOnHand* is sufficient to fill at least a partial order. Otherwise, the customer joins the first-in, first-out (FIFO) queue *WaitQueue* to wait for an update to *InventoryOnHand*. At each review, the order size is based

Table 20.3 Continuous Inventory Review—Primary Events

Event	Description
CustomerSource	Generates customer arrivals
FillOrderFacility	Fill order
ShippingDelay	Inventory restocked

on the inventory position, which reflects backlogged customer orders. However, because there is a four-week delay before an order is received, it is likely that additional customers will join *WaitQueue* during that delay. Consequently, inventory replenishment may not be sufficient to satisfy all waiting customers. A customer who received a partial order waits at the front of *WaitQueue*.

InventoryOnHand, *ReorderLevel*, and *MaximumInventoryLevel* are initialized from the spreadsheet with values of 20, 15, and 20, respectively. Statistical variables are *AvgInvOnHandStatistic* and *AvgOrderSizeStatistic*. Table 20.3 describes the primary events in the model. In addition to a subroutine for each primary event, there are two user-defined subroutines, *ReviewInventoryLevel* and *CheckCustomerQueue*.

The remainder of this section provides an overview of the model procedures and a brief description of the program output.

> *CustomerSourceEvent*: This event handler models arriving customers. In Figure 20.17, attribute *CustomerOrderSize* is *Poisson* (4). The customer joins *WaitQueue*. A call to *CheckCustomerQueue* notifies the system that there has been a state change that may enable another customer to receive service.
>
> *FillOrdersFacilityEvent*: This event handler models the completion of a filled order. In Figure 20.18, the customer disengages the facility. If the customer has a backorder, the backorder becomes the order and the customer is placed in the front of *WaitQueue*. In this example, service time is zero; therefore, restocking will not occur during customer service. As a result, *CheckCustomerQueue* has no effect on the system for this scenario. However, if restocking had occurred, the customer who had not completed service is the next customer serviced. If the customer received a complete order, *CheckCustomerQueue* notifies the system that there has been a state change that may enable another customer to receive service.
>
> *ShippingDelayEvent*: This event handler models the restocking of inventory. In Figure 20.19, after the delay is finished, *InventoryOnHand* is updated and *InventoryOrdered* is changed to *False* to indicate that no outstanding order and *AvgInvOnHandStatistic* are collected. A call to *CheckCustomerQueue* notifies the system that there has been a state change that may enable another customer to receive service.

```
          ' EVENT HANDLERS
Sub CustomerSourceEvent(ByRef Arrival As transaction)
   Arrival.Assign OrderSize, Poisson(4)
   WaitQueue.Join Arrival
   Call CheckCustomerQueue
End Sub
```

Figure 20.17 Continuous inventory review—*CustomerSourceEvent*.

```
Sub FillOrdersFacilityEvent(ByRef Customer As transaction)
   FillOrdersFacility.Disengage Customer
   If Customer.A(CustomerBackOrder) > 0 _
   Then 'partial order => to front of queue
      Customer.assign OrderSize, Customer.A(BackOrderSize)
      Customer.assign BackOrderSize, 0
      WaitQueue.JoinFront Customer
   Else ' order complete
      Set Customer = Nothing
   End If
   Call CheckCustomerQueue
End Sub
```

Figure 20.18   Continuous inventory review—*FillOrdersFacilityEvent*.

```
Sub ShippingDelayEvent(ByRef ReceivedOrder As transaction)
   Set ReceivedOrder = Nothing
   InventoryOnHand = InventoryOnHand + InventoryOnOrder
   InventoryOnOrder = 0
   InventoryOrdered = False
   AvgInvOnHandStatistic.Collect InventoryOnHand
   Call CheckCustomerQueue
End Sub
```

Figure 20.19   Continuous inventory review—*ShippingDelayEvent*.

```
Sub ReviewInventoryLevel()
   If Not InventoryOrdered _
   Then
      If InventoryOnHand < ReorderLevel _
      Then 'order inventory
         InventoryOnOrder = MaximumInventoryLevel - InventoryOnHand
         AvgOrderSizeStatistic.Collect InventoryOnOrder
         ShippingDelay.Start deeds.NewTransaction
         InventoryOrdered = True
      Else
         'inventory adequate
      End If
   End If
End Sub
```

Figure 20.20   Continuous inventory review—sub *ReviewInventoryLevel*.

ReviewInventoryLevel: This program segment determines when to issue a request for restocking inventory. In Figure 20.20, if there is not an outstanding order and *InventoryOnHand* is less than *ReorderLevel*, an order delivery is scheduled and *AvgOrderSizeStatistic* is collected. The order size is determined by

$$InventoryOnOrder = MaximumInventoryLevel - InventoryOnHand$$

CheckCustomerQueue: This subroutine authorizes a customer to be serviced. In Figure 20.21, when there is *InventoryOnHand* and a customer in *WaitQueue*, the customer is removed from *WaitQueue* and authorized for service. If *Attribute (CustomerOrderSize)* is less than or equal to *InventoryOnHand*, the customer receives full service. Otherwise, the customer order is partially filled. In both cases *InventoryOnHand* is updated, inventory level is reviewed, and *AvgInvOnHandStatistic* is collected. For partially filled orders, the attribute *BackOrder* is the unfilled portion of the order and *PartialOrderCount* is incremented by 1. Notice that the entire queue is searched so that an arriving customer at the end of the queue will receive consideration for service.

The reports are shown in Figures 20.22, 20.23, and 20.24. Although not shown, the number of customer orders for this example is 3312; therefore, approximately one-third of the customers

```
Sub CheckCustomerQueue()
  Dim _
    Customer As Transaction
  If InventoryOnHand > 0 And WaitQueue.Length > 0 _
  Then
    Set Customer = WaitQueue.Depart
    If Customer.A(OrderSize) <= InventoryOnHand _
    Then   'send customer to Fill Facility
      Customer.Assign BackOrderSize, 0
    Else ' a partial order to Fill
      PartialOrderCount = PartialOrderCount + 1
      Customer.Assign BackOrderSize, _
                    (Customer.A(OrderSize) -InventoryOnHand)
      Customer.Assign OrderSize, InventoryOnHand
    End If
    InventoryOnHand = InventoryOnHand - Customer.A(OrderSize)
    AvgInvOnHandStatistic.Collect InventoryOnHand
    FillOrdersFacility.EngageAndService Customer, 1
    Call ReviewInventoryLevel
  End If
End Sub
```

Figure 20.21 Continuous inventory review—sub *CheckCustomerQueue.*

Queue	Wait	
Number in	4163	
Number out	4163	
Residing	0	
Destroyed	0	
Removed	0	
Detatched	0	
Minimum	0	
Maximum	41	
	Lq	Wq
Average	5.415	2.601
StdDev	9.234	0.070

Figure 20.22 Continuous inventory review—queue output.

Facility	FillOrders	
Number in	4163	
Number out	4163	
Residing	0	
Interrupted	0	
Destroyed	0	
Minimum	0	
Maximum	1	
	AvgBusy servers	
Average	0.000	
StdDev	0.000	

Figure 20.23   Continuous inventory review—facilities output.

Statistic	(T) AvglnvOnHand	(O) AvglOrderSize
Number in	5373	1211
Min Value	0	6
Max value	20	20
Average	3.324	11.246
StdDev	5.185	3.250

Figure 20.24   Continuous inventory review—statistics output.

received partial orders as implied in Figure 20.22. Notice in Figure 20.23 that utilization of the *FillOrders* facility is 0 because service time to fill the order is 0. Recall that the inventory level is reviewed after each order is filled. In Figure 20.24, the minimum inventory level is 0 and the maximum order size is 20. Because the reorder point is 15, at least one order exceeded the reorder point. Because customer order size is *Poisson* (4), the assumption is that a reorder quantity of 20 did not occur often.

References

1. Pritsker, A. A. B., *Introduction to Simulation and SLAM II*, 3rd ed. New York: Wiley, 1986, pp. 414–424.
2. Schriber, T., *Simulation Using GPSS*. New York: Wiley, 1974, pp. 339–347.

Chapter 21

Scheduling Models

21.1 Introduction

This chapter deals with modeling scheduling problems. In the broadest sense, the purpose of scheduling is to achieve the effective utilization of resources (including labor, machines, and materials) for producing a product. Production-scheduling techniques consider the demand for the item and the limited resources available and then recommend when and in what quantity an item should be produced on a given facility (e.g., machine).

Production processes are generally one of three types: continuous, intermittent, and job shop. The continuous and intermittent systems produce items in mass quantities. In the job shop case, orders mostly are customer tailored, each with different specifications. The simple nature of mass production systems allows the development of reasonably reliable mathematical techniques that can be used to manage the production resources effectively. In contrast, the lack of repetitiveness in the job shop production process usually results in complex situations that cannot be handled by analytic (mathematical) methods. This is where the flexibility of simulation modeling may be used.

21.2 Job Shop Scheduling

In job shop scheduling, the shop includes a number of machines or work centers. Arriving jobs normally are processed on one or more machines, with each job requiring a different route on the available machines. The problem is resolved by specifying an "efficient" rule that can be used to set priorities for processing jobs as they arrive at each machine. This section presents a simulation model of such a situation.

This example is adapted from [1], where it is modeled using SLAM II discrete event scheduling. Its original source is an article by Eilon et al. [2]. For ease of explanation, the version presented here treats the effect of job due dates in a slightly different manner. Also, a generalized model that accommodates any number of machines is developed.

In a shop that includes n different machines, jobs arrive every *Exponential* (25) minutes. The estimated processing time is *Exponential* (20) minutes for all machines. Both interarrival and estimated processing times must be integers with minimum values of 1. The actual processing time is the sum of estimated (integer) processing time and a random component of *Normal* (0, *s*) where *s* = (0.3*estimated processing time). The number of machines per job is estimated as a randomly *k* = *Normal* (4, 1), with each job requiring at least *m* machines (and no more than *n*). The assignment of specific machines to a job is determined randomly with the stipulation that a machine is not used more than once on a job.

The specific scheduling rule evaluated divides jobs waiting at each machine into two groups: priority jobs with past due dates and others with due dates that have not yet been reached. Within each group, the highest priority is given to the job with the shortest processing time. A job due date is estimated as twice the sum of its processing times on the various machines. The metric for this model is the average time a job spends in the shop as well as average job lateness and tardiness. Lateness considers both positive and negative deviations from due dates, whereas tardiness pertains to jobs that are completed after their due dates. The model is applicable to any number of machines *n*. The only change to be made in the model is to specify the appropriate number of attributes. This example is by far the most complex that has been presented in this book.

This program utilizes Visual Basic for Applications (VBA) arrays to represent the machines and their queues. The first five attributes of each transaction (job) are defined as

> 1 = *CompletionDate*
> 2 = *DueDate*
> 3 = *RouteIndex* for routing the job (route = 1, 2,..., *n*)
> 4 = *CurrentMachineID*
> 5 = *CurrentProcessingTime*

Thereafter, pairs of attributes starting with *A(6)* and *A(7)* represent a machine identifier and the corresponding processing time at that machine. The number of pairs is dependent on the number of machines *n* in the shop and *m*, the minimum number of machines for a job. Also, the pairs of attributes reflect the order that the machines visited.

F() is an array of machines and *Q()* is an associated array of queues, one for each machine. *JobLatenessStatistic*, *JobTardinessStatistic*, and *CycleTimeStatistic* are user-defined statistics. *MinOperations* is obtained from the *UserInitial* spreadsheet. Table 21.1 describes the primary events in the model. Because *F()* is used to manage machines, subroutine *Parser* routes all service completions to subroutine *MachineFacilityEvent*.

In addition to the primary events subroutines, the following subroutines are included in the model:

- *Assignments*: Initializes jobs as they enter the model.
- *OrderQueue*: Orders job queue before selecting the next job to be processed.
- *SampleWoReplacement*: Initializes the machine order for each job.

Table 21.1 Job Shop Scheduling—Primary Events

Event	Description
JobOneSourceEvent	Starts the program
MachineFacilityEvent	Machine service completion
SourceLoopDelayEvent	Job arrival

The remainder of this section provides an overview of model procedures and a brief description of program output.

JobOneSourceEvent: In Figure 21.1 the only transaction to originate from *JobOneSource* is processed. Thereafter, subsequent arriving transactions are scheduled by subroutine *SourceLoopDelayEvent*. The first job in the system is *NewTransaction*, and as discussed below, subroutine *Assignments* initializes the first job in the system. The delay time for the next arrival is the integer value of the interarrival time *Exponential* (25) with a minimum value of 1.

SourceLoopDelayEvent: This event handler manages the schedule for arriving jobs. In Figure 21.2 the delay time for each loop is the integer value of the interarrival time *Exponential* (25) with a minimum value of 1. Upon completion of the delay, *NewTransaction* represents a new job entering the system. As discussed later in this section, subroutine *Assignments* initializes each new job in the system and routes the job to the first machine in its job sequence.

MachineFacilityEvent: This is a general procedure that processes service completions for all machines. The procedure is based on the fact that machines are represented by indexed queues *Q()* and facilities *F()*. In Figure 21.3 the procedure begins by using the completed job's attributes and indirect addressing to determine *ThisMachineID*. For jobs in *ThisMachineID* queue, *OrderQueue* ensures that the next job to access *ThisMachineID* satisfies the scheduling criterion.

```
Sub JobOneSourceEvent(ByRef JobLoop As Transaction)
  Dim _
    DelayTime As Integer
  Call Assignments(NewTransaction)
  ' only one arrival sent to a continuous delay loop.
  DelayTime = Int(Exponential(25))
  If DelayTime < 1 _
  Then
    DelayTime = 1
  End If
  SourceLoopDelay.Start JobLoop, DelayTime
End Sub
```

Figure 21.1 Job shop scheduling—*JobOneSourceEvent*.

```
Sub SourceLoopDelayEvent(ByRef NewJob As Transaction)
  Dim _
    DelayTime As Integer
  DelayTime = Int(Deeds.Exponential(25))
  If DelayTime < 1 _
  Then
    DelayTime = 1
  End If
  SourceLoopDelay.Start NewJob, DelayTime
  Call Assignments(NewTransaction)
End Sub
```

Figure 21.2 Job shop scheduling—*SourceLoopDelayEvent*.

```
Sub MachineFacilityEvent(ByRef Job As Transaction)
  Dim _
    ThisMachineID As Integer, _
    NextMachineID As Integer, _
    ServiceTime As Single, _
    JobLateness As Single
  ThisMachineID = Job.A(CurrentMachine)
  F(ThisMachineID).Disengage Job
  Call OrderQueue(ThisMachineID, Job.A(CurrentProcessingTime))
  Job.Assign RouteIndex, Job.A(RouteIndex) + 1
      'Assign the next machine for job
  Job.Assign CurrentMachine, _
    Job.A(2 * (Job.A(RouteIndex) - 1) + 6)
  If Job.A(RouteIndex) > NMachines Or _
    Job.A(CurrentMachine) = Empty _
  Then
    'Job has completed all operations.
    CycleTimeStatistic.Collect Job.SystemTime
    JobLateness = Job.A(DueDate) - Time
    ' Collect lateness stat for all jobs
    JobLatenessStatistic.Collect JobLateness
    ' Collect tardiness stat for late jobs
    If JobLateness < 0 _
    Then
      JobTardinessStatistic.Collect JobLateness
    End If
  Else
    Job.Assign CurrentProcessingTime, _
      Job.A(2 * (Job.A(RouteIndex) - 1) + 7)
    NextMachineID = Job.A(CurrentMachine)
    If F(NextMachineID).NumberOfIdleServers > 0 _
    Then
      Job.Assign CurrentProcessingTime, _
        Job.A(CurrentProcessingTime) + _
          deeds.Normal(0, 0.3 * Job.A(CurrentProcessingTime))
      F(NextMachineID).EngageAndService Job, 1, _
                            Job.A(CurrentProcessingTime)
    Else
      Q(NextMachineID).Join Job
    End If
  End If
  If Q(ThisMachineID).Length > 0 _
  Then
    Set Job = Q(ThisMachineID).Depart
    ServiceTime = Job.A(CurrentProcessingTime) + _
      deeds.Normal(0, 0.3 * Job.A(CurrentProcessingTime))
    F(ThisMachineID).EngageAndService Job, 1, ServiceTime
  End If
End Sub
```

Figure 21.3 Job shop scheduling—*MachineFacilityEvent*.

After the job disengages the machine, *RouteIndex* is incremented and then used to identify the next machine for the current job. The procedure recognizes a finished job when *RouteIndex* exceeds the number of machines in the shop or the next machine identifier is zero. When the job is finished, job statistics *SystemTime and JobLateness* are collected for all jobs. *JobTardiness* is only collected for jobs that are actually late. A completed job exits the system.

Jobs that require additional processing are routed to the next machine. Indirect addressing is used to determine the *NextMachineID* and a preliminary estimate of the corresponding processing time. If *NextMachineID* is idle, the actual processing time is calculated and the job begins service on that machine. Otherwise, the job joins the queue for *NextMachineID*. When *NextMachineID* is idle, there are no jobs waiting on the next machine. The last step is to examine the current machine job queue for a job to schedule for service completion. Sub *OrderQueue* ensures that the appropriate job is scheduled for service.

OrderQueue: This procedure ensures that when a job is completed, the corresponding machine selects and schedules the correct job for the next service completion. In Figure 21.4, the *CompletionDate* attribute is incremented by the *WaitingTime* in the queue while the last job is serviced. *CompletionDate* for those jobs now represents the most optimistic time that the job will finish. Then, all waiting jobs are checked for possible changes in their due dates. As a job becomes overdue, its order in the queue must be adjusted to reflect its new status. Two temporary queues, *LateJobsQueue* and *RegularJobsQueue*, are used to reorder jobs in *Q()*. A job is considered late when either *Time* exceeds *DueDate* or *CompletionDate* exceeds *DueDate*.

ReorderQueue examines each job in the queue and using the above criteria places the job in either the *RegularJobsQueue* or *LateJobsQueue* ordered on the shortest *CurrentProcessingTime* for the machine. Transactions in the temporary queues then join *Q()*. The late jobs join *Q()* first. The result is that a late job with the shortest processing time is at the front of *Q()*.

Notice that updating *CompletionDate* assumes inaccurately that all jobs waiting in *Q()* arrived before the just-completed job started. In this respect, *CompletionDate* will generally be an overestimation of the actual job *CompletionDate*. Because this overestimation represents a more conservative situation regarding the impact of due dates, no attempt is made in the model to rectify the inaccuracy in computing *CompletionDate*.

Assignments: This subroutine initializes job attributes and routes the job to the first machine. As shown in Figure 21.5, this is a lengthy procedure but fairly straightforward.

```
Sub OrderQueue (ByVal MachineID As Integer, _
               ByVal WaitingTime As Single)
  Dim _
    Index As Integer
  For Index = 1 To Q(MachineID).Length
    Q(MachineID).View(Index).Assign CompletionDate, _
      Q(MachineID).View(Index).A(CompletionDate) + WaitingTime
  Next Index
  While Q(MachineID).Length > 0
    If (Time >= Q(MachineID).View(1).A(DueDate) Or _
      Q(MachineID).View(1).A(CompletionDate) >= _
               Q(MachineID).View(1).A(DueDate)) _
    Then
      LateJobsQueue.JoinOnLowAttribute CurrentProcessingTime, _
                             Q(MachineID).Detach(1)
    Else
      RegularJobsQueue.JoinOnLowAttribute CurrentProcessingTime, _
                             Q(MachineID).Detach(1)
    End If
  Wend
  While LateJobsQueue.Length > 0
    Q(MachineID).Join LateJobsQueue.Depart
  Wend
  While RegularJobsQueue.Length > 0
    Q(MachineID).Join RegularJobsQueue.Depart
  Wend
End Sub
```

Figure 21.4   Job shop scheduling—sub *OrderQueue*.

```
Sub Assignments(ByRef Job As Transaction)
  Dim _
    Index As Integer, _
    JobList() As Variant, _
    SampledList() As Variant, _
    NOperations As Integer, _
    ProcessingTime As Integer, _
    ThisMachineID As Integer
  Job.DimAttributes (20)
  Job.Assign DueDate, Time
  ReDim JobList(1 To NMachines)
  ReDim SampledList(1 To NMachines)
  For Index = 1 To NMachines
    JobList(Index) = Index
  Next Index
  SampleWoReplacement JobList, SampledList
  NOperations = Int(deeds.Normal(4, 1))
  If NOperations < MinOperations _
  Then
    NOperations = MinOperations
  End If
  If NOperations > NMachines _
  Then
    NOperations = NMachines
  End If
  For Index = 1 To NOperations
    Job.Assign (5 + 2 * Index - 1), SampledList(Index)
    ProcessingTime = Int(deeds.Exponential(20))
    If ProcessingTime = 0 _
    Then
      ProcessingTime = 1
    End If ' pair machine and processing time
    Job.Assign (6 + 2 * Index - 1), ProcessingTime
      'revise estimated completion date
    Job.Assign CompletionDate, _
        (Job.A(CompletionDate) + ProcessingTime)
      ' revise due date
    Job.Assign DueDate, _
        (Job.A(DueDate) + 2 * ProcessingTime)
  Next Index
  'Set for first machine on the route
  Job.Assign RouteIndex, 1
  Job.Assign CurrentMachine, _
    Job.A(2 * (Job.A(RouteIndex) - 1) + 6)
  Job.Assign CurrentProcessingTime, _
    Job.A(2 * (Job.A(RouteIndex) - 1) + 7)
  ThisMachineID = Job.A(CurrentMachine)
  If F(ThisMachineID).NumberOfIdleServers > 0 _
  Then
    Job.Assign CurrentProcessingTime, _
      Job.A(CurrentProcessingTime) + _
      deeds.Normal(0, 0.3 * Job.A(CurrentProcessingTime))
    F(ThisMachineID).EngageAndService Job, 1, _
                     Job.A(CurrentProcessingTime)
  Else
    Q(ThisMachineID).Join Job
  End If
End Sub
```

Figure 21.5   Job shop scheduling—sub *Assignments*.

Subroutine *SampleWoReplacement* is sent a *JobList*, and a *SampledList* of jobs is returned. *SampledList* is a complete list of jobs, but only the first *NOperations* are used for a job. *NOperations* equals the integer value of *Normal* (4, 1) with a minimum value of *MinOperations* and a maximum value of *NMachines*. *CompletionDate* and *DueDate* are initially equal to *Time*. However, these values are updated as the job progresses through the model.

Starting with attributes 6 and 7, *NOperations* and *SampledList* are used to encode the job sequence and corresponding processing time. *CompletionDate* and *DueDate* are also updated on the basis of estimates of processing time for each machine in the job sequence. After a complete description of the job has been encoded, the *RouteIndex* attribute is set equal to 1 to represent the first operation on the job's route. Attributes *CurrentMachineID* and *CurrentProcessingTime* are also initialized for the first machine on the job route. If that machine is idle, the job engages the facility, revises *CurrentProcessingTime*, and begins service. Otherwise the job joins the queue for that machine.

SampleWoReplacement: This subroutine, shown in Figure 21.6, receives a Variant array, *JobList*, of jobs and randomly samples machines, without replacement, to populate an array *SampledList*. A job is assigned *NOperations* machines so only the first *NOperations* of array *SampledList* is used.

Figures 21.7 and 21.8 present modeling results for six machines with the minimum number of operations per job being equal to three. Typical average utilization of facilities is approximately 50%. Of the 804 jobs that entered the system, 800 have been completed. Because 304 jobs were late, 496 jobs finished on time. The average tardiness of late jobs is 37.968 time units.

```
Sub SampleWoReplacement(ByRef OriginalValue() As Variant, _
                        ByRef SampledValue() As Variant)
Dim _
    Index As Integer, _
    Jndex As Integer, _
    VectorLength As Integer
  VectorLength = UBound(OriginalValue())
  For Index = VectorLength To 1 Step - 1
    Jndex = UniformDiscrete(1, Index)
    SampledValue(Index) = OriginalValue(Jndex)
    OriginalValue(Jndex) = OriginalValue(Index)
  Next Index
End Sub
```

Figure 21.6   Job shop scheduling—sub *SampleWoReplacement*.

Statistic	(O) JobLateness		(O) JobTardiness		(O) CycleTime	
Number in	800		304		800	
Min value	-206.320		-206.320		7.219	
Max value	281.969		-0.088		416.260	
Average	18.727		-37.968		125.878	
StdDev	59.398		35.129		70.117	

Figure 21.7   Job shop scheduling—statistics output.

Facility	1		2	
Number in	598		442	
Number out	597		441	
Residing	1		1	
Interrupted	0		0	
Destroyed	0		0	
Minimum	0		0	
Maximum	1		1	
	AvgBusy servers		AvgBusy servers	
Average	0.566		0.442	
StdDev	0.496		0.497	

Figure 21.8 Job shop scheduling—facilities output.

21.3 PERT Project Scheduling

A normal project consists of a number of activities or jobs with each requiring time (and resources) for completion. Project activities must be interdependent in the sense that a job may not be started until others have been completed. Interdependence among activities may be expressed using network representation. A branch represents an activity. The network nodes define points in time at which an activity or a group of activities can be started or completed. The network representation provides access to information that can assist in the development of an activity-by-activity time schedule for the project. This section presents a project evaluation and review technique (PERT) network in which the times for project activities are described by probabilistic distributions.

Instead of a special-purpose model for the project presented in Figure 21.9, a general formulation is presented that can be applied to any PERT network by simply changing a set of tables that define precedence relationships of the project and the durations of activities. The methodology for this model is in a General Purpose Simulation System (GPSS) formulation by Allen C. Schuerman of Oklahoma State University.

Activities are identified by unique serial numbers given in Table 21.2. For example, the activity joining nodes 1 and 2 is identified by the serial number 1. These serial numbers are used in tables to summarize the entire PERT network.

Specifically, the tables and related information are as follows:

- *SuccessorNodesTable*: Number of successor nodes for a given node.
- *PredecessorNodesTable*: Number of predecessor nodes for a given node.
- *ActivityMapTable*: Returns an activity serial number for a given predecessor and activity index for an activity that originates from the predecessor node.
- *ActivitySuccessorTable*: Returns the successor node for a given activity serial number.
- *ActivityMeanTable*: Returns a mean activity duration for a given activity serial number.
- *ActivitySDTable*: Returns an activity standard deviation for a given activity serial number.

For *ActivityMapTable*, *Index* varies from 1 to the number of activities that originate at node *Kndex*. Each originating activity has a unique serial number, *Jndex*. Using these definitions, *ActivityMapTable* returns *Jndex* for a given *Kndex* and *Index* by condensing the three pieces of data into a single (independent and dependent) table entry given by $Jndex = 10Kndex + Index$. Because $1 < Kndex < 10$ in this example, the entry $10Kndex + Index$ uniquely identifies activity *Jndex* from node *Kndex*. (The multiple 10 should be adjusted to 100 for $10 < Kndex < 100$, etc.).

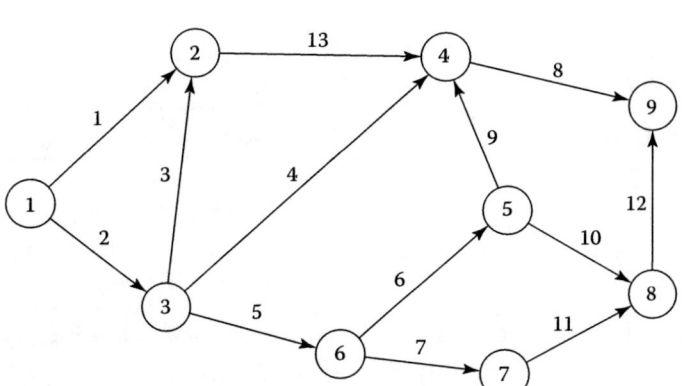

Figure 21.9   Project activity network.

Table 21.2   Activity Means and Standard Deviations

Activity	Predecessor Node	Successor Node	Duration Mean	Duration Standard Deviation
1	1	2	130	30.0
2	1	3	55	11.8
3	3	2	70	10.0
4	3	4	165	25.0
5	3	6	52	8.4
6	6	5	60	10.0
7	6	7	103	16.7
8	4	9	32	5.0
9	5	4	200	33.2
10	5	8	40	7.1
11	7	8	32	8.4
12	8	9	165	11.6
13	2	4	147	13.4

An array, *NodeStatistic()*, of observation-based statistical variables is defined to collect node realization data at each node. An integer array, *NumberRealizedActivities()*, enables the model to track the number of predecessor activities for a node that have been realized. When all predecessor activities for the termination node have been completed, the project is complete. In the model, there are also integer variables *NumberNodes, TerminationNode, SimulationRuns*, and *RunCount*.

Table 21.3   PERT—Primary Events

Event	Description
StartSourceEvent	Begin program
ActivityDelayEvent	Activity completion

Each network transaction flowing through the network has three attributes that identify the activity represented by the transaction. The transaction's *CreationTime* is used for collecting statistics on realization time for each node. The attribute definitions are

1 = *ActivityNumber*
2 = *PredecessorNode*
3 = *SuccessorNode*

Table 21.3 describes the primary event subroutines. The remainder of this section provides an overview of model procedures and a brief description of program output.

StartSourceEvent: This event handler models the transaction that begins the simulation. In Figure 21.10 there is only one arrival that invokes *GenerateActivities(1)* where 1 is the predecessor node for the project.

ActivityDelayEvent: Figure 21.11 represents an activity completion. *NumberRealizedActivities()* for the successor node is incremented by 1. The successor node identifier is from activity attribute *SuccessorNode*. When *NumberRealizedActivities()* is equal to its corresponding *PredecessorNodesTable()*, the observation-based statistical variable *NodeStatistic()* for the node is collected. *NodeStatisitic()* measures the lapsed time from the start of the project to the time in which a node's predecessor activity is completed. When there are unrealized predecessor activities, the sub is finished after recording the activity realization that just occurred.

When all predecessor activities are complete for the termination node and additional runs are needed, *NumberRealizedActivities()* is reset to 0, *RunCount* is incremented by 1, and *GenerateActivities(1)* initiates another project. If *RunCount* is equal to *SimulationRuns*, the model is terminated. The sub *StopSimulation* is used to force stop the simulation because the nature of the model precludes terminating the run with a value for *RunLength*. It is always advisable to use *StopSimulation* to ensure that time-based statistics, if any, are not biased.

```
        ' EVENT HANDLERS
Sub StartSourceEvent(ByRef Arrival As Transaction)
 ' only one source arrival
 Set Arrival = Nothing
 Call GenerateActivities(1) ' predecessor node
End Sub
```

Figure 21.10   PERT—*StartSourceEvent*.

```
Sub ActivityDelayEvent(ByRef Activity As Transaction)
  Dim _
    Index As Integer
      'Update number of activities to a particular node
  NumberRealizedActivities(Activity.A(SuccessorNode)) = _
      NumberRealizedActivities(Activity.A(SuccessorNode)) + 1
  ' When all predecessor node activites are complete,
  ' begin successor node activities; otherwise, wait
  If NumberRealizedActivities(Activity.A(SuccessorNode)) = _
      PredecessorNodesTable.Value(Activity.A(SuccessorNode)) _
  Then
    NodeStatistic(Activity.A(SuccessorNode)).Collect _
                                CurrentTime() - StartTime
    If Activity.A(SuccessorNode) = TerminationNode _
    Then
      Set Activity = Nothing
      RunCount = RunCount + 1
      If RunCount < SimulationRuns _
      Then
        For Index = 1 To NumberNodes
          NumberRealizedActivities(Index) = 0
        Next Index
        StartTime = CurrentTime()
        GenerateActivities (1) ' new run
        Exit Sub
      Else
        StopSimulation
      End If
    Else
      Call GenerateActivities(Activity.A(SuccessorNode))
    End If
  Else
    Set Activity = Nothing 'wait on more activities
  End If
End Sub
```

Figure 21.11   PERT—*ActivityDelayEvent*.

When all predecessor activities are complete for any other node, *GenerateActivities(NodeID)* spawns successor activities for *NodeID*.

GenerateActivities: This program segment spawns a transaction for each successor activity of *NodeID*. In Figure 21.12, attribute *PredecessorNode* now has value *NodeID*. The number of successor nodes for *NodeID* is from *SuccessorNodesTable*. Attribute *SuccessorNode* is determined by using *ActivityMapTable* to identify the activity and *ActivitySuccessorTable* to determine the activity's successor node. A *MapIndex* = 10* *NodeID* + *Index* is passed to *ActivityMapTable*, and *ActivityID* is returned. That value is passed to *ActivitySuccessorNode* and the *SuccessorNode* is returned. *ActivityID* is also used to obtain the mean and standard deviation for the activity duration. The newly created activity then begins its delay.

The summary output of the model is shown in Figure 21.13. The histogram (produced by Excel's data analysis histogram) is based on the distribution of 30 realization times at node 9. Because node 9 is the termination node for the project, the histogram also represents the distribution of project completion time.

```
Sub GenerateActivities(ByVal NodeID)
  Dim _
    Index As Integer, _
    MapIndex As Integer, _
    NextNodeID As Integer, _
    NewActivity As Transaction
  For Index = 1 To SuccessorNodesTable.Value(NodeID)
    Set NewActivity = DEEDS.NewTransaction
    NewActivity.Assign PredecessorNode, NodeID
    MapIndex = 10 * NodeID + Index
    NewActivity.Assign ActivityID, _
                     ActivityMapTable.Value(MapIndex)
    NewActivity.Assign SuccessorNode, _
      ActivitySuccessorTable.Value((NewActivity.A(ActivityID)))
    ActivityDelay.Start NewActivity, _
      Normal(ActivityMeanTable.Value((NewActivity.A(ActivityID))), _
             ActivitySDTable.Value((NewActivity.A(ActivityID)))))
  Next Index
End Sub
```

Figure 21.12   PERT—sub *GenerateActivities*.

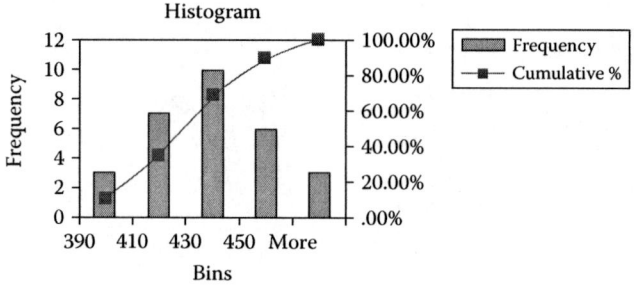

Figure 21.13   Project duration histogram.

21.4 Daily Manpower Allocation

This section presents the problem of periodic allocation of persons to jobs. This type of allocation problem is typical in service centers where workers are dispatched for maintenance or repair service on the customers' premises. The performance metric for this model is the effectiveness of the dispatching rule for allocating workers to jobs. This example is adapted from [3]. Gordon's model uses the advanced "groups" concept in GPSS.

The environment is a company that employs 16 workers to perform maintenance on equipment leased to 20 customers. Service calls arrive daily. For the simulation, a day consists of one eight-hour shift. All times in the model are net values that suppress the effect of the remaining 16 hours of a day. For example, a service time of eight hours represents a full day's work. The time between calls is *UniformContinuous* (1, 3) hours. It requires *UniformContinuous* (4, 12) hours of travel time each way between the company premises and customer sites. The actual allocation of workers to jobs occurs at the beginning of each day when the service supervisor reviews the accumulated list of calls. Unanswered calls due to the unavailability of workers are processed the next day. The dispatch rules are described at the bottom of page 362.

Table 21.4 Manpower Allocation—Primary Events

Event	Description
CallsSourceEvent	Call for service
ReviewSourceEvent	Daily assignment of jobs
JobSiteFacilityEvent	Service completion
TravelDelayEvent	Worker finished travel to/from job site

JobSiteFacility and *WorkersFacility* have 16 and 20 servers, respectively. These facilities maintain information on job sites and worker status. Transactions represent workers and customer calls. The worker is busy when assigned to a call site. Therefore, a busy worker may be working on site or in transit to/from a work site. The transaction attributes for workers are

1 = *CustomerID*
2 = *Destination* ("JobSite," "Office")
3 = *SiteDepartureTime*
4 = *TravelTime*

A job site is "being serviced" when the worker is assigned a call from that site. The site is "being serviced" until the worker departs the job site. Table 21.4 describes each primary event.

In addition to the primary events subroutines, the model includes the following functions:

■ *WorkerOnJob* returns *True* when a worker who is currently on site is assigned a call.
■ *InTransitSameCustomer* returns *True* when a worker returning from the site is assigned the call.
■ *AssignIdleWorker* returns *True* when an idle worker is assigned the call.
■ *InTransit* returns the number of workers in transit to the office.

The remainder of this section provides an overview of model procedures and a brief description of program output.

CallsSourceEvent: Figure 21.14 represents customer calls for service. Attribute *CustomerID* is randomly assigned an integer value from 1 to 20. The call transaction then joins *CallsQueue* to wait for assignment to a worker.

```
        ' EVENT HANDLERS
Sub CallsSourceEvent(ByRef CustomerCall As Transaction)
  CustomerCall.Assign CustomerID, Deeds.UniformDiscrete(1, 20)
  NewCallsQueue.Join CustomerCall
End Sub
```

Figure 21.14 Manpower allocation—*CallsSourceEvent*.

```
Sub ReviewSourceEvent(ByRef DailyReview As Transaction)
  Dim _
     CallIndex As Integer
  For CallIndex = NewCallsQueue.Length To 1 Step -1
    If WorkerOnJob(CallIndex) = False _
    Then
       If InTransitSameCustomer(CallIndex) = False _
       Then
          If AssignIdleWorker(CallIndex) = False _
          Then
                ' perhaps worker returning from another job
          End If
       End If
    End If
  Next CallIndex
  For CallIndex = 1 To InTransit()
    If NewCallsQueue.Length = 0 _
    Then
       Exit Sub
    End If
    NewCallsQueue.View(NewCallsQueue.Length).NewPriority (1)
    AssignedCallsQueue.Join _
                NewCallsQueue.Detach(NewCallsQueue.Length)
  Next CallIndex
End Sub
```

Figure 21.15   Manpower allocation—*ReviewSourceEvent.*

ReviewSourceEvent: This event handler models the daily review and assignment of calls to workers. In Figure 21.15, at the beginning of each day, calls are processed on the basis of the order received. However, *CallsQueue* is ordered in reverse order of arrival (earliest arriving jobs are at the end of *CallsQueue*). Notice that calls are processed from the end of the *CallsQueue* to ensure that earlier calls are dispatched before later calls. The rationale for this approach is to simplify programming for this sub.

Worker status determines how calls are dispatched. The order of consideration for dispatching a worker on a job is

1. A worker is on the call site.
2. A worker is in transit to the call site.
3. A worker is in transit from the call site.
4. A worker is idle.
5. A worker is in transit from a different call site.

For scenarios 3 and 4, calls are removed from *NewCallsQueue* and assigned for immediate processing. For scenarios 1, 2, and 5, calls are detached from *NewCallsQueue* and join *AssignedCallsQueue* to distinguish calls that have been assigned and those yet to be assigned. For scenario 5, subroutine *InTransit* counts the number of workers returning to the office. Because these workers may be dispatched on a job upon arrival at the office, a corresponding number of calls are moved to the *AssignedCallsQueue* with priority equal to 1 to distinguish scenario 5 and scenario 1 and 2 jobs in *AssignedCallsQueue*. Any jobs not immediately dispatched (scenarios 3 and 4) or assigned (scenarios 1, 2, and 5) must wait in *NewCallsQueue* until the next day.

```
Sub JobSiteFacilityEvent(ByRef Worker As Transaction)
  Dim _
    CallIndex As Integer
  JobSiteFacility.Disengage Worker ' job finished
  ServiceCustomerStatistic.Collect Worker.SystemTime
  For CallIndex = 1 To AssignedCallsQueue.Length
    If AssignedCallsQueue.View(CallIndex).A(CustomerID) = _
                                      Worker.A(CustomerID) _
    Then 'company has another call waiting
      WorkersFacility.Disengage Worker
      Set Worker = AssignedCallsQueue.Detach(CallIndex) ' new job
      WorkersFacility.Engage Worker, 1
      JobSiteFacility.EngageAndService Worker, 1
      Exit Sub
    End If
  Next CallIndex
  Worker.Assign Destination, "Office" 'return home
  Worker.Assign SiteDepartureTime, Deeds.CurrentTime
  TravelDelay.Start Worker, Worker.A(TravelTime)
End Sub
```

Figure 21.16   Manpower allocation—*JobSiteFacilityEvent*.

JobSiteFacilityEvent: Figure 21.16 represents the completion of a service call. The worker disengages the *JobSiteFacility* to indicate that the call has been serviced. The *ServiceCustomerStatistic* is collected for *SystemTime* (the time service was completed minus the time the call was received). *AssignedCallsQueue* is checked for additional calls for this site. Before a call is detached from *AssignedJobsQueue*, the previous call (worker) must disengage *WorkersFacility* because the new job engages *WorkersFacility* and *JobSiteFacility* to indicate that the worker is busy and another service completion is scheduled (in effect, the previous call and the next call both represent the same worker). If there are no calls for the site, the worker is destined for the office and *TravelDelay* is started. *CurrentTime* is saved in case the worker returns to the site before arriving at the office. For statistical purposes, the worker is still busy and therefore did not disengage *WorkersFacility*.

TravelDelayEvent: Figure 21.17 represents workers arriving at the "JobSite" or "Office." When arriving at the job site, the worker engages *JobSiteFacility* and begins service. At the "Office," the worker disengages *WorkersFacility* (becomes idle) and checks *AssignedCallsQueue* for calls with priority 1. A priority level 1 job is detached from the queue and now represents the worker assigned to the job site in attribute *CustomerID*. The worker engages *WorkersFacility* and begins *TravelDelay* to the job site. If there are no priority level 1 jobs, the worker remains idle.

WorkerOnJob: Figure 21.18 is a function that returns a Boolean value of *True* to indicate that there is a worker on the job site associated with the call and the call is dispatched by moving it from *NewCallsQueue* to *AssignedCallsQueue*. A value of *False* indicates that the job has not been assigned.

InTransitSameCustomer: Figure 21.19 is a function that returns a Boolean value of *True* to indicate that the call site has a worker in transit to or from the site. When *Destination* is the site, another call from the same customer has been received before the worker has started on the current call. The call is moved from *NewCallsQueue* to *AssignedCallsQueue* and will be serviced at a later date by the same worker being dispatched to the site. If *Destination* is the office, the delay is interrupted and the worker is found and disengaged from *WorkersFacility*. The call is detached from *NewCallsQueue* and becomes the worker who engages *WorkersFacility*.

```
Sub TravelDelayEvent(ByRef Worker As Transaction)
  Dim _
    CallIndex As Integer
  If Worker.A(Destination) = "JobSite" _
  Then 'service call
    JobSiteFacility.EngageAndService Worker, 1
  Else
    WorkersFacility.Disengage Worker
    For CallIndex = 1 To AssignedCallsQueue.Length
      If AssignedCallsQueue.View(CallIndex).Priority = 1 _
      Then
        Set Worker = AssignedCallsQueue.Detach(CallIndex)
        Worker.Assign Destination, "JobSite"
        Worker.Assign TravelTime, Deeds.UniformContinuous(4, 12)
        WorkersFacility.Engage Worker, 1
        TravelDelay.Start Worker, Worker.A(TravelTime)
        Exit Sub
      End If
    Next CallIndex
    Set Worker = Nothing 'hang out
  End If
End Sub
```

Figure 21.17 Manpower allocation—*TravelDelayEvent*.

```
Function WorkerOnJob(ByVal CallIndex As Integer) as Boolean
  Dim _
    Jndex As Integer
  WorkerOnJob = False
  For Jndex = 1 To JobSiteFacility.NumberOfBusyServers
    If NewCallsQueue.View(CallIndex).A(CustomerID) = _
      JobSiteFacility.View(Jndex).A(CustomerID) _
    Then 'worker already on site
      WorkerOnJob = True
      AssignedCallsQueue.Join NewCallsQueue.Detach(CallIndex)
      Exit Function
    End If
  Next Jndex
End Function
```

Figure 21.18 Manpower allocation—function *WorkerOnJob*.

It is routed back to the job site to service the call. *DelayTime* is adjusted for the partial trip back to the job site. A value of *False* indicates that the job has not been assigned.

AssignIdleWorker: Figure 21.20 is a function that returns a Boolean value of *True* to indicate that an idle worker has been dispatched to service the call. An idle server indicates that the server is not associated with any call. The call is detached from *NewCallsQueue*. It becomes the worker who engages *WorkersFacility* and is routed to the job site to service the call. A value of *False* indicates that the job has not been assigned.

InTransit: Figure 21.21 is a function that returns an integer value representing the number of workers destined for the office. Upon arrival at the office, these workers may be immediately dispatched on jobs with priority 1 in *AssignedJobsQueue*. As with previous dispatches, the call becomes the dispatched worker.

```
sitSameCustomer(ByVal CallIndex As Integer) _
                                         As Boolean

eger, _
eger, _
ansaction, _
s Single
ustomer = False
To TravelDelay.Length
eue.View(CallIndex).A(CustomerID) = _
ay.View(Jndex).A(CustomerID) _

SameCustomer = True
Delay.View(Jndex).A(Destination) = "JobSite" _
other call
CallsQueue.Join NewCallsQueue.Detach(CallIndex)
ction

= 1 To WorkersFacility.NumberOfBusyServers
elDelay.View(Jndex).A(CustomerID) = _
ersFacility.View(Kndex).A(CustomerID) _

ersFacility.Disengage WorkersFacility.View(Kndex)
Worker = TravelDelay.Interrupt(Jndex)
  For

  'send employee back
  = Deeds.CurrentTime -_
                Worker.A(SiteDepartureTime)
  = NewCallsQueue.Detach(CallIndex)
ign Destination, "JobSite"
ility.Engage Worker, 1
y.Start Worker, ReturnTime
ion
```

>cation—function *InTransitSameCustomer.*

```
dleWorker(ByVal CallIndex As Integer) as Boolean

ansaction
er = False
lity.NumberOfIdleServers > 0 _

 NewCallsQueue.Detach(CallIndex)
n Destination, "JobSite"
n TravelTime, Deeds.UniformContinuous(4, 12)
Start Worker, Worker.A(TravelTime)
ity.Engage Worker, 1
rker = True
```

>cation—function *AssignIdleWorker.*

```
Function InTransit() As Integer
  Dim _
    Jndex As Integer
  InTransit = 0
  For Jndex = 1 To TravelDelay.Length
    If TravelDelay.View(Jndex).A(Destination) = "Office" _
    Then
      InTransit = InTransit + 1
    End If
  Next Jndex
End Function
```

Figure 21.21   Manpower allocation—function *InTransit*.

Facility	JobSite		Workers	
Number in	994		995	
Number out	988		987	
Residing	6		8	
Interrupted	0		0	
Destroyed	0		0	
Minimum	0		0	
Maximum	13		16	
	AvgBusy servers		AvgBusy servers	
Average	7.838		10.776	
StdDev	1.793		1.969	

Figure 21.22   Manpower allocation—facilities output.

Queue	NewCalls		AssignedCalls	
Number in	1001		474	
Number out	0		0	
Residing	3		3	
Destroyed	0		0	
Removed	0		0	
Detached	998		471	
Minimum	0		0	
Maximum	6		11	
	Lq	Wq	Lq	Wq
Average	1.995	3.982	3.908	16.453
StdDev	1.288	0.073	1.886	0.501

Figure 21.23   Manpower allocation—queues output.

In Figure 21.22, the average number of busy workers for this program was 10.78. During the simulation, 988 of the 1001 calls have been serviced. Also from Figure 21.22, it is clear that upon model termination, eight workers are idle and six of eight busy workers are on site. Therefore, two workers must be in transit. As shown in Figure 21.23, three calls are waiting to be assigned and three calls have been assigned but are waiting for a worker. That occurs when the three assigned calls are from sites that already have a worker either on site or traveling to the site. Figure 21.24 indicates that average service time per call is 31.57.

Statistic	(O) ServiceCustomer	
Number in	988	
Min value	10.04	
Max value	86.55	
Average	31.572	
StdDev	10.519	

Figure 21.24 Manpower allocation—statistics output.

References

1. Pritsker, A. A. B., *Introduction to Simulation and SLAM II*, 3rd ed. New York: Wiley, 1986, pp. 460–478.
2. Ellon, S., Chowdhury, I., and Serghiou, S., Experiments with the *Six* rule in job shop simulation. *Simulation*, 24, 1975, 45–48.
3. Gordon, G., *The Application of GPSS to Discrete System Simulation*. Englewood Cliffs, NJ: Prentice-Hall, 1975, pp. 300–306.

Chapter 22

Maintenance and Reliability Models

22.1 Introduction

Maintenance activities keep a production system operable. Facilities and equipment continually wear out and need repairs and replacement. Maintenance keeps or restores an asset to satisfactory operating condition. Without proper maintenance, production delays can be experienced, excessive idle time may occur, and unsafe working conditions may result. Maintenance costs are lower when the asset is new and increase as the equipment ages.

Maintenance may be centralized, decentralized, or a combination within an organization. Breakdown and preventive maintenance are the two major types of maintenance. Breakdown maintenance occurs when an asset has failed. Preventive maintenance involves activities designed to delay or prevent breakdowns. The best maintenance policy is the one that incurs the lowest total cost. Typical breakdown costs include the repair costs, equipment down time, idle workers, output loss, schedule delays, and customer dissatisfaction. Information needed to conduct a cost analysis includes cost of breakdowns, frequency of breakdowns, and cost of preventive maintenance. This chapter presents simulation models for maintenance planning.

22.2 General Reliability Model

In the block diagram of the system shown in Figure 22.1, the system consists of six components identified as A through F. Components E and F represent parallel units. The model assumes that when the system completely fails, it cannot resume operation until all failed components are repaired and a preventive maintenance check is applied to all system components. The preventive maintenance check is *Exponential* (1) hour.

Using network theory terminology, the entire system will fail if all the components in one *cut set* fail. The cut sets in Table 22.1 represent all failure combinations of the system in Figure 22.1.

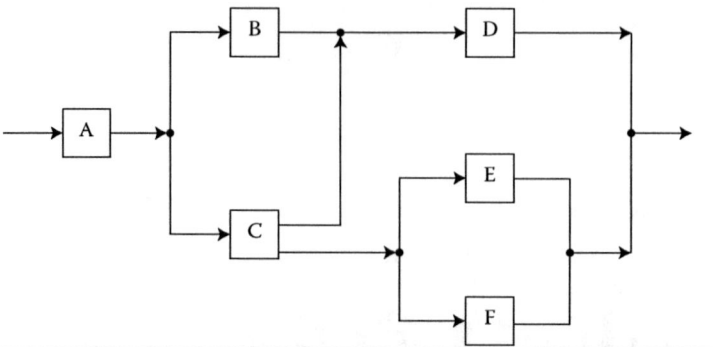

Figure 22.1   General reliability network.

Table 22.1   Reliability Network Cuts

Cut Set Number	Associated Components
1	A
2	B, C
3	C, D
4	D, E, F

To summarize the cut sets in a general format as input data to a simulation model, the information is represented by the special "incidence matrix" in Table 22.2. Each row in the matrix represents a cut set, and each column represents one of the six components. A row element is assigned a value of 1 if its component defines an element of the cut. Otherwise, the value is 0. The incidence matrix provides a complete representation of the system necessary for a reliability study. Notice, however, that the major difficulty in constructing the incidence matrix is the lack of a systematic procedure to enumerate "all" the cut sets from the block diagram. Although some heuristics are available, the task of constructing a complete incidence matrix depends on the complexity of the block diagram. Nevertheless, this is a methodology for translating the block diagram into input data for a general simulation model. Clearly, if cut sets are missed, computed reliability estimates should be regarded as an upper bound on the system's reliability.

This simulation model has arrays to represent *CutSets*, *ComponentFailures*, and *MeanFailureTimes*. Each row of *CutSets* corresponds to a cut set of the reliability system. A model assumption is that the time to failure and the repair time are exponential with different means for each component. *MeanFailureTimes* and *MeanRepairTimes* are arrays for mean time between failures and mean time for repairs of each component. The source of all of the above data is the *UserInitial* spreadsheet.

ComponentFailures is an array of variables that indicate the status of an individual component. Permissible values are

$0 = Operational\ component$
$1 = Failed\ component$

Table 22.2   Incidence Matrix Representation of Cuts

Cut	A	B	C	D	E	F
1	1	0	0	0	0	0
2	0	1	1	0	0	0
3	0	0	1	1	0	1
4	0	0	0	1	1	1

Table 22.3   General Reliability—Primary Events

Event	Description
StartSourceEvent	Starts the program
CheckFacilityEvent	System inspection
SystemFacilityEvent	Component failure
RepairFacilityEvent	Component repair

NumComponents in the system is from the number of columns in the *CutSets* matrix. *SystemFailure* is a Boolean variable that indicates the status of the entire system. *FailureTime* measures lapsed time between failures. *DownTimeStatistic* and *TimeToFailStatistic* are user-defined observation based statistics.

Table 22.3 describes each primary event in the model. In addition to primary events subroutines, the model includes a sub *StartSystem* that resets the system after a system failure.

The remainder of this section provides an overview of model procedures and a brief description of program output.

StartSourceEvent: Figure 22.2 represents initializing the system. Notice that *StartSource* is suspended because the system is subsequently restarted after a system failure. The array *ComponentFailures* is initialized to 0 to indicate that all components are operational. Then, new transactions, one for each component, are scheduled for failure in *SystemFacility* using their respective mean time to failure. The system is now operational for the first time and *StartSourceEvent* is never used again.

CheckFacilityEvent: The completion of system inspection and restarting the system is shown in Figure 22.3. Transaction *PreventiveMaintenance* disengages *CheckFacility* and after *SystemFailure* is set to *False, DownTimeStatistic* is collected and the system is restarted.

RepairFacilityEvent: Figure 22.4 represents the repair of a failed component. The model continues to schedule repairs until the *RepairQueue* is empty. The time to repair a component reflects the individual component's *MeanRepairTime*. What happens after the repair is completed depends on the state of the system. If the system is operational, the repaired component engages the facility and is placed into service. Assigning the respective *ComponentFailures* to 0 indicates that the component is again operational. When the

```
Sub StartSourceEvent(ByRef Begin As Transaction)
  StartSource.Suspend
  Set Begin = Nothing
  'StartSystem
  Dim _
    Arrival As Transaction
  For Index = 1 To NumComponents
   ComponentFailures(Index) = 0
  Next Index
  For Index = 1 To NumComponents
    Set Arrival = Deeds.NewTransaction
    Arrival.Assign 1, Index
    SystemFacility.EngageAndService Arrival, 1, _
           Deeds.Exponential(MeanFailureTimes(Index))
  Next Index
  StartTime = Time
End Sub
```

Figure 22.2   General reliability—*StartSourceEvent*.

```
Sub CheckFacilityEvent(ByRef PreventiveMaintenance As Transaction)
  CheckFacility.Disengage PreventiveMaintenance
  SystemFailure = False
  DownTimeStatistic.Collect (Time - FailureTime)
  StartSource.Restart
End Sub
```

Figure 22.3   General reliability—*CheckFacilityEvent*.

```
Sub RepairFacilityEvent(ByRef RepairedComponent As Transaction)
  Dim _
    NextRepair As Transaction
  RepairFacility.Disengage RepairedComponent
  If RepairQueue.Length > 0 _
  Then
    Set NextRepair = RepairQueue.Depart
    RepairFacility.EngageAndService NextRepair, 1, _
       Deeds.Exponential(MeanRepairTimes(NextRepair.A(1)))
  Else
    If SystemFailure _
    Then
      CheckFacility.EngageAndService RepairedComponent, 1
      Exit Sub
    End If
  End If
  If Not SystemFailure _
  Then
    ComponentFailures(RepairedComponent.A(1)) = 0
    SystemFacility.EngageAndService RepairedComponent, 1, _
           Deeds.Exponential(MeanFailureTimes(Index))
  End If
End Sub
```

Figure 22.4   General reliability—*RepairFacilityEvent*.

system is in failure mode, all repairs must be finished before a preventive maintenance inspection of the system is scheduled.

SystemFacilityEvent: Figure 22.5 represents a component failure. The failed component disengages the facility and the corresponding value in *ComponentFailures* is assigned the value 0 to indicate that the component has failed. *CutSets* is consulted each time a component fails, to ascertain whether the latest failure causes a failure of the entire system. Each row of *CutSets* is compared to *ComponentFailures*. The procedure assumes that a system has failed and then compares each row in *CutSets* to array *ComponentFailure* to verify that at least one critical component has not failed (i.e., the system has not failed). If a system failure occurs, *SystemFailure* is now true, all components are removed from the system, and *TimeToFailStatistic* is collected. Only the failed component is scheduled for repair, with service time equal to its respective repair time, or it joins the repair queue.

```
Sub SystemFacilityEvent (ByRef FailedComponent As Transaction)
  Dim _
    Matches As Integer
  SystemFacility.Disengage FailedComponent
  ComponentFailures(FailedComponent.A(1)) = 1
  SystemFailure = False
  For Index = 1 To NumCutSets
    Matches = 0
    For Jndex = 1 To NumComponents
      If CutSets(Index, Jndex) > ComponentFailures(Jndex) _
      Then
        Exit For ' one critical component operational
      Else
        Matches = Matches + 1
      End If
    Next Jndex
    If Matches = NumComponents _
    Then
      SystemFailure = True
      Exit For
    End If
  Next Index
  If SystemFailure _
  Then
    TimeToFailStatistic.Collect (Time - StartTime)
    FailureTime = Time
    For Kndex = 1 To SystemFacility.NumberOfBusyServers
      SystemFacility.Interrupt 1
    Next Kndex
  End If
  If RepairFacility.NumberOfIdleServers > 0 _
  Then
    RepairFacility.EngageAndService FailedComponent, 1, _
      Deeds.Exponential(MeanRepairTimes(FailedComponent.A(1)))
  Else
    RepairQueue.Join FailedComponent
  End If
End Sub
```

Figure 22.5   General reliability—*SystemFacilityEvent*.

Statistic	(O) DownTime	(O) TimeToFail
Number In	235	236
Min Value	0.141	0.001
Max Value	10.508	9.763
Average	2.869	1.370
StdDev	1.814	1.226

Figure 22.6   General reliability—statistics output.

Facility	Check		System		Repair	
Number In	236		1637		749	
Number Out	235		749		749	
Residing	1		0		0	
Interrupted	0		888		0	
Destroyed	0		0		0	
Minimum	0		0		0	
Maximum	1		6		1	
	AvgBusy	Servers	AvgBusy	Servers	AvgBusy	Servers
Average	0.257		1.676		0.610	
StdDev	0.437		2.470		0.488	

Figure 22.7   General reliability—facilities output.

From Figure 22.6, average *DownTime* is approximately two times the *TimeToFail* for the system. This is obviously an undesirable situation. From simulation output shown in Figure 22.7, *Repair* utilization is not a system bottleneck to performance. The low reliability of the system can be attributed to the system design.

22.3 Maintenance Scheduling

The line-balancing problem in Section 18.2 is modified to demonstrate maintenance scheduling by replacing material replenishment features of the model with maintenance scheduling. A poor maintenance policy will adversely affect the production rate. The system performance metric is *AverageCycleTime*. Of course, the reciprocal of cycle time is the production rate of the system.

The extreme situation is to repair workstations as they fail. However, in a typical maintenance program, the production line is periodically shut down, each workstation is inspected, and worn components are repaired or replaced. After maintenance activities are completed, the line is restarted. However, even with scheduled maintenance there may still be workstation failures between scheduled shutdowns that must be repaired. The lapsed time between shutdowns is dependent upon the mean time between failures of the various workstations.

The details of maintenance procedures are often dictated by the production environment. For example, the disposition of a job being processed when the workstation fails is dependent on the nature of the product. In some instances, when the workstation is restarted, only operations to complete the production unit are required. In other situations, the unit may be scrapped or all previously completed activities must be redone.

As described in Section 18.2, to account for statistical variations in workstation cycle times, each workstation downstream from the first one has a work-in-process buffer with capacity of three units. When an operator completes a unit of production and the downstream buffer is full, the operator becomes blocked, with the completed unit remaining in the workstation. With the exception of the first operator, there must be a unit of production available in the upstream buffer or the operator remains idle.

As part of the scheduled maintenance shutdown procedure for this model, workstations are allowed to complete a unit of production before maintenance begins. If a workstation fails during shutdown, the job remains in the workstation and the remaining processing time is applied to the job when the workstation is restarted. The repair begins immediately but the workstation does not come on line until the entire line is restarted.

Task data is read from worksheet *TaskData* into an array, *TaskTime(1 To StationCount, 1 To 3)*. Columns represent the minimum, middle, and maximum value for each task. Data from Tables 18.3 and 18.4 are read from worksheet *WorkStationData* into array *WSTask(1 To StationCount, 1 To 2)*. Columns represent the first and last task, respectively. Workstations and their associated queues are defined as arrays *WSQ(1 To StationCount)* and *WS(1 To StationCount)*. Of course, the first workstation queue is not used because that workstation never has a waiting job. The workstation states are as follows:

- Blocked when operator has finished a job and the downstream queue is full
- Idle (workstation empty) when the operator has no waiting jobs
- Busy when the operator is working on a job

To simplify program logic, Boolean arrays *WSFailureMode(1 To StationCount)* and *WSBlocked(1 To StationCount)* are defined. When *WSFailureMode()* is *True*, the workstation has failed. Otherwise, *WSFailureMode()* is *False*. When *WSBlocked()* is *True*, the workstation has finished a job but its succeeding workstation's buffer is full. Otherwise, *WSBlocked()* is *False*. *WSSemaphore* is used to count the number of busy workstations. When *WSSemaphore* equals 0, all workstations are idle or blocked. When *WSSemaphore* equals *StationCount*, all workstations are operational. *InShutDownMode* is a Boolean variable that when it equals *True* indicates to operators not to begin another job because maintenance is ready to begin.

Transactions represent units of production and the status of the production line (in operational or maintenance mode). The transaction attributes are

> 1 = *WSID As Integer*
> 2 = *ActivityTime As Integer*

Table 22.4 describes each primary event in the model. As implied in the table, subroutine *Parser* routes all completed workstation jobs to sub *JobCompletionEvent* and all workstation failures and repairs to subs *FailureDelayEvent* and *RepairDelayEvent*, respectively.

In addition to the primary events subroutines, the following subroutines are included in the model:

UpStreamCheck initiates service on jobs at workstations that were previously blocked by its downstream queue.
CreateNewJob at the first workstation initiates a job and begins service.

Table 22.4 Maintenance—Primary Events

Event	Description
JobCompletionEvent	Workstation job completions
FailureDelayEvent	Workstation failure occurs
RepairDelayEvent	Workstation repair completed
OperationsDelayEvent	End of operations activity
MaintenanceDelayEvent	End of maintenance activity

StartNextJob at the current workstation removes a job from the queue and begins service.
InitialWorkStations restarts operations after maintenance completion.
BeginMaintenance initiates maintenance after all workstations have been shut down.
Function *ServiceTime*, in Figure 22.8, determines the time required to process a job in the workstation.

The remainder of this section provides an overview of model procedures and a brief description of program output.

JobCompletionEvent: For a job at the last workstation, cycle time is tabulated and the job exits the system (Figure 22.8). For any other workstation, the workstation disengages the job and moves it to the next workstation or the workstation becomes blocked when the downstream queue is full. When blocked, the job remains in the workstation. The workstation is now ready to begin another job. For the first workstation a new job is initiated. For other workstations, a job must be waiting in the workstation's upstream buffer.

FailureDelayEvent: When a workstation failure occurs (Figure 22.9) and the workstation is busy, the job remains in the workstation but remaining service time is determined. The workstation may be idle but in either case *WSFailure()* is set to *True*. Repair is then started on the failed workstation.

RepairDelayEvent: When the repair has been completed (Figure 22.10) and a job is in the workstation, work begins on that job. Otherwise, assuming a job is in the queue, the job is placed into service. *WSFailure()* is set to *False*.

OperationsDelayEvent: At the end of operations (Figure 22.11) the system is prepared for shutdown. *InMaintenanceMode* is set to *True*. *WSSemaphore* is set to *StationCount* and then decremented for each workstation in repair mode (repair operations are also terminated). As workstations finish a job, *WSSemaphore* is decremented.

MaintenanceDelayEvent: When maintenance has been completed (Figure 22.12) each workstation is restarted. If a job is in the workstation, work begins on that job. Otherwise, assuming a job is in the queue, the job is placed into service. *InMaintenanceMode* is set to *False*.

UpStreamCheck: When a blocking condition (Figure 22.13) has been removed, *UpStreamCheck* begins by removing the blocking condition at the previous workstations and continues until either the first workstation is reached or the workstation is not blocked. For each workstation, the blocked transaction is moved to the downstream queue or workstation because the

```
     ' SECONDARY EVENTS FOR EACH PRIMARY EVENT
Sub JobCompletionEvent(ByRef Job As Transaction)
  Dim _
    CurrentWS As Integer, _
    Index As Integer
  CurrentWS = Job.A(WSID)
  If CurrentWS = StationCount _
  Then
    WS(CurrentWS).Disengage Job
    CycleTimeStatistic.Collect Job.SystemTime
    Set Job = Nothing
  Else
    If WSQ(CurrentWS + 1).Full _
    Then
      WSBlocked(CurrentWS) = True
    Else
      WS(CurrentWS).Disengage Job
      Job.Assign WSID, CurrentWS + 1
      Job.Assign ActivityTime, ServiceTime(CurrentWS + 1)
      WSQ(CurrentWS + 1).Join Job
      If Not InShutDownMode And Not WSFailureMode(CurrentWS + 1) _
      Then
        StartNextJob (CurrentWS + 1)
      End If
    End If
  End If
  If InShutDownMode _
  Then
    WSSemaphore = WSSemaphore - 1
    Call BeginMaintenance
    Exit Sub
  End If
  If Not WSFailureMode(CurrentWS) _
  Then
    If CurrentWS = 1 _
    Then
      Call CreateNewJob
    Else
      StartNextJob (CurrentWS)
    End If
  End If
End Sub
```

Figure 22.8   Maintenance—*JobCompletionEvent*.

downstream station may have been blocked with an empty upstream buffer. For the first workstation, a new job is created when the blocking condition is removed. For all other workstations, a job is pulled from the upstream queue with service time from Figure 22.18, or the workstation becomes idle.

CreateNewJob: When the first workstation is idle and parts are available (Figure 22.14), work begins on a new job.

StartNextJob: When the workstation server is idle (Figure 22.15), its queue has a job, and parts are available for the next job, work begins on the job removed from the queue. Because a job has been removed from the queue, beginning at the previous workstation, the system is checked for upstream workstations that may have become unblocked.

```
Sub FailureDelayEvent(ByRef FailedWorkStation As Transaction)
  Dim _
    Index As Integer, _
    Job As Transaction, _
    Repair As Transaction
  Index = FailedWorkStation.A(WSID)
  If Not WSBlocked(Index) _
  Then
    If WS(Index).NumberOfBusyServers > 0 _
    Then
      Set Job = WS(Index).Interrupt(1)
      Job.Assign ActivityTime, Job.TimeStamp - CurrentTime
      WS(Index).Engage Job, 1
    End If
  End If
  If InShutDownMode _
  Then
    WSSemaphore = WSSemaphore -1
  Else
    WSFailureMode(Index) = True
    Set Repair = NewTransaction
    Repair.Assign WSID, Index
    RepairDelay.Start Repair
  End If
End Sub
```

Figure 22.9   Maintenance—*FailureDelayEvent*.

```
Sub RepairDelayEvent(ByRef RepairedWorkStation As Transaction)
  Dim _
    Index As Byte, _
    Job As Transaction, _
    NextFailure As Transaction
  Index = RepairedWorkStation.A(WSID)

  If WS(Index).NumberOfBusyServers > 0 _
  Then
    Set Job = WS(Index).View(1)
    WS(Index).Service Job, Job.A(ActivityTime)
  Else
    If Index = 1 _
    Then
      Call CreateNewJob
    Else
      StartNextJob (Index)
    End If
  End If
  WSFailureMode(Index) = False
  Set NextFailure = NewTransaction
  NextFailure.Assign WSID, Index
  FailureDelay.Start NextFailure
End Sub
```

Figure 22.10   Maintenance—*RepairDelayEvent*.

```
Sub OperationsDelayEvent(ByRef OperationsTimer As Transaction)
   Dim _
      Index As Integer
   Set OperationsTimer = Nothing
   WSSemaphore = StationCount
   InShutDownMode = True
   While RepairDelay.Length > 0
      Index = RepairDelay.View(1).A(WSID)
      WSFailureMode(Index) = False
      RepairDelay.Interrupt 1
      WSSemaphore = WSSemaphore - 1
   Wend
End Sub
```

Figure 22.11   Maintenance—*OperationsDelayEvent*.

```
Sub MaintenanceDelayEvent(ByRef MaintenanceTimer As Transaction)
   Dim _
      Index As Integer, _
      NextFailure As Transaction
   Set MaintenanceTimer = Nothing
   InShutDownMode = False
   For Index = 2 To StationCount
      Set NextFailure = NewTransaction
      NextFailure.Assign WSID, Index
      FailureDelay.Start NextFailure
      WSFailureMode(Index) = False
      If Not WSBlocked(Index) _
      Then
         StartNextJob (Index)
      End If
   Next Index
   Set NextFailure = NewTransaction
   NextFailure.Assign WSID, 1
   FailureDelay.Start NextFailure
   If Not WSBlocked(1) _
   Then
      Call CreateNewJob
   End If
   OperationsDelay.Start NewTransaction ' Operations activity
End Sub
```

Figure 22.12   Maintenance—*MaintenanceDelayEvent*.

InitialWorkStations: When maintenance has been completed (Figure 22.16) each workstation is restarted. If a job is in the workstation, work begins on that job. Otherwise, assuming a job is in the queue, the job is placed into service. Failure delays to affect workstation failures are initiated for each workstation.

BeginMaintenance: When all workstations (Figure 22.17) are inactive and *WSSemaphore* is zero, maintenance delay begins.

```
Sub UpStreamCheck(ByVal CurrentWS As Integer)
  Dim _
    Job As Transaction, _
    Index As Integer
  Do While (CurrentWS > 0)
    If WSBlocked(CurrentWS) _
    Then
      WSBlocked(CurrentWS) = False
      Set Job = WS(CurrentWS).View(1)
      WS(CurrentWS).Disengage Job
      Job.Assign WSID, CurrentWS + 1
      WS(CurrentWS + 1).NextFacility Job, 1, WSQ(CurrentWS + 1)
      If WSQ(CurrentWS).Length > 0 _
      Then
        WS(CurrentWS).EngageAndService WSQ(CurrentWS).Depart, 1, _
                                          ServiceTime(CurrentWS)
      End If
      If CurrentWS = 1 _
      Then
        Call CreateNewJob
      Else
        CurrentWS = CurrentWS - 1
      End If
    Else
      Exit Do ' no additional blocking conditions
    End If
  Loop
End Sub
```

Figure 22.13   Maintenance—sub *UpStreamCheck*.

```
Sub CreateNewJob()
  Dim _
    Job As Transaction
  If WS(1).NumberOfIdleServers > 0 _
  Then
    Set Job = NewTransaction
    Job.Assign WSID, 1
    Job.Assign ActivityTime, ServiceTime(1)
    WS(1).EngageAndService Job, 1, Job.A(ActivityTime)
  End If
End Sub
```

Figure 22.14   Maintenance—sub *CreateNewJob*.

The model reports are shown in Figures 22.19 and 22.20. Notice that *AverageCycleTime* is 2194.56 as compared to 579.589 for the model in Figure 18.9. Figure 22.20 shows workstation statistics for workstations with utilization that exceeds 80%. As expected for this model, facility utilization should remain essentially the same and cycle time will show a significant increase.

```
Sub StartNextJob(ByVal CurrentWS As Integer)
  Dim _
    Job As Transaction
  If WS(CurrentWS).NumberOfIdleServers > 0 _
  Then
    If WSQ(CurrentWS).Length > 0 _
    Then
      Set Job = WSQ(CurrentWS).Depart
      WS(CurrentWS).EngageAndService Job, 1, Job.A(ActivityTime)
      Call UpStreamCheck(CurrentWS - 1)
    End If
  End If
End Sub
```

Figure 22.15   Maintenance—sub *StartNextJob*.

```
Sub InitialWorkStations()
  Dim _
    index As Integer, _
    NextFailure As Transaction
  For index = 2 To StationCount
    Set NextFailure = NewTransaction
    NextFailure.Assign WSID, index
    FailureDelay.Start NextFailure
    WSFailureMode(index) = False
    If Not WSBlocked(index) _
    Then
      StartNextJob (index)
    End If
  Next index
  Set NextFailure = NewTransaction
  NextFailure.Assign WSID, 1
  FailureDelay.Start NextFailure
  If Not WSBlocked(1) _
  Then
    Call CreateNewJob
  End If
End Sub
```

Figure 22.16   Maintenance—sub *InitialWorkStations*.

```
Sub BeginMaintenance()
  If WSSemaphore = 0 _
  Then
    While FailureDelay.Length > 0
      FailureDelay.Interrupt 1
    Wend
    MaintenanceDelay.Start NewTransaction ' maintenance activity
  End If
End Sub
```

Figure 22.17   Maintenance—sub *BeginMaintenance*.

```
Function ServiceTime(ByVal CurrentWS As Integer) As Single
  Dim _
    Index As Integer
  ServiceTime = 0
  For Index = WSTask(CurrentWS, First) To WSTask(CurrentWS, Last)
    ServiceTime = ServiceTime + _
      Distribution.Triangle(TaskTime(Index, 1), TaskTime(Index, 3),
TaskTime(Index, 2))
  Next Index
End Function
```

Figure 22.18 Maintenance—function *ServiceTime*.

Statistic	(O) Cycle Time	
Number In	168	
Min Value	234.311	
Max Value	7568.755	
Average	2194.561	
StdDev	2392.395	
LCL		
UCL		
Observation	Mean	S
0		
1	2194.561	2392.395

Figure 22.19 Assembly line maintenance—statistics output.

Facility	F1		F2		F3		F4	
Number In	181		180		173		172	
Number Out	180		176		172		168	
Residing	1		1		1		1	
Interrupted	0		3		0		3	
Destroyed	0		0		0		0	
Minimum	0		0		0		0	
Maximum	1		1		1		1	
	AvgBusy	Servers	AvgBusy	Servers	AvgBusy	Servers	AvgBusy	Servers
Average	0.988		0.989		0.833		0.849	
StdDev	0.111		0.105		0.373		0.358	
LCL								
UCL								
Observation	Avg	StdDev	Avg	StdDev	Avg	StdDev	Avg	StdDev
0								
1	0.988	0.111	0.989	0.105	0.833	0.373	0.849	0.358

Figure 22.20 Assembly line maintenance—facilities output.

Chapter 23

Quality Control Models

23.1 Introduction

Quality control is much more than the application of statistical methods to process improvement. However, statistical process control (SPC) methods are the basis for measuring quality. The methodology is the same if the processes are in disparate areas such as manufacturing or health care. Within an organization, these methods may also be applied to process development, management, engineering design, finance, accounting, and marketing.

This chapter presents two examples that demonstrate the use of discrete event simulation to model problems in quality control. In contrast to problems presented previously, for these problems the lapsed time between primary events is implied in the problem, that is, the sampling interval may be constant or variable. The focus is on the SPC aspects of the problem. Both problems were suggested by James R. Smith at Tennessee Tech University.

23.2 Costing Inspection Plans

This problem illustrates the use of simulation to evaluate an inspection strategy for an assembly activity that consists of an axle and two wheels. Specifically, this model is used to determine the effects of tolerances and inspection errors on the cost of producing an assembly. The problem of tolerances on assemblies and components is as follows: given an assembly characteristic Y that is a function of n component characteristics, X_i, according to some function

$$Y = f(X_1, X_2, \ldots, X_n)$$

and tolerances on Y, what should be the tolerances on the X_is? The problem is complicated by inspection errors.

The performance metric for this simulation is how tolerances on the axles and wheels (X_is for a specific assembly) and inspection errors affect the cost of an inspection program. Inspection costs include the cost of inspection, rework, and scrap of components and assemblies. The decision variables are (L_i, U_i), $i = 1, \ldots, n$, the specifications for the n components ($n = 2$ for this model). Nominal values for the X_i are assumed (nominal $= (L_i + U_i)/2$) for the analysis. The distributions

X_i are assumed *Normal* with means $E(X_i)$ and variances σ_i^2. Although a linear function f and normal distribution of the X_is are assumed, a simulation analysis easily permits nonlinear functions and non-normal distributions to be considered.

The combinations of (L_i, U_i) that may be considered are

$$\text{Nominal } X_i \pm \sigma_i; \pm 2\sigma_i; \pm 3\sigma_i$$

Assuming that each $E(X_i)$ is the nominal value, each process is centered at the nominal value. The specifications on the assembly are (L, U).

For the axle-and-wheel assembly shown in Figure 23.1, the clearances between the wheel and axle are $Y_1 = X_{21} - X_1$ and $Y_2 = X_{22} - X_1$, respectively. The axle is assumed to be uniform in diameter. There are two production lines, one for wheels and one for axles.

The effects of sensitivity and specificity are included in the model. *Sensitivity* is defined as the probability of correctly classifying a defective component. *Specificity* is defined as the probability of correctly classifying a nondefective component. The value for specificity and sensitivity for component inspection is 0.80. Assembly inspection is assumed to be perfect.

The assumption is that there is a 100% inspection on all parts when $U - L < 6\sigma$. When $U - L \geq 6\sigma$ the assumption is that parts are not inspected because virtually all parts will be within specification if the process is in control. The nominal dimensions are $E(X_1) = 1.000$ for the axles and $E(X_2) = 1.003$ for the wheels. Y_1 and Y_2 clearances must be between 0.001 and 0.005. Otherwise, the costs incurred are $2.00 if either $Y < 0.001$ (rework) or $4.00 if either $Y > 0.005$ (scrap).

The wheel and axle lines have a normally distributed variation where $\sigma_{X_1} = \sigma_{X_2} = .001$. The specifications on axles and wheels are (L_1, U_1) and (L_2, U_2), respectively. The costs of out-of-specification parts are as follows:

$$\$0.75 \quad \text{if } X_1 < L_1(\text{scrap}) \quad \text{or} \quad \$0.25 \quad \text{if } X_1 > U_1(\text{rework})$$
$$\$0.50 \quad \text{if } X_2 > L_2(\text{scrap}) \quad \text{or} \quad \$0.30 \quad \text{if } X_2 < L_2(\text{rework})$$

The costs of inspecting an axle, wheel, and assembly are $0.10, $0.05, and $0.25, respectively.

The source data for this model is described in Section 17.4. A String array *InspectionStatus(1 To 3)* is used to maintain inspection results for each component. Permissible values are *Passed* and *Reworked*. Transactions represent a unit of production that consists of an axle and two wheels. The transaction attributes are

1 = *AxleDiameter*
2 = *WheelADiameter*
3 = *WheelBDiameter*

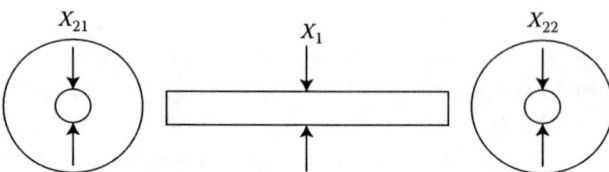

Figure 23.1   Axle and wheel assembly.

The only primary event in the model corresponds to an arrival of a unit of production. Therefore, *RunLength* represents the number of units produced. In addition to the primary event subroutine, sub *NextMaterialSet* details information on a unit of production. The following functions are included in the model:

- *NextMaterial* specifies production parameters for individual components.
- *FailedAxleInspection* returns *False* when an axle passes inspection. Otherwise it returns *True*.
- *FailedWheelInspection* returns *False* when a wheel passes inspection. Otherwise it returns *True*.
- *FailedAssemblyInspection* returns *False* when an assembly passes inspection. Otherwise it returns *True*.
- *Calculations* summarizes the end-of-run inspection, failure, and cost data.

The remainder of this section provides an overview of model procedures and a brief description of program output.

PartsSourceEvent: As shown in Figure 23.2, after obtaining component parameters, the axle is inspected. Components are inspected separately. However, if at any time during the inspection process any component fails an inspection, then the entire production unit, axle and both wheels, are scrapped and a new production unit is started. It is only after all parts pass inspection and meet tolerance requirements that an assembly is produced. However, a component may pass an inspection after rework.

```
' SECONDARY EVENTS FOR EACH PRIMARY EVENT
Sub PartsSourceEvent(ByRef MaterialSet As Transaction)
  Dim _
    Index As Integer
  For Index = 1 To 3
    InspectionStatus(Index) = ""
  Next Index
  Call NextMaterialSet(MaterialSet)
  If FailedAxleInspection(MaterialSet.A(AxleDiameter)) _
  Then  ' bad axle in set
    NumberAxlesScraped = NumberAxlesScraped + 1
    NumberWheelsScraped = NumberWheelsScraped + 2
    Exit Sub
  End If
  For Index = 2 To 3 ' each wheel
    If FailedWheelInspection(Index, MaterialSet.A(Index)) _
    Then  ' bad wheel
      NumberAxlesScraped = NumberAxlesScraped + 1
      NumberWheelsScraped = NumberWheelsScraped + 2
      Exit Sub
    End If
  Next Index
  If FailedAssemblyInspection(MaterialSet) _
  Then
    NumberAxlesScraped = NumberAxlesScraped + 1
    NumberWheelsScraped = NumberWheelsScraped + 2
    NumberAssembliesScraped = NumberAssembliesScraped + 1
  Else
    AssemblyCount = AssemblyCount + 1
  End If
End Sub
```

Figure 23.2   Inspection—*PartsSourceEvent*.

NextMaterialSet: In Figure 23.3, production parameters for individual components are detailed.
FailedAxleInspection: In Figure 23.4, the number of axle inspections is incremented by 1. *False* is returned when an axle passes inspection or is reworked. Otherwise, *True* is returned. *InspectionStatus(1)* will be either *Passed* or *Reworked*. The three scenarios are

1. Axle diameter within specification limits: The axle passes inspection. However, because of a *specificity* error, the axle may be incorrectly classified as needing rework.
2. Axle diameter above upper specification limits: The axle should be reworked. However, because of a *sensitivity* error, the axle may be scrapped.
3. Axle diameter below lower specification limits: The axle fails inspection. However, because of a *sensitivity* error, the axle is reworked before being scrapped.

```
Sub NextMaterialSet(ByRef Job As Transaction)
  Job.Assign AxleDiameter, Normal(AxleMean, AxleStdDev, 1)
  Job.Assign WheelADiameter, Normal(WheelMean, WheelStdDev,1)
  Job.Assign WheelBDiameter, Normal(WheelMean, WheelStdDev, 1)
End Sub
```

Figure 23.3 Inspection—sub *NextMaterialSet*.

```
Function FailedAxleInspection(ByVal AxleDiameter As Single) As Boolean
  NumberAxleInspections = NumberAxleInspections + 1
  FailedAxleInspection = False
  InspectionStatus(1) = "Passed"
  If AxleDiameter < UpperAxleTolerance And _
    AxleDiameter > LowerAxleTolerance _
  Then ' good axle
    If Random(1) > Specificity _
    Then ' reworked good axle
      InspectionStatus(1) = "Reworked"
      NumberAxlesReworked = NumberAxlesReworked + 1
    End If
    Exit Function
  End If
  If AxleDiameter > UpperAxleTolerance _
  Then 'axle should be reworked
    If Random(2) > Sensitivity _
    Then ' should be reworked
      FailedAxleInspection = True
    Else
      InspectionStatus(1) = "Reworked"
      NumberAxlesReworked = NumberAxlesReworked + 1
    End If
    Exit Function
  End If
  FailedAxleInspection = True
  If AxleDiameter < LowerAxleTolerance _
  Then 'should be scraped
    If Random(3) > Sensitivity _
    Then ' reworked and scraped
      NumberAxlesReworked = NumberAxlesReworked + 1
    End If
  End If
End Function
```

Figure 23.4 Inspection—function *FailedAxleInspection*.

```
Function FailedWheelInspection(ByVal Index As Integer, _
                               ByVal WheelDiameter As Single) _
                               As Boolean
   FailedWheelInspection = False
   InspectionStatus(Index) = "Passed"
   NumberWheelInspections = NumberWheelInspections + 1
   If WheelDiameter < UpperWheelTolerance And _
      WheelDiameter > LowerWheelTolerance _
   Then 'good wheel
      If Random(4) > Specificity _
      Then  ' good wheel reworked
         InspectionStatus(Index) = "Reworked"
         NumberWheelsReworked = NumberWheelsReworked + 1
      End If
      Exit Function
   End If
   If WheelDiameter < LowerWheelTolerance _
   Then 'wheel should be reworked
      If Random(5) > Sensitivity _
      Then  ' scraped good wheel
         FailedWheelInspection = True
      Else
         InspectionStatus(Index) = "Reworked"
         NumberWheelsReworked = NumberWheelsReworked + 1
      End If
      Exit Function
   End If
   FailedWheelInspection = True
   If WheelDiameter > UpperWheelTolerance _
   Then 'wheel should be scraped
      If Random(6) > Sensitivity _
      Then  ' kept bad wheel
         NumberWheelsReworked = NumberWheelsReworked + 1
      End If
   End If
End Function
```

Figure 23.5   Inspection—function *FailedWheelInspection*.

FailedWheelInspection: In Figure 23.5, the number of wheel inspections is incremented by 1. *False* is returned when a wheel passes inspection or is reworked. Otherwise, *True* is returned. *InspectionStatus(2)* will be either *Passed* or *Reworked*. The three scenarios are

1. Wheel diameter within specification limits: The wheel passes inspection. However, because of a *specificity* error, the wheel may be incorrectly classified as needing rework.
2. Wheel diameter above upper specification limits: The wheel should be scrapped. However, because of a *sensitivity* error, the wheel may be incorrectly reworked before being scrapped.
3. Wheel diameter below lower specification limits: The wheel should be reworked. However, because of a *sensitivity* error, the wheel may be scrapped.

FailedAssemblyInspection: As shown in Figure 23.6, the final assembly inspection is assumed to be without error. Also, a reworked axle is automatically assumed to be within tolerance, and therefore a good assembly is produced. When a wheel has been reworked, it is also assumed to fit properly. When the assembly clearance is less than the lower assembly clearance or exceeds upper assembly clearance, the assembly fails inspection.

Calculations: As shown in Figure 23.7, end-of-run summary information on inspections, failures, and unit cost of production is produced.

Output from the above model is presented in Section 17.4. In Section 25.3, the inspection program is used to demonstrate design of experiments as an analysis tool for simulation models.

```
Function FailedAssemblyInspection(ByRef MaterialSet As Transaction)
  Dim _
    Clearance As Single, _
    Index As Integer
  FailedAssemblyInspection = False
  NumberAssemblyInspections = NumberAssemblyInspections + 1
  If InspectionStatus(1) = "Reworked" _
  Then
    Exit Function
  End If
  For Index = 2 To 3
    If InspectionStatus(Index) = "Passed" _
    Then
      Clearance = MaterialSet.A((Index)) -MaterialSet.A(AxleDiameter)
      If Clearance < LowerAssemblyClearance _
      Then
        NumberAssembliesReworked = NumberAssembliesReworked + 1
      End If
      If Clearance > UpperAssemblyClearance _
      Then
        FailedAssemblyInspection = True
        Exit Function
      End If
    End If
  Next Index
End Function
```

Figure 23.6   Inspection—function *FailedAssemblyInspection*.

```
Sub Calculations()
  TotalInspectionCostAxle = NumberAxleInspections * AxleInspectionCost
  AverageInspectionCostAxle = TotalInspectionCostAxle / AssemblyCount
  TotalReworkCostAxle = NumberAxlesReworked * AxleReworkCost
  AverageReworkCostAxle = TotalReworkCostAxle / AssemblyCount
  TotalScrapCostAxle = NumberAxlesScraped * AxleScrapCost
  AverageScrapCostAxle = TotalScrapCostAxle / AssemblyCount
  TotalCostAxles =
    TotalInspectionCostAxle + TotalReworkCostAxle + TotalScrapCostAxle
  AverageCostAxle = TotalCostAxles / AssemblyCount
  TotalInspectionCostWheel = NumberWheelInspections * WheelInspectionCost
  AverageInspectionCostWheel = TotalInspectionCostWheel / AssemblyCount
  TotalReworkCostWheel = NumberWheelsReworked * WheelReworkCost
  AverageReworkCostWheel = TotalReworkCostWheel / AssemblyCount
  TotalScrapCostWheel = NumberWheelsScraped * WheelScrapCost
  AverageScrapCostWheel = TotalScrapCostWheel / AssemblyCount
  TotalCostWheels =
    TotalInspectionCostWheel + TotalReworkCostWheel + TotalScrapCostWheel
  AverageCostWheel = TotalCostWheels / AssemblyCount
  TotalInspectionCostAssembly =
    NumberAssemblyInspections * AssemblyInspectionCost
  AverageInspectionCostAssembly = TotalInspectionCostAssembly / AssemblyCount
  TotalReworkCostAssembly = NumberAssembliesReworked * WheelReworkCost
  AverageReworkCostAssembly = TotalReworkCostAssembly / AssemblyCount
  TotalScrapCostAssembly = NumberAssembliesScraped * AssemblyScrapCost
  AverageScrapCostAssembly = TotalScrapCostAssembly / AssemblyCount
  TotalCostAssemblies =
    TotalInspectionCostAssembly + TotalReworkCostAssembly +
                                            TotalScrapCostAssembly
  AverageCostAssembly = TotalCostAssemblies / AssemblyCount
  AverageCost =
    (TotalCostAxles + TotalCostWheels + TotalCostAssemblies) / AssemblyCount
  Yield = AssemblyCount / Time
End Sub
```

Figure 23.7   Inspection—sub *Calculations*.

23.3 Monitoring Control Charts

Sample statistics such as x, $\bar{x}$, s, p, and so on are the basis for plotting control charts. When individual values of x are plotted as a time-ordered sequence, all observations should be within a $\mu_x \pm 3\sigma_x$ interval in a pattern consistent with the parent distribution, assuming independence of the x_i. The intervals are referred to as the upper and lower control limits and are defined as

$$UCL = \mu_x + 3\sigma_x \quad \text{and} \quad LCL = \mu_x - 3\sigma_x$$

μ_x and σ_x are either target results or estimates based on historical data. A value of x outside either limit signals a "special cause" that requires an explanation. From the *Western Electric Handbook* (1956), the rules for detecting a signal are

1. One point is outside the 3σ control limits.
2. Two of three consecutive points plot beyond the 2σ warning limits on the same side of the center line.
3. Four of five consecutive points plot at a distance of 1σ or beyond from the center line.
4. Eight consecutive points plot on one side of the center line.

With values of μ and σ from a known distribution, $P(x > UCL)$ and $P(x < UCL)$ can be determined. For a large positive shift in μ_x to μ_x', $P(x > UCL)$ will also increase. In a similar manner, a positive shift in standard deviation from σ_x to σ_x' will cause $P(x > UCL)$ to also increase. These probabilities may be easily estimated for samples from a known distribution.

The average run length (ARL) is a measure of lapsed time for the chart to reflect a change (detect a signal) in process mean ($\mu_x => \mu_x'$) or process standard deviation ($\sigma_x => \sigma_x'$). For a sample size $1 < n \leq 10$, $\bar{x}$ and R charts are used. The $\bar{x}$ chart monitors process mean, and the R chart monitors process standard deviation. The control limits for $\bar{x}$ are

$$LCL = \mu_{\bar{x}} - 3\sigma_{\bar{x}}/\sqrt{n} \quad \text{and} \quad UCL = \mu_{\bar{x}} + 3\sigma_{\bar{x}}/\sqrt{n}$$

For R, the control limits are

$$LCL = \mu_R - 3\sigma_R \sim D_3\bar{R} \quad \text{and} \quad UCL = \mu_R + 3\sigma_R \sim D_4\bar{R}$$

where $\bar{R}$ is the average range of all samples and D_3 and D_4 are from quality control tables for range charts. The estimate of μ_x is $\bar{\bar{x}}$, which is the average of all sample $\bar{x}$'s and the estimate of σ_x is $\bar{R}/d/d$. The value of d_2 is also from a range chart table.

The issue of an R chart ARL is investigated for the normal distribution to address the sensitivity of changes to σ. The procedure is the same when the distribution is unknown and begins by computing control limits on the basis of assumed values for μ and σ. Then, use a simulation model to vary the actual value of σ and determine the average number of observations needed to detect a signal. Of course, the average is the ARL for the process.

Source data from spreadsheet *UserInitialed* includes the process mean and variance, the sample size for the range chart and a revised population variance. D_3 and D_4 control chart parameters are available to the program from table definitions. Model *RunLength* is the total number of observations observed by the model. The only model output is the observation-based statistical variable *ARL*.

The only primary event, *ObservationSourceEvent*, in the model is a source arrival that represents a range chart observation. *SampleSize* attributes are each assigned an observation of the random variable and are the basis for determining the sample range. In addition to the primary event subroutine, the following are included in the model:

- *PopulationData* initializes the range chart parameters.
- *AboveCenterLine* processes a range chart observation above the range center line.
- *BelowCenterLine* processes a range chart observation below the range center line.
- *UpdateChart* includes the observation in the chart and monitors the control chart rules described above.
- *RecordARL* records the ARL and resets the range chart to determine the next ARL.

The remainder of this section provides an overview of model procedures and a brief description of program output.

ObservationSourceEvent: In Figure 23.8, *SampleSize* random variables based on the population mean and revised population variance are recorded in attributes. A case structure is used with the sample range to determine where the sample range occurs in the range chart. A sample range outside of the 3σ control limit is a signal that the population variance has changed. Otherwise the chart is updated and then the other signals are considered.

PopulationData: Initializes the control chart on the basis of the process mean, process variance, and the sample size for the range chart (Figure 23.9). Thirty observations of *SampleSize*

```
' SECONDARY EVENTS FOR EACH PRIMARY EVENT
Sub ObservationSourceEvent(ByRef Observation As Transaction)
  Dim _
    index As Integer
  RunLength = RunLength + 1
  Observation.DimAttributes (SampleSize)
  For index = 1 To SampleSize
    Observation.Assign (index), distribution.Normal(Mean, RevisedStdDev, 1)
  Next index
  Max = Observation.A(1)
  Min = Observation.A(1)
  For index = 1 To SampleSize
    If Observation.A(index) > Max Then
      Max = Observation.A(index)
    End If
    If Observation.A(index) < Min Then
      Min = Observation.A(index)
    End If
  Next index
  SampleRange = Max - Min
  Select Case SampleRange
    Case Is > CL(3)
      Call RecordARL
    Case Is < CL(-3)
      Call RecordARL
    Case Is > RangeMean
      Call AboveCenterLine
    Case Else
      Call BelowCenterLine
  End Select
  Call UpdateChart(UpperObservationSequence())
  Call UpdateChart(LowerObservationSequence())
End Sub
```

Figure 23.8 Quality control—*ObservationSourceEvent*.

```
Sub PopulationData()
  Dim _
    Job As Transaction, _
    index As Integer, _
    Jndex As Integer, _
    SumOfRanges As Single, _
    AverageRange As Single, _
    SigmaEstimate As Single, _
    D3 As Single, _
    D4 As Single
  Set Job = simulator.NewTransaction
  Job.DimAttributes (SampleSize)
  For index = 1 To 30
    For Jndex = 1 To SampleSize
        Job.Assign (Jndex), distribution.Normal(Mean, StdDev, 1)
    Next Jndex
    Max - Job.A(1)
    Min = Job.A(1)
    For Jndex = 1 To SampleSize
      If Job.A(Jndex) > Max Then
        Max = Job.A(Jndex)
      End If
      If Job.A(Jndex) < Min Then
        Min = Job.A(Jndex)
      End If
    Next Jndex
    SampleRange = Max - Min
    SumOfRanges = SumOfRanges + SampleRange
  Next index
  AverageRange = SumOfRanges / 30
  D3 = D3RChartTable.F(SampleSize)
  D4 = D4RChartTable.F(SampleSize)
  RangeMean = AverageRange
  SigmaEstimate = (D4 * AverageRange - AverageRange) / 3
  For index = 1 To 3
    CL(index) = AverageRange + SigmaEstimate * index
  Next index
  SigmaEstimate = (D3 * AverageRange - AverageRange) / 3
  For index = -3 To -1
    CL(index) = AverageRange + SigmaEstimate * Abs(index)
  Next index
  RunLength = 0
  For index = 1 To 2
    For Jndex = 1 To 5
      UpperObservationSequence(index, Jndex) = 0
      LowerObservationSequence(index, Jndex) = 0
    Next Jndex
  Next index
  UpperObservationSequence(1, 0) = 5
  UpperObservationSequence(2, 0) = 3
  LowerObservationSequence(1, 0) = 5
  LowerObservationSequence(2, 0) = 3
  UpperObservationCount = 0
  LowerObservationCount = 0
End Sub
```

Figure 23.9   Quality control—sub *PopulationData*.

random variables are used to estimate the population range. A range standard deviation estimate is calculated from the estimate of the average population range and values from a design environment for event-driven simulation (DEEDS) table for D_3 and D_4. The control chart limits and arrays used to monitor chart signals are initialized. Notice that the zero position in the *UpperObservationSequence* and *UpperObservationSequence* array indicates the number of observations for the corresponding control limit signal.

RecordARL: As shown in Figure 23.10, the number of observations since the last signal is the run length. It is recorded in the *ARL* statistic, and then the control signal arrays are reinitialized.

AboveCenterLine: As shown in Figure 23.11, each time an observation is above the average range, *LowerObservationCount* is reset to 0 and *UpperObservationCount* is incremented by 1. When *UpperObservationCount* is 8, the signal indicates a change in population mean. Otherwise, the observation is mapped to the appropriate control limit interval by incrementing the appropriate interval counter.

BelowCenterLine: Figure 23.12 is the mirror image of the *AboveCenterLine* sub.

```
Sub RecordARL()
  Dim _
    index As Integer, _
    Jndex As Integer
  ARLStatistic.TimeBetweenArrivals
  RunLength = 0
  For index = 1 To 2
    For Jndex = 1 To 5
      UpperObservationSequence(index, Jndex) = 0
      LowerObservationSequence(index, Jndex) = 0
    Next Jndex
  Next index
  UpperObservationCount = 0
  LowerObservationCount = 0
End Sub
```

Figure 23.10 Quality control—sub *RecordARL*.

```
Sub AboveCenterLine()
  Dim _
    index As Integer, _
    Jndex As Integer, _
    TempCounter As Integer
  LowerObservationCount = 0
  UpperObservationCount = UpperObservationCount + 1
  If UpperObservationCount = 8 _
  Then
    Call RecordARL
    Exit Sub
  End If
  Select Case SampleRange
    Case Is > CL(2)
      index = UpperObservationSequence(2, 0)
      UpperObservationSequence(2, index) = 1
    Case Is > CL(1)
      index = UpperObservationSequence(1, 0)
      UpperObservationSequence(1, index) = 1
  End Select
End Sub
```

Figure 23.11 Quality control—sub *AboveCenterLine*.

```
Sub BelowCenterLine()
  Dim _
    index As Integer, _
    Jndex As Integer, _
    TempCounter As Integer
  UpperObservationCount = 0
  LowerObservationCount = LowerObservationCount + 1
  If LowerObservationCount = 8 _
  Then
    Call RecordARL
    Exit Sub
  End If
  Select Case SampleRange
    Case Is < CL(-2)
      index = LowerObservationSequence(2, 0)
      LowerObservationSequence(2, index) = 1
    Case Is < CL(-1)
      index = LowerObservationSequence(1, 0)
      LowerObservationSequence(1, index) = 1
  End Select
End Sub
```

Figure 23.12   Quality control—sub *BelowCenterLine*.

```
Sub UpdateChart(ByRef ObservationSequence() As Byte)
  Dim _
    index As Byte, _
    Jndex As Byte, _
    TempCounter As Byte
  For index = 1 To 2
    TempCounter = 0
    For Jndex = 1 To ObservationSequence(index, 0)
      TempCounter = TempCounter + ObservationSequence(index, Jndex)
    Next Jndex
    If TempCounter = ObservationSequence(index, 0) -1 _
    Then
      Call RecordARL
      Exit Sub
    End If
  Next index
  For index = 1 To 2
    For Jndex = 1 To ObservationSequence(index, 0) -1
      ObservationSequence(index, Jndex) = _
                          ObservationSequence(index, Jndex -1)
    Next Jndex
    ObservationSequence(index, Jndex) = 0
  Next index
End Sub
```

Figure 23.13   Quality control—*UpdateChart*.

UpdateChart: This program segment tests the Western Electric condition for $n-1$ of n observations outside the control limits. As shown in Figure 23.13, either *UpperObservationSequence* or *LowerObservationSequence* becomes *ObservationSequence*. For each control limit, *ObservationSequence* is evaluated for a variance change signal. If a signal is detected, ARL is recorded. Otherwise, *ObservationSequence* is updated and is ready for the next observation. The last step is to remove an observation from the *ObservationSequence* to make room for the next observation.

Statistic	(O) ARL
Number In	215
Min Value	1.000
Max Value	375.000
Average	46.419
StdDev	46.455

Figure 23.14 Quality control—statistics output.

As shown in Figure 23.14, there were 215 runs. A total of 10,000 observations were recorded for these runs. The minimum run length was one observation and the maximum run length was 375 observations. ARL was 46.1 with a standard deviation of 46.5. Of course, other distributions may be analyzed in an identical manner. For situations where the population parameters are unknown, raw data can be used to estimate the population range.

Reference

1. Montgomery, D. C., *Introduction to Statistical Quality Control*, 5th ed. New York: Wiley, 2005.

Chapter 24

Supply Chain Models

24.1 Introduction

Supply-chain management integrates supply and demand management within and across companies. It encompasses the planning and management of all sourcing and procurement, conversion activities, as well as logistics management activities. It also includes coordination and collaboration with suppliers, intermediaries, third-party service providers, and customers.

Logistics management is that part of supply-chain management that plans, implements, and controls the efficient flow and storage of goods, services, and related information between the point of origin and the point of consumption to meet customers' requirements. Logistics activities typically include inbound and outbound transportation management, fleet management, warehousing, materials handling, order fulfillment, logistics network design, inventory management, supply/demand planning, and management of third-party providers of logistics services. This chapter demonstrates the application of simulation modeling to a variety of logistics models.

24.2 Port Operation

This example was first proposed by Schriber [1]. A simulation model of the same problem using the SLAM II network was given by Pritsker [2]. An embellished version using SIMSCRIPT was provided by Law and Larmcy [3]. This presentation is similar to the Law and Larmcy version.

Three classes of tankers arrive at a port to be loaded with crude oil. The port's loading facilities can accommodate a maximum of three tankers. Tankers arrive randomly every *UniformContinuous* (4, 18) hours and loading times are dependent upon the tanker's size. Table 24.1 summarizes loading time requirements.

The port operates with only one tug to berth and deberth tankers. An arriving tanker waits outside the harbor until the tug is available to move the tanker to the loading facility. After a tanker is loaded, the tug moves it out of the harbor. Moving a tanker into or out of the harbor requires approximately one hour. Priority access to the tug is given to a deberthing operation. If no tankers are waiting to be deberthed, the tug embarks on a berthing operation if any tankers are waiting. Otherwise, the tug is idle near the berths. The time for a tug to reach a waiting tanker or to return to the harbor (empty) after completing a deberthing operation is 0.25 hours.

Table 24.1 Loading Times for Tankers

Category	Percentage of Occurrence	Loading Time (Hours)
1	25	UniformContinuous (16, 20)
2	25	UniformContinuous (21, 27)
3	50	UniformContinuous (32, 40)

The port operation is hampered by frequent storms that occur every *Exponential* (48) hours and a storm lasts *UniformContinuous* (2, 6) hours. The tug may not start a berthing or deberthing operation during a storm. If a storm develops while the tug is traveling to a waiting tanker outside the harbor, it must return to the berths. In contrast, the tug must always complete an ongoing berthing and deberthing operation regardless of weather conditions.

The run length for the model is 9480 hours. Performance metrics output must include the time a tanker spends in port, as well as information on port utilization, the average number of tankers waiting for berthing or deberthing, and tug usage. Tankers and the tug are represented by transactions. In addition to creation time, which is used to calculate time in the system for the tanker, the following two attributes define an arriving tanker:

1 = *Class* (1, 2, or 3)
2 = *ServiceTime* (loading time)

Both *Class* and *ServiceTime* for an arriving tanker are from user-defined PDFs. Other random variables are modeled from distribution functions. The model includes queues for arriving tankers, tankers ready to be deberthed, and an idle tug. There is also a statistical variable for each class of tanker to collect data on time in the system.

User-defined Boolean variables are *TugBusy, StormConditions*, and *TugDestination. TugBusy* is *True* when the tug is berthing or deberthing a tanker. *TugDestination* is a Visual Basic (VB) String variable assigned values *Bay, Port*, and *Idle*. Table 24.2 describes the primary events in the model.

In addition to subroutines for each of the primary events, there are two subroutines, *ToBayDelay* and *ToPortDelay*, that detail events related to a tug arriving in the bay and in the port, respectively.

The remainder of this section provides an overview of model procedures and a brief description of program output.

StormSourceEvent: This event handler models an arriving storm. As shown in Figure 24.1, upon arrival of a storm, *StormSource* is suspended to ensure the proper interarrival time between storms. The storm is initiated by *StormDelay.Start Storm*. The Boolean variable *StormConditions* is set to *True* and the status of the tug is considered. If *TugDestination* is "Port," the tug continues the trip. Of course, if the tug is "Idle," it remains idle. If *TugDestination* is "Bay," the tug either returns to port or continues the deberthing operation, as appropriate. *TugBusy* is equal to *True* when the tug is moving a tanker. To immediately return to the port, the tug's trip to the bay is interrupted (removed from the primary events list, or PEL). The time for the return trip to port is calculated and the statement

TugDelay.Start Tug, DelayTime

begins the tug's journey back to the port.

Table 24.2   Port Operations—Primary Events

Event	Description
StormSource	Storm arrival
TankerSource	Tanker arrival
PortFacility	Loaded tanker
TugDelay	Finished travel to the bay or the port
StormDelay	End of the storm

```
Sub StormSourceEvent(ByRef Storm As Transaction)
  Dim _
    DelayTime As Single
  StormSource.Suspend
    'A tug transporting a tanker continues
    'otherwise, it returns to port.
  StormDelay.Start Storm
  StormConditions = True
  If TugDestination = "Port" _
  Then
     Exit Sub 'continue activity
  End If
  If Not TugBusy And TugDestination = "Bay" _
  Then
    Set Tug = TugDelay.Interrupt(1)
    DelayTime = 0.25 -(Tug.TimeStamp - Deeds.CurrentTime)
    TugDestination = "Port"
    TugDelay.Start Tug, DelayTime
  End If
End Sub
```

Figure 24.1   Port operation—*StormSourceEvent*.

TankerSourceEvent: As shown in Figure 24.2, attribute *ServiceTime* is assigned a value that is dependent on tanker attribute *Class*. The tanker then joins *ArrivalQueue*. The following mutually exclusive scenarios are then considered in the order presented:

1. If any one or more of the conditions tug is busy, port facility has no idle berths, or storm conditions exist, the tanker waits in the queue.
2. If the tug is in *TugQueue* (idle), it is dispatched to the bay to begin berthing of the arriving tanker.
3. If the tug is destined to port without a tanker and no tankers are waiting for deberthing, the tug returns to the bay to berth the tanker with an appropriate dispatch time. Recall that when *TugBusy* equal *False* it indicates the tug is not moving a tanker.

It is important that scenario 1 is considered first. For example, if scenario 2 is considered before scenario 1, the tug may be dispatched during storm conditions. Also, there is no reason to dispatch the tug if all berths are occupied. The order of consideration of scenarios 2 and 3 is not critical.

```
Sub TankerSourceEvent(ByRef Tanker As Transaction)
  Dim _
    DelayTime As Single
  Tanker.Assign Class, TankerClassPDF.RV
  Select Case Tanker.A(Class)
   Case 1
     Tanker.Assign ServiceTime, UniformContinuous(16, 20)
   Case 2
     Tanker.Assign ServiceTime, UniformContinuous(21, 27)
   Case Else
     Tanker.Assign ServiceTime, UniformContinuous(32, 40)
  End Select
  ArrivalQueue.Join Tanker
  If TugBusy Or _
     PortFacility.NumberOfIdleServers = 0 Or _
     StormConditions _
  Then
     Exit Sub
  End If
  If TugQueue.Length > 0 _
  Then  'Tug idle in port queue
    TugBusy = False
    TugDestination = "Bay"
    TugDelay.Start TugQueue.Depart, 0.25
    Exit Sub
  End If
  If TugDestination = "Port" And _
     Not TugBusy And DepartureQueue.Length = 0 _
  Then ' Inbound tug w/o tanker turns around
    Set Tug = TugDelay.Interrupt(1)
    DelayTime = 0.25 - (Tug.TimeStamp - CurrentTime)
    TugDestination = "Bay"
    TugDelay.Start Tug, DelayTime
  End If
End Sub
```

Figure 24.2   Port operation—*TankerSourceEvent*.

PortFacilityEvent: As shown in Figure 24.3, if the tug is in *TugQueue* when the loading operation is finished, the tug departs its queue and becomes busy deberthing the tanker. Deberthing begins as the tanker disengages the port and begins the trip to the bay. Otherwise, the tanker joins *DepartureQueue* while maintaining its position in the port. The following mutually exclusive scenarios are then considered:

1. If the tug is busy or storm conditions exist, the tanker's deberthing is now dependent on other primary event(s).
2. If the tug is destined to port, the tanker's deberthing is now dependent on other primary event(s).
3. If the tug is destined to the bay to begin a berthing operation (i.e., idle), the tug is interrupted and returns to the bay for a deberthing operation. Notice that this program segment gives priority to deberthing.

From a logical perspective, the order that these scenarios are considered is unimportant. For program efficiency, the scenarios should be ordered by frequency of occurrence.

TugDelayEvent: The tug is delayed for a trip to the bay or port. As shown in Figure 24.4, upon completion of the delay, this procedure determines where the tanker has arrived and calls

```
Sub PortFacilityEvent(ByRef Tanker As Transaction)
  Dim _
    DelayTime As Single
  If TugQueue.Length > 0 _
  Then  'Tug is at the port
    TugDestination = "Bay"
    TugBusy = True
    Set Tug = TugQueue.Depart
    Set Tug = Tanker
    PortFacility.Disengage Tanker
    TugDelay.Start Tug, 1
    Exit Sub
  End If
  DepartureQueue.Join Tanker
  If TugBusy Or StormConditions _
  Then  'storm or transporting tanker
    Exit Sub
  End If
  If TugDestination = "Port" _
  Then
    Exit Sub ' Tug already inbound
  End If
  If TugDestination = "Bay" And Not TugBusy _
  Then  'Outbound tug w/o tanker returns to deberth tanker
    Set Tug = TugDelay.Interrupt(1)
    DelayTime = 0.25 - (Tug.TimeStamp - CurrentTime)
    TugDestination = "Port"
    TugDelay.Start Tug, DelayTime
  End If
End Sub
```

Figure 24.3   Port operation—*PortFacilityEvent*.

```
Sub TugDelayEvent(ByRef Tug As Transaction)
  If TugDestination = "Bay" _
  Then
    Call ToBayDelay(Tug)
  Else
    Call ToPortDelay(Tug)
  End If
End Sub
```

Figure 24.4   Port operation—*TugDelayEvent*.

the corresponding sub to process related secondary events. These related events could be included in the event handler. They are separated to facilitate development of the model.

ToBayDelay: The tug has arrived in the bay. The destination is now the port. If, as shown in Figure 24.5, *TugBusy* is *True*, a deberthing operation is complete and cycle time data for the tanker is collected. The following mutually exclusive return scenarios are considered:

1. If storm conditions exist, the tug must return to port.
2. If a tanker is waiting for berthing, remove the tanker from *ArrivalQueue*, indicate that the tug is busy, and begin the berthing operation. Otherwise the idle tug begins the return trip to the port.

The order of consideration for scenarios in this procedure is critical.

```
Sub ToBayDelay(ByRef Tug As Transaction)
  TugDestination = "Port"
  If TugBusy _
  Then
    TankerCycleTimeStatistic(Tug.A(Class)).Collect _
                                        Tug.SystemTime
  End If
  If StormConditions _
  Then
    TugBusy = False
    TugDelay.Start Tug, 0.25
    Exit Sub
  End If
  If ArrivalQueue.Length > 0 _
  Then ' tanker in arrival queue
    Set Tug = ArrivalQueue.Depart
    TugBusy = True
    TugDelay.Start Tug, 1
    Exit Sub
  End If
  TugBusy = False    ' back to port idle
  TugDelay.Start Tug, 0.25
End Sub
```

Figure 24.5   Port operation—sub *ToBayDelay*.

ToPortDelay: The tug has arrived in the port. The destination is now the bay. As shown in Figure 24.6, if *TugBusy* is *True*, a tanker is being berthed and is ready to begin the loading operation. Notice that a new transaction is created to separate the tanker and tug. It is important that the tanker keep the original transaction because its *CreationTime* is needed to collect tanker *SystemTime*. If storm conditions exist, the tug joins *TugQueue* to wait for the passing of the storm. Otherwise, the following mutually exclusive scenarios are considered in the order presented:

1. If a tanker is in *DepartureQueue*, remove the tug from *TugQueue*, indicate *TugBusy* is *True* and *TugDestination* is "Bay." Next, depart the tanker from *DepartureQueue*, disengage it from the port, and begin the deberthing operation.
2. If there is a tanker in *ArrivalQueue*, and the *PortFacility* has an available berth, dispatch the tug to a waiting tanker in the bay. Recall that the tug is idle whenever it is not moving a tanker.
3. If both queues are empty, then the tug becomes idle and joins the *TugQueue*.

It is critical that scenario 1 be considered first to ensure that priority is given to deberthing. The order of consideration for scenarios 2 and 3 is unimportant.

StormDelayEvent: As shown in Figure 24.7, after a storm ends, *StormConditions* is set to *False* and a *NewStormEvent* is scheduled. If *TugQueue* is empty, the tug will be unable to return to "idle" before the storm ends. If the tug is "idle" (i.e., in *TugQueue*), the following mutually exclusive alternatives are considered:

1. If a tanker is in *DepartureQueue*, remove the tug from *TugQueue*, and indicate *TugBusy* is *True* and *TugDestination* is "Bay." Next, depart the tanker from *DepartureQueue*, disengage it from the port, and begin the deberthing operation.
2. If there is a tanker in *ArrivalQueue* and the *PortFacility* has an available berth, dispatch the tug to a waiting tanker in the bay.
3. The tanker has no work to do and remains in *TugQueue*.

Notice the similarities between *ToPortDelay* and *StormDelayEvent*.

```
Sub ToPortDelay(ByRef Tug As Transaction)
  TugDestination = "Bay"
  If TugBusy _
  Then  'Transporting a tanker
    PortFacility.EngageAndService Tug, 1, Tug.A(ServiceTime)
    Set Tug = NewTransaction
  End If
  If StormConditions _
  Then
    TugQueue.Join Tug
    TugDestination = ""
    TugBusy = False
    Exit Sub
  End If
  If DepartureQueue.Length > 0 _
  Then
    TugBusy = True
    Set Tug = DepartureQueue.Depart
    PortFacility.Disengage Tug
    TugDelay.Start Tug, 1
    Exit Sub
  End If
  TugBusy = False
  If ArrivalQueue.Length > 0 And _
     PortFacility.NumberOfIdleServers > 0 _
  Then
    TugDelay.Start Tug, 0.25
    Exit Sub
  End If
  TugQueue.Join Tug
  TugDestination = ""
End Sub
```

Figure 24.6   Port operation—sub *ToPortDelay*.

```
Sub StormDelayEvent(ByRef Storm As Transaction)
  StormSource.ScheduleNextArrival
  StormConditions = False
  If TugQueue.Length = 0 _
  Then
    Exit Sub
  End If
  If DepartureQueue.Length > 0 _
  Then
    Set Tug = TugQueue.Depart
    Set Tug = DepartureQueue.Depart
    PortFacility.Disengage Tug
    TugDelay.Start Tug, 1
    TugBusy = True
    TugDestination = "Bay"
    Exit Sub
  End If
  If ArrivalQueue.Length > 0 And _
       PortFacility.NumberOfIdleServers > 0 _
  Then
    TugBusy = False
    TugDelay.Start TugQueue.Depart, 0.25
    TugDestination = "Bay"
  End If
    ' no work for the tug
End Sub
```

Figure 24.7   Port operation—*StormDelayEvent*.

Queue	Arrival		Departure		Tug	
Number In	870		115		1000	
Number Out	866		115		999	
Residing	4		0		1	
Destroyed	0		0		0	
Removed	0		0		0	
Detached	0		0		0	
Minimum	0		0		0	
Maximum	12		2		1	
	Lq	Wq	Lq	Wq	Lq	Wq
Average	1.606	17.475	0.011	0.859	0.803	7.602
StdDev	2.198	0.780	0.102	0.095	0.398	0.195

Figure 24.8   Port operation—queues output.

Statistic	(O) TankerCycleTime1	(O) TankerCycleTime2	(O) TankerCycleTime3
Number In	212	221	430
Min Value	18.26	23.36	34.26
Max Value	133.95	134.92	146.38
Average	37.60	42.92	56.00
StdDev	23.30	23.77	22.93

Figure 24.9   Port operation—statistics output.

Facility	Port	
Number In	866	
Number Out	863	
Residing	3	
Interrupted	0	
Destroyed	0	
Minimum	0	
Maximum	3	
	AvgBusy	Servers
Average	2.611	
StdDev	0.562	

Figure 24.10   Port operation—facilities output.

Before presenting simulation results, it is important to clarify the assumptions that affect the performance characteristics of the system. The most important assumption in this model is the definition of a busy berth. In this implementation, the berth is not occupied until the tug completes its one-hour trip to the port. In addition, the tanker is "removed" from the berth when the tug begins the trip to the bay. In contrast, the berth could be considered "occupied" and utilization begins when the tug begins a trip to the port with a tanker. In a similar manner, the port could be considered "occupied" until the tanker is returned to the bay. Clearly, berth utilization is much lower for the first case. Simulation results are shown in Figures 24.8 through 24.10.

- Average number of tankers waiting for berthing = 1.61 tankers.
- Average number of tankers waiting for deberthing = 0.01 tankers.
- An average *TugQueue* length of 0.80 indicates that tug utilization is approximately 0.20.
- Average number of busy berths = 2.61.

24.3 Automatic Warehouse Operation

This example, adapted from [4], is the simulation of an automatic warehouse that stores goods for later redistribution to different outlets. A schematic diagram of the operation of the warehouse is shown in Figure 24.11.

Goods arrive at a receiving bay on standard-size pallets. The pallets feed automatically into a lower-level conveyor and are lifted by an elevator to an upper-level conveyor. The upper level then feeds into three corridors that house vertically stacked bins for storing the pallets. When a request for items arrives, the order, on its pallet, is retrieved from a designated corridor and placed on the upper conveyor that moves it to the lower conveyor using a down elevator. The lower conveyor then transports the pallet to a shipping bay.

A computer controls the warehouse operations so that the conveyor is accessible only when a pallet is in front of an entry or exit point. Each of the three corridors has an incoming buffer area that accommodates at most two pallets. If a pallet designated for a given corridor finds the inbound buffer full, the pallet is not allowed to leave the conveyor and waits for the next cycle before a new attempt is made to enter the buffer area.

Requested orders are moved from their bins in the bay to an outbound buffer before being loaded onto the (upper-level) conveyor. The outbound buffer can accommodate only one pallet. In a similar manner, loading from the outbound buffer to the conveyor occurs only if an empty space is available in front of the buffer area.

Pallets are moved to and from the buffers of each corridor by a single overhead crane. The crane alternates between inbound and outbound buffers. After moving a pallet from an inbound

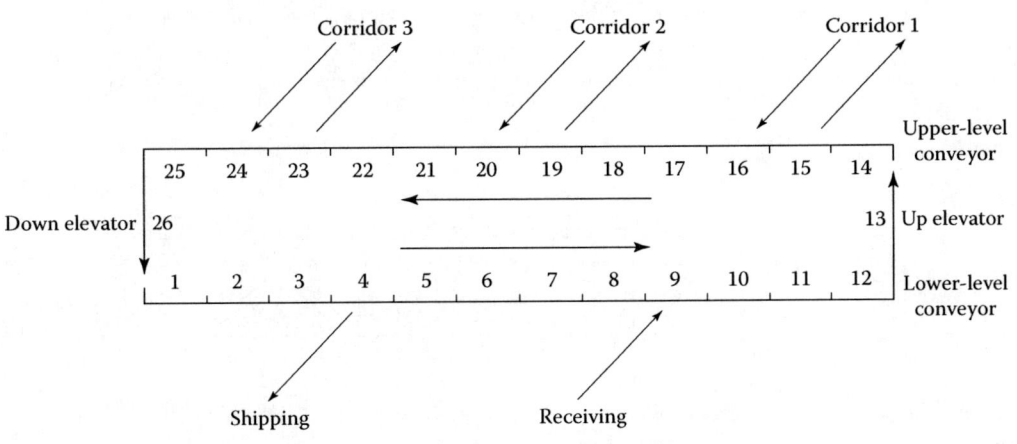

Figure 24.11 Warehouse layout.

buffer to a storage bin, the crane fills an order request by moving it to the outbound buffer. The latter occurs only when the outbound buffer is empty. Otherwise, the crane continues to move any pallets from the inbound buffer to the bins. Pallets arrive at the shipping bay in a Poisson stream at a rate of two pallets per minute. Requests for items also follow a Poisson distribution with mean one pallet per minute. Received pallets are destined to any one of the three corridors with equal probabilities. The upper and lower conveyors, which are 24 meters long, move at a speed of 20 meters per minute. A pallet occupies a two-meter space on the conveyor. Additionally, each of the two end elevators accommodates one pallet at a time.

For the simulation, an elevator is considered a two-meter conveyor space. As a result, the entire conveyor may be represented by 26 two-meter segments moving at a speed of ten segments per minute. As shown in Figure 24.12, conveyor segment 4 is the shipping bay and the receiving bay is segment 9. The up and down elevators are segments 16 and 26, respectively. The inbound and outbound buffers of the three corridors are the respective segments (15, 16), (19, 20), and (23, 24). Each corridor is 20 meters long and 8 meters high. The crane has horizontal and vertical speeds of 10 and 1 meters per second, respectively. Both horizontal and vertical motions occur simultaneously. The time to pick or release a pallet is negligible.

An inbound pallet picked by the crane is placed in any of the bins with equal probabilities. In a similar manner, an outbound pallet is also picked from any of the bins with equal probabilities. The model consists of (1) pallets arriving at the shipping area; (2) operation of the corridor's inbound buffer and the link to the conveyor and crane; (3) receipt of requests and the linkage between the crane, outbound buffer, and conveyor; and (4) conveyor movement and linkage with the shipping and receiving bays.

The conveyor, including the end elevators, is represented by an array, *Conveyor()*, of 26 transaction variables. Pallet transactions are assigned to a *Conveyor()* element that represents a conveyor segment. Transaction attribute *CorridorIndex* represents the segment where the pallet leaves the conveyor. For example, when an empty *Conveyor()* element arrives at the receiving area (segment 9), the *Conveyor()* element is assigned a transaction with attribute *CorridorIndex* equal to 1, 2, or 3 (with equal probabilities). The assignment is equivalent to specifying destination segment 15, 19, or 23, respectively. In a similar manner, a pallet (transaction) on the conveyor from one of the outbound buffers (*Conveyor()* segment 16, 20, or 24) will have *Destination* attribute equal to 4, indicating that the pallet is destined for *Conveyor()* segment 4, the shipping bay. In this convention, a conveyor segment is free only when the corresponding *Conveyor()* element is *Nothing*. In the model, the conveyor is advanced in discrete steps representing the length of one segment (two meters). With a speed of 20 meters per minute, the *ConveyorTimer* for forward movement of the conveyor is $2/20 = 0.1$ minute. At the end of each *ConveyorTimer* the model tests for pallets entering and leaving the conveyor.

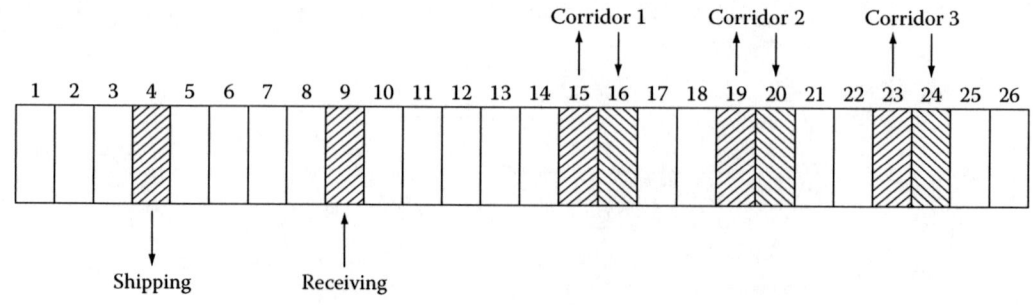

Figure 24.12   Warehouse corridor layout.

Table 24.3 Warehouse Primary Events

Event	Description
ReceiveSource	Warehouse destined pallet arrived
RequestSourceEvent	Shipment request arrived
GeneralCraneFacility	Crane completed service
ConveyorDelay	Conveyor stopped

To demonstrate modeling flexibility, the details of the model exceed the requirements of the problem. For example, pallets in the warehouse are maintained in *CorridorQueue()* when variables for transaction counts would be sufficient. Also, integer values for *Conveyor()* elements would be sufficient to model the system. Because of the warehouse symmetry, the following arrays are used to represent corridors 1, 2, and 3:

- *CorridorQueue()*: Pallets (transactions) in the warehouse
- *CraneQueue()*: Requests (transactions) waiting for processing
- *InBufferQueue()*: Shipping pallet (transaction) waiting for crane access (capacity = 2)
- *OutBufferQueue()*: Shipping pallet (transactions) waiting for conveyor access (capacity = 1)
- *CraneFacility()*: Corridor crane

The transaction attributes for each pallet are

1 = *Destination*
2 = *CorridorIndex*

ShippingBay and *ReceivingBay* are both Visual Basic for Applications (VBA)-defined Const variables. Table 24.3 describes the primary events in the model.

In addition to a subroutine for each primary event, there are four additional subroutines, *ShipOrders, ReceiveArrival, CorridorIn*, and *CorridorOut*. The remainder of this section provides an overview of model procedures and a brief description of program output.

ReceiveSourceEvent: This event handler models an arriving pallet destined for the warehouse. As shown in Figure 24.13, an arrival is randomly assigned a corridor index from *ArrivalsPDF* and the corresponding destination segment using *InCraneCorridorTable*. The arrival then joins *ReceivingQueue* to wait for conveyor access. *ConveyorDelay* controls access to the conveyor.

```
Sub ReceiveSourceEvent(ByRef Pallet As Transaction)
  Dim _
    Index As Integer
  Index = ArrivalsPDF.RV
  Pallet.Assign CorridorIndex, Index
  Pallet.Assign Destination, InCraneCorridorTable.Value(Index)
  ReceivingQueue.Join Pallet
End Sub
```

Figure 24.13 Warehouse—*ReceiveSourceEvent*.

RequestSourceEvent: This event handler models a received order to be shipped. The order, as shown in Figure 24.14, is randomly assigned a corridor index using *DeparturesPDF*. Of course, the destination is *ShippingBay*. The following three conditions for the corresponding corridor are necessary for the order to be processed:
1. The item is in the warehouse.
2. The crane is idle.
3. The out buffer is empty.
When these three conditions are satisfied, a pallet is removed from *CorridorQueue()* and engages the crane for processing. Otherwise, the order joins *CraneQueue()* to wait for processing.

GeneralCraneFacilityEvent: Figure 24.15 models the crane moving a pallet from the inbound buffer to the warehouse or from the warehouse to the outbound buffer. In general, the crane alternates between trips to and from the warehouse. However, when the next "destination" is blocked by buffer capacity, the crane skips that portion of the cycle. As shown in Figure 24.15, upon completion of service, the pallet disengages the crane. Other activities are dependent on the destination of the pallet.

When the crane has finished moving a pallet, the pallet's *Destination* attribute is also the basis for locating the crane. If the pallet destination is *ShippingBay*, the crane has moved from the warehouse to the *OutBuffer()* and the pallet joins *OutBuffer()*. If *InBuffer()* has a pallet, the crane engages the pallet and moves it to the warehouse. Otherwise, the crane is idle because the corresponding *OutBuffer()* (with capacity of 1) is *Full*.

When the crane has moved from *InBuffer()* to the warehouse the same three conditions in *ReceiveSourceEvent* must be satisfied and there must be an order residing in *CraneQueue()*. If all of these conditions are satisfied, an order is removed from *CraneQueue()* and a pallet from *CorridorQueue()* engages the crane for processing. If one or more of the conditions is not satisfied and a pallet is in *InBufferQueue()*, it becomes the next pallet to be processed. An inbound pallet departs *InBufferQueue()* and engages the crane for transporting to the warehouse. The crane is idle if it is unable to find a waiting pallet. Notice that before the crane moves the pallet to the warehouse, the *Destination* attribute is changed to *ShippingBay*.

```
Sub RequestsSourceEvent(ByRef Order As Transaction)
  Dim _
    ID As Integer
  ID = DeparturesPDF.RV
  Order.Assign CorridorIndex, ID
  If CorridorQueue(ID).Length > 0 And _
    CraneFacility(ID).NumberOfIdleServers > 0 And _
    Not (OutBufferQueue(ID).Full) _
  Then
    CraneFacility(ID).EngageAndService _
                        CorridorQueue(ID).Depart, 1, _
      WorksheetFunction.Max((UniformDiscrete(1, 20) / 10), _
                            UniformDiscrete(0, 8))
  Else
    CraneQueue(ID).Join Order
  End If
End Sub
```

Figure 24.14 Warehouse—*RequestSourceEvent*.

```
Sub GeneralCraneFacilityEvent(ByRef Pallet As Transaction)
  Dim _
    Index As Integer
  Index = Pallet.A(CorridorIndex)
  CraneFacility(Index).Disengage Pallet
  Select Case Pallet.A(Destination)
   Case ShippingBay
    OutBufferQueue(Index).Join Pallet
    If InBufferQueue(Index).Length > 0 _
    Then ' to warehouse moves preceed from warehouse moves
      CraneFacility(Index).EngageAndService _
        InBufferQueue(Index).Depart, 1, _
          WorksheetFunction.Max((UniformDiscrete(1, 20) / 10), _
                                      UniformDiscrete(0, 8))
    End If ' outbound buffer full => crane becomes idle
   Case Else
      CorridorQueue(Index).Join Pallet         ' now in warehouse
      Pallet.Assign Destination, ShippingBay ' next destination
      If OutBufferQueue(Index).Length = 0 And _
        CraneQueue(Index).Length > 0 _
      Then ' move from warehouse
        CraneQueue(Index).Remove 1, 1
        CraneFacility(Index).EngageAndService _
          CorridorQueue(Index).Depart, 1, _
            WorksheetFunction.Max((UniformDiscrete(1, 20) / 10), _
                                        UniformDiscrete(0, 8))
      Else ' or from inbuffer
        If InBufferQueue(Index).Length > 0 _
        Then
          CraneFacility(Index).EngageAndService _
            InBufferQueue(Index).Depart, 1, _
              WorksheetFunction.Max((UniformDiscrete(1, 20) / 10), _
                                          UniformDiscrete(0, 8))

        End If
      End If
  End Select
End Sub
```

Figure 24.15 Warehouse—*GeneralCraneFacilityEvent*.

Crane transport time to move a pallet from the warehouse to the outbound buffer or from the inbound buffer to the warehouse is the maximum of the simultaneous horizontal and vertical movements of the crane. The time is calculated using Excel's *WorksheetFunction. Max()* method. For simplicity, the time for the crane to move between the warehouse and the conveyor without a pallet is negligible. To process an order, a pallet is removed from warehouse *CraneQueue()* and an order departs *CorridorQueue()* and engages the crane for a trip from the warehouse. Service time is calculated using Excel's *WorksheetFunction.Max()* method.

ConveyorDelayEvent: This model segment represents the movement of the conveyor. As shown in Figure 24.16, after the conveyor delay is finished, the relative position of each pallet on the conveyor is updated. The transaction in location 26 is moved to a temporary location then each transaction from 1 to 25 is advanced one position. Finally, the transaction in the temporary position is moved to position 1. The conveyor is checked for a pallet at the shipping bay or receiving bay and then for a pallet in the warehouse corridors ready to be moved to/from the conveyor. After a pallet moves to/from the conveyor, the delay advances the conveyor by one position. Subroutines *CorridorOut* and *CorridorIn* describe the details of warehouse access to and from the conveyor.

```
Sub ConveyorDelayEvent(ByRef ConveyorTimer As Transaction)
  Dim _
    Index As Integer, _
    Temp As Transaction
  Set Temp = Conveyor(26)
  For Index = 25 To 1 Step -1
    Set Conveyor(Index + 1) = Conveyor(Index)
  Next Index
  Set Conveyor(1) = Temp
    'Check for Pallet at predetermined slots
  Call ShipOrders
  Call ReceiveArrival
  For Index = 1 To 3
   Call CorridorIn(Index)
   Call CorridorOut(Index)
  Next Index
  ConveyorDelay.Start ConveyorTimer
End Sub
```

Figure 24.16 Warehouse—*ConveyorDelayEvent*.

```
Sub ShipOrders()
  If Not Conveyor(ShippingBay) Is Nothing _
  Then
    If Conveyor(ShippingBay).A(Destination) = ShippingBay _
    Then
      Set Conveyor(ShippingBay) = Nothing
    End If
  End If
End Sub
```

Figure 24.17 Warehouse—sub *ShipOrders*.

```
Sub ReceiveArrival()
  If ReceivingQueue.Length > 0 And _
     Conveyor(ReceivingBay) Is Nothing _
  Then
    Set Conveyor(ReceivingBay) = ReceivingQueue.Depart
  End If
End Sub
```

Figure 24.18 Warehouse—sub *ReceiveArrival*.

ShipOrders: This program segment examines the shipping bay location for orders to ship. In Figure 24.17, a pallet with destination *ShippingBay* is removed from the conveyor and the *ShippingBay* location is set equal to *Nothing* to indicate an empty location.

ReceiveArrival: This program segment examines *ReceivingQueue* for an order destined to the warehouse. If, in Figure 24.18, there is a warehouse-destined pallet and the receiving bay conveyor segment is empty, the pallet is moved from *ReceivingQueue* to *ReceivingBay*.

CorridorIn: This program segment links the conveyor to inbound buffers for pallets destined for the warehouse. *CorridorIn* is invoked three times, once for each corridor. In Figure 24.19, conveyor *SegmentID* is obtained from the *InCraneCorridorTable*. If there is a pallet destined to the conveyor *SegmentID* and *InBufferQueue()* space is available, the pallet is moved either to *InBufferQueue()*

```
Sub CorridorIn(ByVal Index As Integer)
   Dim _
      SegmentID As Integer
   SegmentID = InCraneCorridorTable.Value(Index)
   If Not (Conveyor(SegmentID) Is Nothing) _
   Then
      If Conveyor(SegmentID).A(Destination) = SegmentID And _
         Not InBufferQueue(Index).Full _
      Then
         If CraneFacility(Index).NumberOfIdleServers > 0 _
         Then
            CraneFacility(Index).EngageAndService _
                              Conveyor(SegmentID), 1, _
            WorksheetFunction.Max((UniformDiscrete(1, 20) / 10), _
                              UniformDiscrete(0, 8))
         Else
            InBufferQueue(Index).Join Conveyor(SegmentID)
         End If
         Set Conveyor(SegmentID) = Nothing
      End If
   End If
End Sub
```

Figure 24.19 Warehouse—sub *CorridorIn*.

or directly to the warehouse, depending on whether the crane is idle. When the pallet is removed from the conveyor, the corresponding conveyor segment is changed to *Nothing*. If the crane is idle, the pallet engages the crane for a trip to the warehouse. As described in *GeneralCraneFacilityEvent*, service time is calculated using Excel's *WorksheetFunction.Max()* method.

CorridorOut: This program segment links the conveyor to inbound buffers for pallets destined for the *ShippingBay*. In a similar manner to *CorridorIn*, *CorridorOut* is invoked once for each corridor, and conveyor *SegmentID* is obtained from the *OutCraneCorridorTable*. In Figure 24.20, if space is available in conveyor *SegmentID* and *OutBufferQueue()* has a pallet, the pallet departs *OutBufferQueue()* and moves to the conveyor space. Because space is now available in *OutBufferQueue()*, the crane processes an order if the following conditions are satisfied:

1. The crane is idle.
2. An order is in *CraneQueue*.
3. A pallet is in *CorridorQueue* (warehouse).

In Figure 24.21, the average output buffer length is less than 0.05 for each buffer; therefore, an outbound buffer of one pallet does not represent a bottleneck in the warehouse operation. On the contrary, the average input buffer length for each buffer in Figure 24.22 is greater than 1.90 (capacity of 2.0) and the number of pallets waiting in the receiving bay (140) seems to be increasing. Notice also in Figure 24.23 that crane utilization exceeds 0.99 for each crane. The possible causes for this situation may be attributed to the following factors: (1) the in buffers may be too small and (2) the conveyor or crane speed may be too slow. More experimenting is necessary to assess the effects of these factors. Also, including a time-based statistical variable in the model can provide additional information critical to improving performance characteristics of the system. Recall that in *CorridorIn* and *CorridorOut* descriptions, the model assumes that travel time to the warehouse or buffer is negligible when the crane is idle. However, as shown in the *InBuffer()* and *Outbuffer()*, the crane does alternate in and out of the warehouse with a pallet. Therefore, crane utilization estimates are not based on the instantaneous response assumption.

```
Sub CorridorOut(ByVal Index As Integer)
Dim _
    SegmentID As Integer
  SegmentID = InCraneCorridorTable.Value(Index)
  If Conveyor(SegmentID) Is Nothing And _
     OutBufferQueue(Index).Length > 0 _
  Then
     Set Conveyor(SegmentID) = OutBufferQueue(Index).Depart
  End If
  ' check for busy crane
  If CorridorQueue(Index).Length > 0 And _
     CraneFacility(Index).NumberOfIdleServers> 0 And _
     CraneQueue(Index).Length > 0 _
  Then
     CraneQueue(Index).Remove 1, 1
     CraneFacility(Index).EngageAndService _
       CorridorQueue(Index).Depart, 1, _
         WorksheetFunction.Max((UniformDiscrete(1, 20) / 10), _
                                   UniformDiscrete(0, 8))
  End If
End Sub
```

Figure 24.20   Warehouse—sub *CorridorOut*.

Queue	OutBuffer1		OutBuffer2		OutBuffer3		Receiving	
Number In	13		11		11		189	
Number Out	13		11		11		45	
Residing	0		0		0		144	
Destroyed	0		0		0		0	
Removed	0		0		0		0	
Detached	0		0		0		0	
Minimum	0		0		0		0	
Maximum	1		1		1		144	
	Lq	Wq	Lq	Wq	Lq	Wq	Lq	Wq
Average	0.040	0.285	0.017	0.141	0.052	0.433	71.253	37.502
StdDev	0.196	0.084	0.129	0.042	0.222	0.135	40.779	1.822

Figure 24.21   Warehouse—out buffer queues output.

Queue	InBuffer1		InBuffer2		InBuffer3	
Number In	16		13		13	
Number Out	14		11		11	
Residing	2		2		2	
Destroyed	0		0		0	
Removed	0		0		0	
Detached	0		0		0	
Minimum	0		0		0	
Maximum	2		2		2	
	Lq	Wq	Lq	Wq	Lq	Wq
Average	1.949	11.465	1.978	14.129	1.909	13.636
StdDev	0.242	1.8162	0.160	1.597	0.339	1.678

Figure 24.22   Warehouse—in buffer queues output.

Facility	Crane1		Crane2		Crane3	
Number In	28		24		24	
Number Out	27		23		23	
Residing	1		1		1	
Interrupted	0		0		0	
Destroyed	0		0		0	
Minimum	0		0		0	
Maximum	1		1		1	
	AvgBusy Servers		AvgBusy Servers		AvgBusy Servers	
Average	0.998		0.999		0.988	
StdDev	0.045		0.032		0.109	

Figure 24.23   Warehouse—facilities output.

24.4 Cross Dock Operations

The cross dock program was proposed by the logistics group at Averitt Express, Cookeville, Tennessee. In the cross docking environment, arriving freight, measured in handling units, is moved from a trailer at a stripping door to a destination trailer at a loading door. An arriving trailer is assigned one of several stripping doors; similarly, loading doors are permanently assigned a destination city. Depending on the total number of handling units from all sources to a given destination, there may be more than one loading door for a destination. Freight arriving at a stripping door is parsed to handling units destined to different loading doors, as determined by the destination.

The decision variables are x_{ik} and x_{jl} (which doors are stripping doors and which loading doors are assigned to which destinations).

$$\in I \; origin \; doors$$
$$j \in J \; destination \; doors$$
$$k \in K \; origins$$
$$l \in L \; destinations$$
$$x_{ik} \text{ - } origin \; door \; i \; assigned \; to \; origin \; k$$
$$x_{ji} \text{ - } destination \; door \; j \; assigned \; to \; destination \; l$$
$$d_{ij} \text{ - } distance \; from \; door \; i \; to \; door \; j$$
$$a_{kl} \text{ - } number \; of \; pallets \; from \; origin \; k \; to \; destination \; l$$

The feasibility constraints are resolved by ensuring that the dock configurations satisfy the first four equations and workload data satisfies the last two.

$$\sum_k x_{ik} = 1 \; for \; i \in I, each \; origin \; door \; assigned$$
$$\sum_i x_{ik} = 1 \; for \; k \in K, each \; origin \; assigned \; a \; door$$
$$\sum_l x_{jl} = 1 \; for \; j \in J, each \; destination \; door \; assigned$$

$$\sum_j x_{jl} = 1 \ for\ l \in L,\ each\ destination\ assigned\ a\ door$$

$$\sum_l a_{kl} = S_k \ for\ k \in K,\ all\ pallets\ have\ an\ origin$$

$$\sum_k a_{kl} = D_l \ for\ l \in L,\ all\ pallets\ have\ a\ destination$$

The performance metric for the model is the total distance that handling units are moved. Of course, the "best" configuration is determined by a workload that is expressed as a time-dependent arrival of freight. Additional constraints reflected in the simulation model are as follows:

1. Handling units processing is based on the dock configuration for arriving and departing trailers.
2. The time to move handling units between doors is a function of traffic density as reflected in *TowMotorVelocityTable* and *TowMotorTimerTable*.
3. The number of tow motors on the dock as reflected in *TowMotorCountTable*.

An array *StripDoorID()* maps strip door facilities to an actual dock location. *StripDoorFacility()* is an array of facilities used to monitor strip door activity. *YardQueue* maintains a time-ordered list of waiting trailers. *TowMotorFacility* monitors all tow motors on the dock. The maximum number of tow motors allowed in the system is controlled by altering the number of servers in *TowMotorFacility*. At any time, the tow motor count is controlled by *TowMotorTimerTable*. *AllTrailorsArrived* is a Boolean variable that indicates when all trailers in the script have been processed. The model terminates when *AllTrailorsArrived* is *True* and *YardQueue* is empty. *HandlingUnitsPDF* determines the number of handling units for each trip to a loading door.

Distances between dock doors are in spreadsheet *DistanceMatrix*. As shown in Figure 24.24, information on trailer arrival day and time, as well as the corresponding shipments, is in spreadsheet *ShipmentData*. Day and time are used to schedule an arrival. The remaining data on each row details the work requirements for the associated arrival. *ScriptRow* is a row counter to facilitate reading the script. In sub *Initial* of *EventsManager* is a program segment that calls the primary event subroutine to read shipment data and schedule the first scripted arrival as a primary event. A transaction represents a tow motor with the following attributes:

Day	Time*	TrailorID	Dest	Units	Dest	Units
1	7:05 AM	534183	BHM1	0	KNX1	0
1	12:11 PM	511735	BHM1	0	KNX1	0
1	1:11 PM	400021	BHM1	0	KNX1	2
1	4:22 PM	534180	BHM1	0	KNX1	0
1	4:47 PM	Z171	BHM1	17	KNX1	10
1	4:58 PM	534783	BHM1	0	KNX1	0
1	4:58 PM	535765	BHM1	0	KNX1	0
1	5:14 PM	534283	BHM1	4	KNX1	24
1	5:49 PM	483284	BHM1	3	KNX1	3
1	6:33 PM	534816	BHM1	12	KNX1	3
1	6:38 PM	534072	BHM1	1	KNX1	4
1	6:46 PM	535262	BHM1	5	KNX1	3
1	6:52 PM	533642	BHM1	9	KNX1	2
1	7:24 PM	2821	BHM1	6	KNX1	17
2	6:19 AM	536047-2	BHM1	0	KNX1	0

Figure 24.24   Trailer arrival script.

1 = *TotalDelay*
2 = *FacilityIndex*
3 = *OriginDoor*
4 = *DestinationDoor*
5 = *HandlingUnits*
6 = *Distance*
7 = *DataRow*
8 = *DataColumn*

Table 24.4 describes primary events in the model. In addition to subroutines for each of the primary events, there are additionally two subs and two functions. *ScheduleScriptArrival* and *ScheduleStripDoor* are the subs, and the functions are *NowEmpty* and *Door*.

The remainder of this section provides an overview of model procedures and a brief description of program output.

ScriptArrivalEvent: In Figure 24.25, an arriving trailer joins *YardQueue*. *ScriptRow* is updated so that data for the arriving trailer is available from the *ShipmentData* spreadsheet. The next trailer arrival is scheduled. The number of tow motors never exceeds the number of strip doors. Therefore, if a tow motor is idle a trailer is ready to be stripped. The following program segment in sub *Initial* schedules the first trailer arrival:

```
Set ScriptData = Worksheets("ShipmentData").Range("A2")
ScriptRow = 0
ScheduleScriptArrival
```

Sub *Parser* uses the following code segment to recognize an arriving transaction associated with the script:

```
Case "ScriptArrival"
    ScriptArrivalEvent CurrentEvent
```

TimerDelayEvent: In Figure 24.26, the value of *TimerIndex* is the basis for allocating tow motors. *TimerIndex* cycles from 1 to the number of time cycles for the dock. Time intervals are enumerated in *TowMotorTimeTable*. When a timer delay is complete, it is time to change the number of tow motors and tow motor velocity on the dock. *TimerIndex* is updated and the timer delay

Table 24.4   Cross Dock Primary Events

Event	Description
ScriptArrivalEvent	Generates trailer arrivals.
TimerDelayEvent	Maintains the clock for changing the number of tow motors on the dock.
TowMotorDelayEvent	Tow motor completes a round trip between stripping and load door.
DoorSwitchDelay	Prepares strip door for next trailer.

```
Sub ScriptArrivalEvent(ByRef Trailer As Transaction)
  ' Process script arrival
  YardQueue.Join Trailer
  ScriptRow = ScriptRow + 1
  ScheduleScriptArrival
  If TowMotorFacility.NumberOfIdleServers > 0 _
  Then
    ScheduleStripDoor
  End If
End Sub
```

Figure 24.25 Cross dock—*ScriptArrivalEvent*.

```
Sub TimerDelayEvent(ByRef Timer As Transaction)
  Dim _
    Index As Integer, _
    NumberOfTowMotors As Integer
  TimerIndex = TimerIndex + 1
  If TimerIndex > NumberTimeIntervals _
  Then
    TimerIndex = 1
  End If
  TowMotorVelocity = TowMotorVelocityTable.F(TimerIndex) * 60
  NumberOfTowMotors = TowMotorFacility.NumberOfServers
  TimerDelay.Start Timer, TowMotorTimerTable.F(TimerIndex) * 60
  TowMotorFacility.NewNumberOfServers _
                            TowMotorCountTable.F(TimerIndex)
  ScheduleStripDoor
End Sub
```

Figure 24.26 Cross dock—*TimerDelayEvent*.

is restarted. With information from *TowMotorVelocityTable* and *TowMotorCountTable* the dock is reconfigured. When the number of tow motors increases, subroutine *ScheduleStripDoor* allocates any newly available tow motors to trailers in *YardQueue*.

TowMotorDelayEvent: In Figure 24.27, the delayed transaction is the tow motor that has completed a round trip delay, beginning at the stripping door and returning from the loading door. However, the tow motor transaction also represents the trailer being stripped. Initially, the transaction has engaged a server in *TowMotorFacility* and *StripDoorFacility*. Although not shown, the transaction is never "serviced" by either facility. The purpose of both facilities is simply to control access to resources.

If all handling units to the current destination have been moved, function *NowEmpty* checks for shipments to another destination. When the trailer is empty, *TowMotor* disengages the *TowMotorFacility* (tow motor is now idle), and *DoorSwitchDelay* uses the trailer transaction to replace the current trailer at the stripping door with one waiting to be stripped. *ScheduleStripDoor* checks for another strip door that may be waiting for an idle tow motor. When *YardQueue* is empty and all tow motors are idle, then all trailers are stripped and the simulation is terminated.

With handling units equal to zero and additional shipments on the trailer, the tow motor begins work on shipments for the next destination. To reduce data requirements, the distance from origin *i* to destination *j* is the same as the distance between origin *j* and

```
Sub TowMotorDelayEvent(ByRef TowMotor As Transaction)
  Dim _
    Trailor As Transaction, _
    NextHandlingUnits As Integer
  Set Trailor = TowMotor
  If Trailor.A(HandlingUnits) = 0 _
  Then 'finished current destination
    If NowEmpty(Trailor) _
    Then
      TowMotorFacility.Disengage TowMotor
      If AllTrailorsArrived And YardQueue.Length = 0 And _
        TowMotorFacility.NumberOfBusyServers = 0 _
      Then
        StopSimulation ' ALL SHIPMENTS PROCESSED !!!!
        Exit Sub
      End If
      DoorSwitchDelay.Start Trailor
      ScheduleStripDoor
      Exit Sub ' wait for door to clear
    End If
    ' ready for next destination door
    Origin = TowMotor.A(OriginDoor)
    Destination = TowMotor.A(DestinationDoor)
    If Origin > Destination _
    Then
      TowMotor.Assign Distance, _
             2 * DistanceData.Offset(Destination, Origin)
    Else
      TowMotor.Assign Distance, _
             2 * DistanceData.Offset(Origin, Destination)
    End If
    TowMotor.Assign TotalDelay, 2 'handling at each end of trip
    TowMotor.Assign TotalDelay, _
      TowMotor.A(TotalDelay) + _
                         TowMotor.A(Distance) / TowMotorVelocity
  End If
  NextHandlingUnits = HandlingUnitsPDF.RV
  If NextHandlingUnits > Trailor.A(HandlingUnits) _
  Then
    NextHandlingUnits = Trailor.A(HandlingUnits)
    Trailor.Assign HandlingUnits, 0
  Else
    Trailor.Assign HandlingUnits, _
               Trailor.A(HandlingUnits) - NextHandlingUnits
  End If
  P = P + TowMotor.A(Distance)
  TowMotorDelay.Start TowMotor, TowMotor.A(TotalDelay)
End Sub
```

Figure 24.27   Cross dock—*TowMotorDelayEvent*.

destination *i*. Because the delay is round trip, the total is twice the one-way distance. A standard two-minute delay is used for picking up and dropping off handling units. The tow motor delay time is incremented by *Distance/TowMotorVelocity*.

Notice that the distance calculations are omitted for a subsequent trip to a loading door. The remaining program segment applies whether *TowMotor* is working on a new destination or making another trip for the current destination. The number of handling units is the minimum of a random variable from *HandlingUnitsPDF* and the remaining number of units to that destination. The total tow motor distance traveled is incremented for another tow motor round trip and the tow motor begins another delay.

DoorSwitchDelayEvent: As shown in Figure 24.28, after the delay, the trailer transaction disengages *StripDoorFacility*. If a tow motor is available and a trailer is in *YardQueue*, a strip door is scheduled. Again, an available tow motor implies an available strip door.

ScheduleScriptArrival: As shown in Figure 24.29, no more arrival events are scheduled when all rows of shipment data have been read. A transaction is created to represent the next trailer. The number of attributes is incremented to accommodate parameters needed to track shipment data. A negative interarrival time indicates a new day of shipments. For example, a 1:15 a.m. arrival that follows an 11:30 p.m. arrival will produce a negative interarrival time. When that condition occurs, *IAT* is incremented by 1440 time units to maintain the sequential simulation time. *NewEventType* and *NewTimeStamp* are set for the new arrival and the transaction is inserted into the primary events list. After each scripted arrival, the procedure schedules the next arrival. After the last scripted arrival occurs, neither the *ScriptArrivalEvent* or *ScheduleScriptArrival* subroutine is called again.

```
Sub DoorSwitchDelayEvent (ByRef Trailor As Transaction)
    StripDoorFacility(Trailor.A(FacilityIndex)).Disengage Trailor
    Set Trailor = Nothing
    If TowMotorFacility.NumberOfIdleServers > 0 And _
      YardQueue.Length > 0 _
    Then
        ScheduleStripDoor
    End If
End Sub
```

Figure 24.28 Cross dock—*DoorSwitchDelayEvent*.

```
Sub ScheduleScriptArrival(Arrival As Transaction)
    ' Process script arrival
    Dim _
      ScriptEvent As Transaction, _
      IAT As Integer
    If (ScriptData.Offset(ScriptRow, 0)) = "" _
    Then
        AllTrailorsArrived = True
        Exit Sub
    End If
    Set ScriptEvent = NewTransaction
    ScriptEvent.Assign DataRow, ScriptRow
    ScriptEvent.Assign DataColumn, 4
    PreviousArrivalTime = CurrentArrivalTime
    CurrentArrivalTime = Worksheets("Sheet1").Range("A2")
    IAT = DateDiff("n", CurrentArrivalTime, PreviousArrivalTime)
    If IAT < 0 _
    Then
        IAT = IAT + 1440
    End If
    ScriptEvent.NewEventType "ScriptArrival"
    ScriptEvent.NewTimeStamp CurrentTime + IAT
    Simulator.InsertInPEL ScriptEvent
End Sub
```

Figure 24.29 Cross dock—sub *ScheduleScriptArrival*.

DateDiff is a VBA function that determines the lapsed time between arrivals, or the interarrival time. The *n* option requests that the difference be measured in minutes. A negative lapsed time indicates a next day arrival. By adding 1440 (minutes in 24 hours), the lapsed time reflects the actual time since the last arrival.

NewTimeStamp and *NewEventType* are transaction procedures that enable the user to model primary events outside the scope of classes *Source*, *Facility*, and *Delay*. The *InsertInPEL* procedure enables the modeler to schedule a user-defined primary event.

ScheduleStripDoor: As shown in Figure 24.30, to schedule a strip door, there must be an available tow motor and a waiting trailer. An available tow motor implies an available strip door. Strip doors are scheduled until all tow motors are allocated. The trailer transaction is removed from *YardQueue* and the transaction engages *StripDoorFacility* and *TowMotorFacility* to indicate both resources are being used. The transaction is delayed by *TowMotorDelay* with zero delay time and zero handling units. It is therefore ready for processing by *TowMotorDelayEvent*. Function *Door* returns the dock location of the strip door being scheduled.

NowEmpty: Pairs of cells for a given trailer are examined to find the end of the shipments list. In Figure 24.31, *True* is returned when the trailer is empty. Otherwise, *Destination* and *HandlingUnits* attributes are updated for the next destination.

MapDoors: In Figure 24.32, using an insertion sort, *MapDoor*, a VBA collection of *DoorData* instances ordered on strip door identifier, is created. *DoorData* is a class definition with dock location and strip door identifier.

```
Sub ScheduleStripDoor()
  Dim _
    Index As Integer, _
    AssignedDoor As Boolean, _
    Trailor As Transaction
  Do While (TowMotorFacility.NumberOfIdleServers > 0 And _
            YardQueue.Length > 0)
    AssignedDoor = False
    For Index = 1 To NumberStripDoors
      If StripDoorFacility(Index).NumberOfIdleServers > 0 _
      Then
        Set Trailor = YardQueue.Depart
        Trailor.Assign FacilityIndex, Index
        StripDoorFacility(Index).Engage Trailor, 1
        Trailor.Assign OriginDoor, Door("S" & Index)
        Trailor.Assign HandlingUnits, 0
        TowMotorFacility.Engage Trailor, 1
                    ' trailor is also the TowMotor
        TowMotorDelayEvent Trailor
        AssignedDoor = True
        Exit For
      End If
    Next Index
    If Not AssignedDoor _
    Then
      Exit Sub ' no strip door
    End If
  Loop
End Sub
```

Figure 24.30 Cross dock—sub *ScheduleStripDoor*.

```
Function NowEmpty(Trailor As Transaction) As Boolean
  Dim _
    RowIndex As Integer, _
    ColumnIndex As Integer
  RowIndex = Trailor.A(DataRow)
  ColumnIndex = Trailor.A(DataColumn)
  NowEmpty = True
  Do While (ScriptData.Offset(RowIndex, ColumnIndex) <> "")
    If (ScriptData.Offset(RowIndex, ColumnIndex + 1) <> 0) _
    Then
      NowEmpty = False
      Trailor.Assign DestinationDoor, _
              Door(ScriptData.Offset(RowIndex, ColumnIndex))
      Trailor.Assign HandlingUnits, _
              ScriptData.Offset(RowIndex, ColumnIndex + 1)
      Trailor.Assign DataColumn, ColumnIndex + 2
      Exit Function ' next destination
    End If
    ColumnIndex = ColumnIndex + 2
  Loop
End Function
```

Figure 24.31   Cross dock—sub *NowEmpty*.

```
Sub MapDoors()
  Dim _
    Index As Integer, _
    Jndex As Integer
  Index = 0
  With Worksheets("UserInitialed").Range("I4")
    Do While (.Offset(Index, 0) <> "")
      Set DoorData = New DoorData
      DoorData.Location = .Offset(Index, 0)
      DoorData.StripDoor = .Offset(Index, 1)
      For Jndex = 1 To DoorMap.Count
        If DoorData.StripDoor < DoorMap.Item(Jndex).StripDoor _
        Then
          DoorMap.Add Item:=DoorData, before:=Jndex
          Exit For
        End If
      Next Jndex
      If Jndex > DoorMap.Count _
      Then
       DoorMap.Add Item:=DoorData
      End If
      Index = Index + 1
    Loop
  End With
End Sub
```

Figure 24.32   Cross dock—sub *MapDoors*.

DoorSearch: In Figure 24.33, a binary search of *DoorMap* is used to find the dock location of a strip door.

Analysis of the cross dock problem is presented in Section 25.5, to demonstrate genetic algorithms as a model analysis tool.

```
Function DoorSearch(ByRef Name As String)As Integer
    Dim _
       Index As Integer, _
       Upper As Integer, _
       Middle As Integer, _
       Lower As Integer
    Lower = 1
    Upper = DoorMap.Count
    For Index = 1 To DoorMap.Count
       Middle = (Upper + Lower) / 2
       If DoorMap.Item(Middle).StripDoor = Name _
       Then
          DoorSearch = DoorMap.Item(Middle).Location
          Exit Function
       End If
       If DoorMap.Item(Middle).StripDoor < Name _
       Then
          Lower = Middle + 1
       Else
          Upper = Middle - 1
       End If
    Next Index
End Function
```

Figure 24.33   Cross dock—function *DoorSearch*.

Problems

24.1. Industrial bolts are produced in two sizes: a 10-inch long bolt that weighs 15 pounds and a 7-inch bolt that weighs 7 pounds. These bolts are produced at the rate of one unit every *Exponential* (9) seconds. Small bolts account for 45% of the production. The remaining 55% are large bolts. In both cases, rejects comprise about 10% of the production.

Bolts are fed down a conveyor belt that moves at a constant speed. Rejectors are placed one-third of the way down from the start of the belt. Rejected bolts are reworked and it takes *UniformContinuous* (20, 45) seconds to do the reworking, after which time the bolt is sent back to the conveyor. Accepted bolts must be oriented head down for proper pickup by a robot. Selectors properly placed along the belt are used to reorient the bolts, if necessary. The belt stops when a bolt breaks a sensor at the robot pickup point. The sensor determines the size of the bolt and sends a signal to the proper robot. The robot picks up the bolt and checks a sensor to see if the loading ramp is full. If the ramp is full, the robot places the bolt on another conveyor belt that feeds another loading ramp. Otherwise, the robot sets the bolt on the ramp and retracts to a waiting position. *Exponential* (7) seconds are needed for the robot to dispose of a bolt. The capacity of each loading ramp is five bolts. Loading ramps feed into loaders (one for each type bolt). *Exponential* (20) seconds are needed to package each bolt before it is sent to shipping. Simulate the system for different conveyor speeds to determine bottlenecks.

24.2. A shipping/receiving dock with six bays is used in conjunction with a production facility. Trucks arrive with supplies and leave with finished products. The interarrival time of trucks with loaded trailers is *Normal* (360, 30) minutes. A truck that does not find an empty bay in the dock waits until one is available. The truck departs after being hooked to a loaded trailer. *Exponential* (5) minutes is needed to hook or unhook a trailer.

A full trailer load of the final product is produced every *Exponential* (360) minutes and is loaded as empty trailers become available. Two workers load and unload the trailers. The hours are from 8:00 a.m. to 8:00 p.m. daily. However, a job, once started, must be completed. *Triangular* (56, 60, 80) minutes are needed to load or unload a trailer. Simulate the environment to determine any bottlenecks in the system. Assume that the simulation starts with one (incoming) full and two empty trailers.

24.3. Consider the flow of shirt factory garments from the cutting room to the final storage area. Cut fronts arrive from the cutting room every *Exponential* (0.0192) hours. The right and left fronts of a shirt are sent to two separate machines for hemming that requires *Exponential* (0.025) hour per shirt. Hemmed left fronts are then sent to one of two machines for buttonholes. *Exponential* (0.028) hours are needed for this task. The right fronts are sent to one of two machines that sew on buttons in *Exponential* (0.04) hours per shirt. The two fronts are then sent to one of four machines that attach pockets on each side. *Exponential* (0.05) hours are needed for each pocket.

Cut backs arrive every *Exponential* (0.0192) hours. *Exponential* (0.071) hours are needed to join the front and back of a shirt. At this point, collars that arrive from the cutting room every *Exponential* (0.0192) hours are attached to the shirt. This process requires *Exponential* (0.1) hours per shirt. The remaining processes of joining the sleeves, attaching the cuffs, and sewing the inseams down the inside of a shirt are performed in *Exponential* (0.5) hours. There are two machines dedicated for these jobs.

Inspection is done by seven inspectors in *Exponential* (0.125) hour per shirt. In 3% of the cases, the shirt must be reworked in *Exponential* (0.1) hours. A shirt is discarded when it does not pass the second inspection. Accepted shirts are folded and bagged at the rate of one every *Exponential* (0.21) hours. The final stage involves boxing three dozen shirts to a box, a process that takes *Exponential* (0.02) hours. Write a simulation model that will determine the utilization of the different manufacturing stages.

24.4. Multiprogramming is used in most computer operating systems to increase Central Processing Unit utilization and jobs processed. Primary memory is segmented into several regions or partitions. Each of these partitions is loaded with a job ready for execution. A job continues to execute until it issues an input/output (I/O) request. While the job waits for completion of its I/O request, the operating system allocates the CPU to another job. This pattern continues every time a job has to wait for an I/O request.

Figure 24.34 provides a schematic diagram of a multprogrammed computing system composed of a CPU, a CPU scheduler, 50 partitions (pages) of primary memory, a job scheduler, four disk drives, and a data (I/O) channel.

The computer operates two eight-hour shifts. Jobs are from three user classes: interactive, batch, and special. Interactive and batch jobs can be submitted to the system at any time. Special jobs can be submitted during the second shift only. Table 24.5 provides interarrival times for the three classes during the two shifts. It also indicates the corresponding processing times, memory requirements, and number of I/O requests.

The CPU processing time consists of bursts, each of which is followed by an I/O request. Each job starts and ends with a CPU burst. The length of each burst is inversely related to the number of I/O requests for that job and is calculated as follows:

$$\text{Length of burst} = \text{CPU time}/(\text{number of I/O requests} + 1)$$

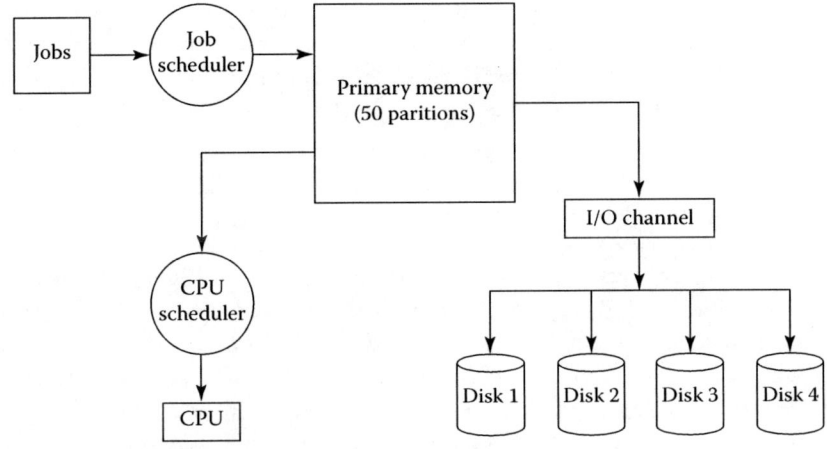

Figure 24.34 Computer description.

Upon arrival at the system, a job is assigned a priority according to its class, with interactive jobs having the highest priority and special jobs the lowest. I/O operations are assigned to a specific disk upon arrival. The job scheduler loads a job into the primary memory using the highest priority scheduling algorithm. However, not more than two special jobs can be in memory at any one time. Also, special jobs that are not completed during the second shift are held until the second shift of the following day.

Processor time is divided into small intervals known as *time quantums* or *time slices*. The time slices for shifts 1 and 2 are one and four minutes, respectively. The CPU scheduler allocates the CPU to one of the jobs in memory on a first-in, first-out (FIFO) basis. A job continues to execute until an I/O request is issued or until its time slice expires. If an I/O request is issued, the job must wait for its I/O completion. However, upon expiration of the time slice, the operating system will take note of the status of the job so that when the CPU returns to process the job, it will start from where it last terminated. In either case, the CPU is allocated to the next job that has been waiting the longest.

After a CPU burst, an I/O request is automatically made on the disk assigned to the job. The seek time to locate the proper position on any disk is *UniformContinuous*

Table 24.5 Job Service Time

	Interarrival Times (Minutes)[a]		Processing Time (Minutes)	Memory (Pages)	I/O Requests
Class	Shift1	Shift2			
Interactive	NO (3, .33)	NO (5, 0.67)	UN (1, 2.5)	NO (4, .67)	UN (0, 3)
Batch	NO (13, 1.67)	NO (18, 1.67)	UN (2, 7)	NO (7, 1.67)	UN (1, 3)
Special	No arrivals	NO (90, 1.67)	UN (20, 25)	NO (15, 1.67)	UN (5, 10)

[a] *NO* and *UN* denote *Normal* and *UniformContinuous* distributions.

(0, 0.01) minutes. Only one seek operation per disk can be performed at a time. Following the seek, a data transfer through the channel is performed. Transfer time is *Exponential* (0.002) minutes. After the transfer, the I/O request is complete. Simulate the system to determine the following performance metrics:
   a. CPU utilization
   b. Disk drive utilization
   c. Waiting time for resources
   d. Response time = sum of all waiting times + CPU execution time + I/O execution time
24.5. A farmer makes a delivery route every day to deliver milk and cheese, and in the process collects orders for the next day. Orders are filled and packaged very early in the morning and must be ready in time for the start of the daily delivery trip. The farmer is considering purchasing equipment to replace the manual system. If the average time between packaged orders is less than the time to do the job manually, the equipment will be bought.
   The following information summarizes the packaging/delivery process:

1. Milk is filled in one-quart bottles and cheese is wrapped in one-pound blocks. *UniformContinuous* (20, 30) seconds are needed to fill (and seal) a quart of milk, and *UniformContinuous* (18, 32) seconds are needed to wrap one pound of cheese.
2. The boxes used to pack an order have six slots for milk and another six for cheese. Placing items in the box requires five seconds per item for milk and four seconds for cheese. Remaining slots must be filled with cardboard to ensure safety during transportation. The time to package each empty slot is three seconds.
3. Each order can have up to six quarts of milk and six packages of cheese according to the probability distributions shown in Table 24.6. Notice that according to the distributions an order may consist entirely of milk or cheese.
4. The equipment manufacturer claims that although the machines are very reliable, they require service every 120 minutes and 100 minutes for milk and cheese, respectively. The time to service the machinery is 12 minutes for milk and 10 minutes for cheese.
5. Statistics show that, on average, one of every 50 quarts of milk and one of every 60 pounds of cheese are packaged improperly and require an extra 45 seconds to repackage when filling the box.
6. Approximately 1% of the boxes must be completely repackaged.

Determine the average time for an order from the moment it starts until it is completely packaged. Also, determine the average time between successive order completions.
24.6. Two types of aircraft (light and heavy) arrive at an intermediate range airport with the interarrival times of *Exponential* (60) and *Exponential* (65) minutes, respectively. The airport has two runways, one for landing and the other for takeoff. There are two exits from the runways. The first one is for light planes and the second is reserved for heavy planes. The two exits intersect and merge into a taxiway that leads to the apron at the terminal's gate.

Table 24.6 Order Distribution

Units	0	1	2	3	4	5	6
Milk	0.08	0.10	0.20	0.20	0.17	0.17	0.08
Cheese	0.05	0.10	0.15	0.25	0.20	0.15	0.10

It takes four minutes to pass the exits or taxiways. The light aircraft can pass the intersection only if the heavy plane is at least two minutes away from the intersection. At the terminal, there are two gates: one for light planes and the other for heavy aircraft. Each plane must use its designated gate. If the gate is busy, the aircraft must wait in line until it becomes available. Inspection and maintenance of aircraft are performed during passenger loading and unloading.

The time each plane waits at the boarding gate is *UniformContinuous* (20, 30) minutes for light planes and *UniformContinuous* (40, 50) minutes for heavy planes. When the aircraft is ready, it moves into an entrance taxiway that leads to the takeoff runway.

Aircraft wait at the end of the taxiway until the runway becomes available. In 10% of the cases, departing planes passing through the entrance taxiway are delayed for additional inspection. These planes remain in the holding apron until inspection is completed. Inspection requires *UniformContinuous* (20, 40) minutes. Upon completion of inspection, these planes are given priority for takeoff. Takeoff time is *UniformContinuous* (4, 7) minutes. Air traffic control prefers that the number of aircraft in the holding stack not exceed eight. If that limit is exceeded, the takeoff runway is cleared and used for landing until the number of planes in the holding stack is back to eight. Develop a simulation model to determine the utilization of the runways, the taxiways, and the boarding facilities.

24.7. A camel rental facility operates in the shadow of the Great Pyramid of Cheops on the outskirts of Cairo. Tourists arrive at the site approximately every *Exponential* (3) minutes between the hours of 8:00 a.m. and 2:00 p.m. After 2:00 p.m., the interarrival time slows to *Exponential* (4.5) minutes. The tour guides allow tourists to stay on the Giza Pyramid plateau for only 30 minutes; after all, their agenda for the day includes not only the pyramids but also a great many other attractions of Ancient Egypt. Any tourist who has begun a camel ride within the allotted 30-minute time limit is allowed to finish. Those still waiting have to leave without riding a "ship" of the desert. A ride on the colossal beasts can last anywhere from 9 to 45 minutes (uniformly distributed) at a cost of between $1 and $10, depending on the bargaining power of the tourist and the cleverness of the camel owner. Upon dismounting from a camel, the rider must leave immediately, if the departure time was reached during the ride; otherwise, the tourist is free to gaze at the ancient wonders (or count the sands of the Sahara). Approximately 40% of all tourists who can stay for a little while after their rides will buy some souvenirs. The salesperson can usually attend to any number of customers at once, and each transaction is completed in about six minutes. Again, a tourist is allowed to complete the transaction if it is started before departure time.

When the arrival rate of tourists tapers down in the afternoon, the owner allows half of the camels to return to the stables for lunch. It takes 6–12 minutes to reach the stables, 18–30 minutes to eat, and 6–18 minutes to return up the incline after lunch. Naturally, a tourist's ride is completed before camels are sent to the stables. When the first shift of camels returns from lunch, they immediately relieve all the remaining camels so that the second lunch shift can begin. On the day when an odd number of camels are on duty, the extra camel is with the second shift.

At eight o'clock every evening, the camels retire for the night. Any ride started before 8:00 p.m. is completed, but no new rides begin after 8:00 p.m. If a camel works past 8:00 p.m., it is brought to work two minutes late for every minute of overtime. The rental facility has five camels. Determine whether this number is sufficient to serve the tourists adequately.

24.8. In an automotive parts store, a counter serves as the entrance and exit points for all transactions in the facility. The entire operation includes auto parts and accessories sales, hydraulic hose assembly, brake drum and rotor turning, machine shop service, and garage service.

The hose assembly and brake drum turning are done by the salespersons at the counter. Service times in the shop and garage depend on the various jobs that are done in those departments. Customers arrive every *Exponential* (9) minutes. The customer checks the number of people waiting in line at the counter. If six persons are already waiting, the customer drives down the block to another store. If fewer than six persons are waiting, the customer enters the store and waits for service. There are five classes of customers:

1. About 83% of the customers want to purchase a part or accessory for their automobiles. The service time per customer in this category is *Normal* (7, 1) minutes.

2. Another 3% of the customers purchase a hydraulic hose, which must be assembled by the counter salesperson. *UniformContinuous* (7, 12) minutes are needed to complete this task.

3. Another 2% of the customers request the turning of a brake drum or a generator rotor. This task requires *Triangle* (10, 20, 30) minutes and must be done by the counter salesperson.

4. Customers needing machine shop service represent 9% of the total. They must enter through the counter area, where a salesperson writes a work order in about *Exponential* (4) minutes. The job is then sent to the shop for processing, which lasts *Exponential* (120) minutes. After the shop work is completed, the customer picks up the finished job and pays the bill at the counter. *Exponential* (4) minutes are needed to pay the bill.

5. The remaining customers (3%) require garage service. They too must be processed at the counter, where a salesperson writes up a work order in *Exponential* (5) minutes before sending the job to the garage. The garage has four bays and ten parking spaces for waiting jobs. If all the bays are busy when the customer arrives, the salesperson will advise the customer to return at a later time. Garage service lasts *UniformContinuous* (120, 960) minutes per job. Garage jobs requiring eight or more hours, usually engine overhauls, also require shop service. For these jobs, half of the garage time is spent in dismantling the engine, which is sent to the shop for the necessary work. It takes *Exponential* (400) minutes to complete this work. The engine is then returned to the garage to be mounted on the car. The mounting time is the remaining garage time. Such jobs receive priority shop service. After garage work is completed, the customer picks up the car and pays the bill at the counter in *Exponential* (5) minutes.

Develop a simulation model to test the effectiveness of these operations, given that the counter is served by one, two, or three salespersons.

References

1. Schriber, T., *Simulation Using GPSS*. New York: Wiley, 1974.
2. Pritsker, A. A. B., *Introduction to Simulation and SLAM II*, 3rd ed. New York: Wiley, 1986, pp. 195–200.
3. Law, A. M. and C. S. Larmcy, *An Introduction to Simulation Using SIMSCRIPT 11.5*. Los Angeles: CACI, 1984, Chapter 6.
4. Bobillier, P., B. Kahan, and A. Probst, *Simulation with GPSS and GPSS V*. Englewood Cliffs, NJ: Prentice-Hall, 1976, pp. 259–291.

Chapter 25

Analysis of Large Scale Models

25.1 Introduction

In Chapter 6 the implication was that simply because there are fewer runs the *subinterval* method is much more efficient than the *replication* method for estimating system parameters. However, the replication method is important for "what if" analysis described in Chapter 1. Also, system specification concepts from Chapter 17 are the basis for performing "what if" analysis. The analysis consists of changing workload or system specifications and then using the subinterval method to estimate performance metrics for the new configuration. The new metrics are compared with previous estimates to assess the effects of changes to the system.

This chapter focuses on design environment for event-driven simulation (DEEDS) as a research tool for discrete event simulation. A more sophisticated approach to analysis than those presented in Chapter 13 is to use experimental design, simulated annealing, genetic algorithms, and other search techniques to systematically search for improved system configurations. Problems from previous chapters are used to demonstrate how to conduct an experimental design and how to integrate a search algorithm into the simulation project.

25.2 Evaluation of Alternatives

As presented in Chapter 17, the performance evaluation problem is used to compare the ability of different systems (S_{1t}, S_{2t}, ..., S_{nt}) to process a given workload, W_t. To facilitate data collection, DEEDS has a special feature, the *New Replicate* option shown on the *Initial Model* page of Figure 25.1 that enables the user to archive simulation results for each experimental observation (workload/system configuration). DEEDS archives all output spreadsheets as well as the *InitialModel* and *UserInitial* spreadsheets with a replication number appended to the name of each spreadsheet. New versions of spreadsheets *InitialModel*, *UserInitial*, and *UserOutput* with previous values of W_t and S_t parameters, or previous user output format, as the basis for the next configuration are then produced. The archived results can be summarized and analyzed using statistical tools embedded in Excel.

As discussed in Chapter 11, DEEDS has 50 random number sequences available to the model. Unless the seed is changed, the sequence of random numbers produced by a pseudo-random number

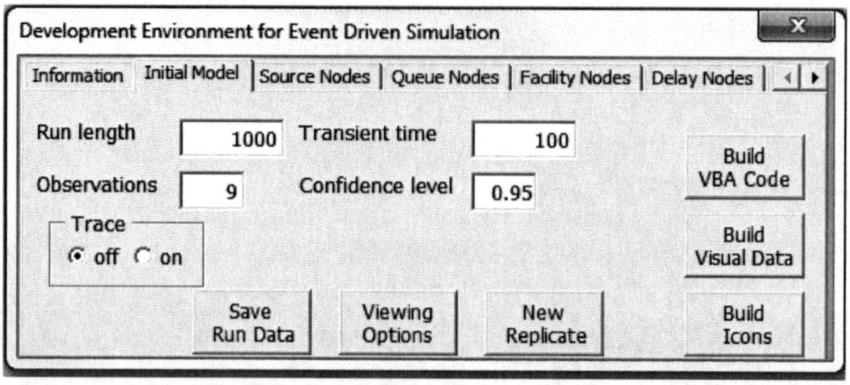

Figure 25.1 *New Replicate* **option for DEEDS.**

is identical each time the program is executed. To ensure that the workload is "identical" for each system configuration, a random number sequence can be reserved to produce only the workload.

For example, preserving the service time sequence for a given workload is problematic when comparing a current system with a new system designed to perform identical tasks in less time. In such a case, the workload must be characterized using tables to define tasks. Tasks are then translated to system service times. However, the same random number sequence preserves the basic features of service time.

25.3 Design of Experiments

The basis for the comment in Section 13.1, that simulation is a statistical experiment, is that inferential statistics are used to estimate performance measures that are inherently random. The conceptual expression for P, the performance metric, in Section 17.2 is another view of the statistical predictor model. If W_t and S_t form a set of independent variables that affect dependent variable P, an analyst can develop regression or experimental design models to predict P for given values of W_t and S_t.

Because the expression for P is only a conceptual model, the user must selectively develop alternatives for W_t and S_t to obtain as much information as possible on masked relationships that affect P. In effect, the alternatives become "what if" observations. Simply stated, the questions are, "What is the value of P for selected values of W_t and S_t?" and "What values of W_t and S_t will produce the best performance metric for the system?"

The inspection problem introduced in Section 17.3 and modeled in Section 23.2 is used to demonstrate an experimental design using a simulation model for obtaining performance metrics of the system. Assuming that the decision variables are sensitivity, specificity, and production variance, the question is "Which of these variables, or combination of variables, has the most impact on unit cost of production?" Changes to any of these variables are in effect a change to the inspection program. Using system terminology from Chapter 17, management can compare costs of affecting changes to decision variables with savings that result from those changes. Table 25.1 shows the level settings of factors (decision variables) for the experiment.

A full factorial design required a total of $2^3 = 8$ scenarios to test the individual factors and all their interactions. A scenario corresponds to a simulation run, and two observations were collected for each scenario. To produce the two observations, the program segment shown below was added to sub *Parser*. The two call statements were also added to sub *ShutDown*. Additional observations

Table 25.1   Inspection Model—Three-Factor Design

Level	Sensitivity	Specificity	Variance
−	70	70	0.001
+	95	90	0.007

may be obtained by changing the *Number of Observations* parameter on spreadsheet *InitialModel*. For this analysis, *Transient Time* was 0.

Sub *DOEObservation*, presented in Figure 25.2, calculates unit cost and places the result in spreadsheet *UserInitial*. After each run, the *New Replicate* option, in Figure 25.1, is used to prepare

```
Case "NewObservation"
     Call DOEObservation
```

spreadsheets for the next run. The replication archives all output spreadsheets. However, only spreadsheet *UserInitial* is kept because it has the information needed for the factorial experiment. Summary results of the experiment are presented in Figure 25.3, a Pareto chart. Variability in production and sensitivity are the most important factors. Notice that there is also an important interaction effect between these factors.

```
Sub DOEObservation()
  DOEOutputIndex = DOEOutputIndex + 1
  Call Calculations
  DOEOutput.Offset(DOEOutputIndex, 0).Value = DOEOutputIndex
  DOEOutput.Offset(DOEOutputIndex, 1).Value = AverageCost
End Sub
```

Figure 25.2   Inspection model—program segment for data collection.

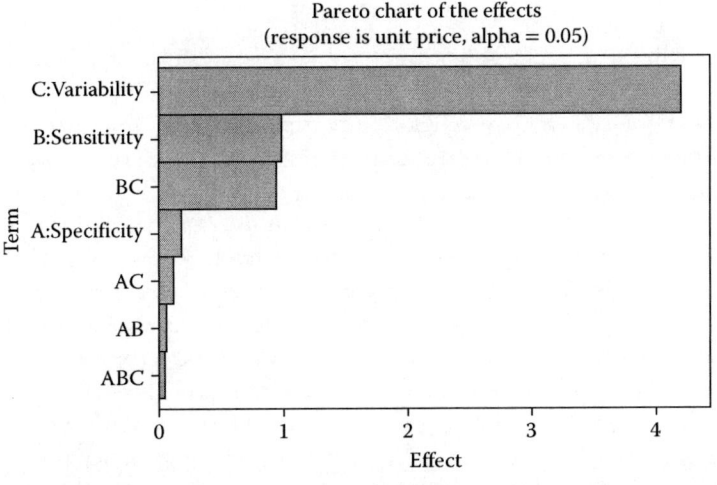

Figure 25.3   Inspection model—Pareto chart of decision variables.

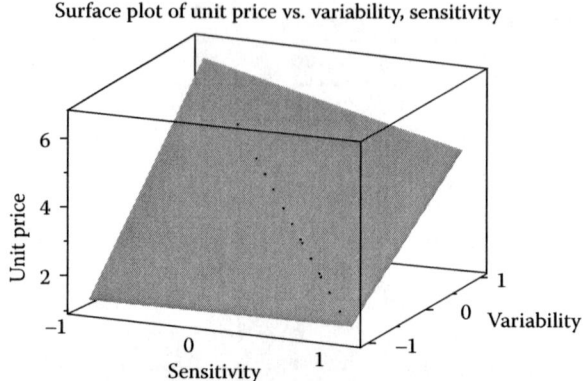

Surface plot of unit price vs. variability, sensitivity

Figure 25.4 Inspection model—surface plot.

Figure 25.4 provides another view of the effects of changes in production variability and sensitivity on unit price of production. The surface plot is shown as a plane; however, a regression analysis with data from additional simulation runs can be used to refine the surface shape. Information from the experimental design reduces the number of simulation runs necessary to assess the effects of changes in process variability and sensitivity.

25.4 Simulation and Search Algorithms

For optimization problems where the objective function or constraints may be nonlinear, solving for an optimum solution becomes problematic. Search algorithms are helpful for finding an improved but not necessarily optimum solution. Simulation may be used because precise relationships between W_t and S_t cannot be expressed algebraically or when precise estimates of how interactions between W_t and S_t affect P are a problem. Search algorithms from optimization theory that are much more flexible than regression and design of experiments models may be applied to discrete event simulation models. Integrating search algorithms with discrete event simulation offers a powerful research tool that performs a systematic search for a "better" system configuration, S_t, for a fixed workload, W_t.

Simulated annealing and genetic algorithms are search algorithms that have been successfully applied to simulation problems. Similar to traditional optimization algorithms, these algorithms start with a feasible solution and search for improved solutions while maintaining feasibility. Simulated annealing is based on the concept of a neighborhood search that begins with the starting solution. To enable backtracking, a simulated annealing algorithm selectively allows for accepting a configuration with an objective function value worse than the current solution's objective function.

In genetic search algorithms, a population of *chromosomes* represents a generation of feasible system configurations. Each chromosome represents a different configuration, and *genes* on the chromosome represent details of the configuration. Permissible values of a chromosome are *alleles*. By selectively mutating genes and splicing chromosomes, successive generations of alternative feasible chromosome configurations are developed. A selection process is used to determine which population and offspring chromosomes pass to the next generation. Rather than a neighborhood search, genetic algorithms consider "similar" solutions to achieve better results. The following sections demonstrate the flexibility of combining a genetic search algorithm with discrete event simulation in DEEDS.

25.5 Cross Dock Problem

The cross dock problem was introduced in Section 24.4. Using the conventional approach to this problem, the modeler enumerates several configurations, uses the simulation model in Section 24.4 to test each configuration on a given workload, and then selects the best configuration. The approach in Section 25.2 uses archiving of simulation results for each configuration to reduce the effort necessary for perform comparisons.

In Section 24.4, the performance metric for the model is the total distance that handling units are moved. An objective function for constraints in Section 24.4 is to minimize P where

$$P = \sum_i \sum_j \sum_k \sum_l x_{ik} x_{jl} a_{kl} d_{ij}$$

Because of the number of decision variables and the fact that the decision variables are affected by a time variant system, predicting P using concepts of experimental design would prove to be unwieldy. Discrete event simulation is combined with a genetic search algorithm to demonstrate a simulation-optimization model as a viable approach to this problem.

The relationship between the search algorithm and simulation model is shown in Figure 25.5. The communications link is the spreadsheet *UserInitialed*. Both programs retrieve data from and present results to spreadsheets. The search algorithm referenced in Figure 25.5 is applied to the cross dock problem.

A chromosome representation of the dock layout for the genetic algorithm is as follows:

Knx	Atl	Me	S1	S2	Cin	Lex

Each cell represents a dock door. Destination doors have destination names assigned to the cell, and an S_i in the cell indicates that the door is a stripping door. In this example, Knoxville, Atlanta, Memphis, Cincinnati, and Lexington are destinations. Permissible values of a chromosome are alleles. A valid assignment of doors ensures a starting feasible solution.

A splicing operator to produce a new chromosome is demonstrated using the following parent chromosomes:

Parent 1	Knx	Atl	Me	S1	S2	Cin	Lex

Parent 2	Me	S1	Atl	Knx	Cin	S2	Lex

The first step is to identify common genes in the parent. *Lex* occupies gene number 7 in both parents and therefore occupies gene 7 in the offspring chromosome.

Common	*	*	*	*	*	*	Lex

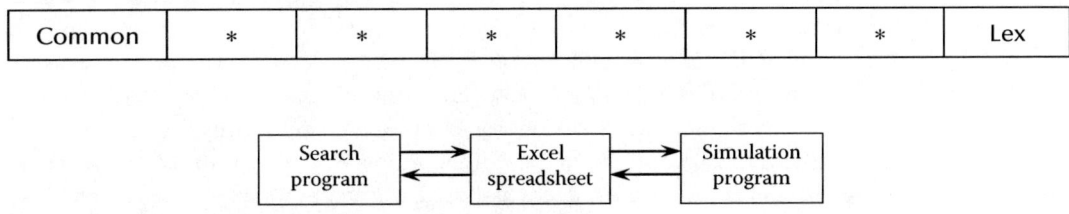

Figure 25.5 **Simulation—genetic algorithm model.**

Beginning with gene 1 in the parent chromosomes, randomly select a name from a parent. A leftover list is maintained during the gene selection process, and the final step is to move names from the leftover list to empty genes on the offspring chromosome. The sequence of operations is as follows:

Gene 1: Select *Knx* and add allele *Mem* to the leftover list.
Gene 2: Select allele *Atl* and add allele *S1* to the leftover list.
Gene 3: Select allele *Atl*, but because it is already on the chromosome, leave gene empty. Allele *Mem* is already on the leftover list.
Gene 4: Select allele *S1* but because allele *Knx* is already on the chromosome, do not add it to the leftover list. Remove allele *S1* from the leftover list.
Gene 5: Select allele *Cin* and add allele *S2* to the leftover list.
Gene 6: Select allele *S2*, but because allele *Cin* is already on the chromosome, do not add it to the leftover list. Remove allele *S2* from the leftover list.

After the above operations, the random choice for the offspring chromsome is

Random Choice	Knx	Atl	*	S1	Cin	S2	Lex

Allele *Mem* is the only location on the leftover list. It is therefore moved to gene 3 and the new offspring chromosome is

Offspring	Knx	Atl	Me	S1	Cin	S2	Lex

An example of a transpose chromosome mutation is

Chromosome	Knx	Atl	Me	S1	S2	Cin	Lex

The alleles of gene 2 through gene 4 are simply transposed. The starting and ending genes are chosen randomly. The result is

Mutation	Knx	S1	Me	Atl	S2	Cin	Lex

Notice that procedures for splicing and mutation ensure that feasibility is maintained for each system configuration. Visual Basic (VB) procedures for producing offspring by splicing and mutation are in Figures 25.15 and 25.19.

Sub *ProduceOffspring* produces new chromosomes by mutating and splicing the current population of chromosomes. Before selecting a new generation of chromosomes from the previous generation and the newly produced chromosomes, the algorithm considers termination condition(s). Typically, the termination criterion is number of generations produced by the algorithm. The above genetic algorithm also produces a short list of configurations ranked on their corresponding value of *P*. The algorithm also ensures that each feasible solution is embedded in the chromosome representation of the configuration.

Design issues for the above algorithm include:

- A chromosome structure to represent the system configuration
- The number of chromosomes in the population
- The number of chromosomes to splice for the next generation
- Which chromosomes to splice
- The number of chromosome mutations
- Which chromosomes to mutate for the next generation
- Selection criteria for the next generation

These issues affect the execution time and effectiveness of the program. Generally, the execution time on a PC is measured in hours.

VB data definitions for a genetic algorithm are presented in Figure 25.6. *PopulationData* and *SimulationData* reference spreadsheet cells with configuration data. The starting solution configurations are from *PopulationData*, and the next configuration to be evaluated by the simulation program is in *SimulationData*. VB Collections are used to maintain *Population, Offspring*, and *BestSolution*. To facilitate the processing, these collections are ordered on the value of *P*. A *NewChromosome* is the result of chromosome splicing or gene mutations.

The major Visual Basic for Applications (VBA) search program procedures are presented below. Several that deal with data entry and search program initialization have been omitted but are included on the companion CD.

Search, in Figure 25.7, is the search program referenced in Figure 25.5. After evaluating the initial population of chromosomes, the procedure loops on sub *NextGeneration* until the number of generations specified has been evaluated. Thereafter, the best solutions are written to spreadsheet *BestSolutions* and the search algorithm is terminated.

EvaluateP in Figure 25.8 writes a chromosome (dock layout configuration) to spreadsheet *UserInitial* and then initiates the simulation program. Sub *ShutDown* of the simulation program writes the corresponding value of *P* to spreadsheet *UserInitialed* and returns control to sub *EvaluateP*. Whenever a "better" chromosome is found, the *BestSolution* Collection is updated for that chromosome. For a maximization (minimization) problem, a "better" solution is when a new value of *P* is smaller (larger) than one or more of the current "best" solutions.

```
Private _
    MaximumGenerations As Integer, _
    PopulationData As Range, _
    SimulationData As Range, _
    SolutionData As Range, _
    NumberGenes As Integer, _
    MutationParent As Chromosome, _
    SpliceParent1 As Chromosome, _
    SpliceParent2 As Chromosome, _
    NewChromosome As Chromosome, _
    Population As New Collection, _
    OffSpring As New Collection, _
    BestSolution As New Collection, _
    PopulationSize As Integer, _
    OffSpringSize As Integer, _
    GenerationIndex As Integer
```

Figure 25.6 Genetic algorithm—data definitions.

```
Sub Search()
  Call Distribution.InitialPRNG
  Set PopulationData = Worksheets("Population").Range("A5")
  Set SimulationData = Worksheets("UserInitialed").Range("I3")
  Set SolutionData = Worksheets("BestSolutions").Range("A1")
  MaximumGenerations = Worksheets("UserInitialed").Range("E4")
  PopulationParameters
  GenerationIndex = 1
  Worksheets("UserInitialed").Range("E5") = GenerationIndex
  For Index = 1 To PopulationSize
    InitialChromosome Index   ' initial population chromosome
    EvaluateP
    InsertionSort Population
  Next Index
  Do While (GenerationIndex < MaximumGenerations)
    GenerationIndex = GenerationIndex + 1 ' next generation
    Worksheets("UserInitialed").Range("E5") = GenerationIndex
    ProduceOffspring
    NextGeneration ' from Population and OffSpring
  Loop
  PresentResults
  Do While (Population.Count > 0)
      Population.Remove 1
  Loop
  Do While (OffSpring.Count > 0)
    OffSpring.Remove 1
  Loop
  Do While (BestSolution.Count > 0)
    BestSolution.Remove 1
  Loop
End Sub
```

Figure 25.7   Genetic algorithm—sub *Search*.

```
Sub EvaluateP()
    ' send data to simulator
  Dim _
    Index As Integer, _
    Jndex As Integer
  With SimulationData
    Jndex = 1
    For Index = 1 To NumberGenes
      If NewChromosome.Allele(Index) <> "XX" _
      Then
        .Offset(Jndex, 0) = Index
        .Offset(Jndex, 1) = NewChromosome.Allele(Index)
        Jndex = Jndex + 1
      End If
    Next Index
    Controller
    NewChromosome.P = .Offset(-1, 5)
  End With
  InsertionSort BestSolution
  If BestSolution.Count > 5 _
  Then
      BestSolution.Remove 6
  End If
End Sub
```

Figure 25.8   Genetic algorithm—sub *EvaluateP*.

NextGeneration, in Figure 25.9, discards the weakest population chromosomes and then uses an insertion sort to merge chromosomes from the previous generation with new chromosomes that result from combining and mutating previous-generation chromosomes.

The *Chromosome* class definition is shown in Figure 25.10. An instance of the user-defined VB class *Chromosome* represents a specific system configuration and is represented as a collection of genes. Notice that the limit on gene count from this chromosome is 80. So for the cross dock problem, only 80 doors may be assigned. As previously mentioned, VB Collections are used to accumulate chromosomes in *Population, Offspring,* and *BestSolution.*

ProduceOffspring, as shown in Figure 25.11, develops a probability distribution function used to estimate the probability of a particular chromosome to spawn or mutate. For each generation, a chromosome's probability is based on its contribution to the sum of *P*s for all chromosomes in the population. The purpose of this approach is to ensure that "weak" chromosomes participate in the

```
Sub NextGeneration()
  Dim _
    Mu As Integer, _
    ReplacedPopulation As Integer
  ' select chromosomes
  Mu = CInt(Population.Count * 0.25)
  If Mu < OffSpring.Count _
  Then
      ReplacedPopulation = Mu
  Else
      ReplacedPopulation = OffSpring.Count
  End If
  For Index = 1 To ReplacedPopulation
    Population.Remove Population.Count
  Next Index
  For Index = 1 To ReplacedPopulation
      Set NewChromosome = OffSpring(1)
      OffSpring.Remove 1
      InsertionSort Population
  Next Index
  Do While (OffSpring.Count > 0)
    OffSpring.Remove 1
  Loop
End Sub
```

Figure 25.9   Genetic algorithm—sub *NextGeneration*.

```
Public _
  P As Single, _
  GenerationIndex As Integer, _
Private _
  Gene(1 To 80) As String
Public Sub InitialGene(ByVal Index As Integer, ByRef Allele As String)
    Gene(Index) = Allele
  End Sub
Public Function Allele(ByVal Index As Integer) As String
  Allele = Gene(Index)
End Function
```

Figure 25.10   Genetic algorithm—class *Chromosome*.

```
Sub ProduceOffspring()
  Dim _
    TotalP As Double, _
    Probability As Single, _
    CumulativeProbability As Single
  TotalP = 0
  For Index = 1 To Population.Count
    TotalP = TotalP + Population.Item(Index).P
  Next Index
  CumulativeProbability = 0
  For Index = 1 To Population.Count
    Population.Item(Index).Probability = _
                  Population.Item(Index).P / TotalP
    CumulativeProbability = _
        CumulativeProbability + Population.Item(Index).Probability
    Population.Item(Index).CumulativeProbability = Cumulative Probability
  Next Index
  Do While (1)
    Spawn
    ProcessChromosome
    If OffSpring.Count = 5 _
    Then
      Exit Sub
    End If
    Mutate
    ProcessChromosome
    If OffSpring.Count = 5 _
    Then
      Exit Sub
    End If
  Loop
End Sub
```

Figure 25.11   Genetic algorithm—sub *ProduceOffspring*.

```
Sub ProcessChromosome()
  If Not DuplicateChromosome(Population) And _
     Not DuplicateChromosome(OffSpring) And _
     Not DuplicateChromosome(BestSolution) _
  Then
      EvaluateP
      InsertionSort OffSpring
  End If
End Sub
```

Figure 25.12   Genetic algorithm—sub *ProcessChromosome*.

spawning process before being excluded from the population. As indicated, a total of five offspring are produced by either spawns or mutations.

ProcessChromosome, in Figure 25.12, is a procedure used by sub *ProduceOffspring* to ensure that a newly produced chromosome does not currently exist somewhere in the system.

DuplicateChromosome, presented in Figure 25.13, is a function that receives a chromosome and collection from sub *ProcessChromosome* and returns *True* when the chromosome resides in the collection.

Spawn, shown in Figure 25.14, uses the probability distribution function developed in sub *ProduceOffspring* to select two chromosomes and then invokes sub *SpliceChromosome* to perform the splicing operation.

```
Function DuplicateChromosome(Temp As Collection) As Boolean
  Dim _
    DuplicateCount As Integer
  DuplicateChromosome = False
  DuplicateCount = Temp.Count
  For Index = 1 To Temp.Count
    For Jndex = 1 To NumberGenes
      If Population.Item(Index).Allele(Jndex) <> _
                             NewChromosome.Allele(Jndex) _
    Then
        DuplicateCount = DuplicateCount - 1
        Exit For
      End If
    Next Jndex
  Next Index
  If DuplicateCount > 0 _
  Then
    DuplicateChromosome = True
  End If
End Function
```

Figure 25.13   Genetic algorithm—function *DuplicateChromosome*.

```
Sub Spawn()
  Dim _
    ParentIndex(1 To 2) As Integer
  Do While (ParentIndex(1) = ParentIndex(2))
    Rn(1) = Random(10)
    Rn(2) = Random(11)
    Index = 1
    Do While (1)
      If Rn(1) > Population.Item(Index).CumulativeProbability Or _
        Index = Population.Count _
      Then
        ParentIndex(1) = Index
        Exit Do
      End If
      Index = Index + 1
    Loop
    Do While (1)
      If Rn(2) > Population.Item(Index).CumulativeProbability Or _
        Index = Population.Count _
      Then
        ParentIndex(2) = Index
        Exit Do
      End If
      Index = Index + 1
    Loop
  Loop
  Set SpliceParent1 = Population(ParentIndex(1))
  Set SpliceParent2 = Population(ParentIndex(2))
  SpliceChromosomes ' SpliceParent1, SpliceParent1
End Sub
```

Figure 25.14   Genetic algorithm—sub *Spawn*.

SpliceChromosomes, presented in Figure 25.15, implements the previously described splicing operation.

AlleleOnChromosome, presented in Figure 25.16, is a support function to assist with implementation of the splicing procedure. *NewChromosome* is the spliced chromosome, and *True* is returned when an allele is found on *NewChromosome*.

```
Sub SpliceChromosomes()
 Dim _
   AlleleToUse As String, _
   AlleleToSave As String, _
   AlleleList As New Collection, _
   Gene As Gene
 Set NewChromosome = New Chromosome
 NewChromosome.GenerationIndex = GenerationIndex
 For Index = 1 To NumberGenes
   If SpliceParent1.Allele(Index) = SpliceParent2.Allele(Index) _
   Then
     NewChromosome.InitialGene Index, SpliceParent1.Allele(Index)
   End If
 Next Index  ' found same alleles
 For Index = 1 To NumberGenes
   If NewChromosome.Allele(Index) = "" _
   Then
     If Random(1) < 0.5 _
     Then
       AlleleToUse = SpliceParent1.Allele(Index)
       AlleleToSave = SpliceParent2.Allele(Index)
     Else
       AlleleToUse = SpliceParent2.Allele(Index)
       AlleleToSave = SpliceParent1.Allele(Index)
     End If
     If Not AlleleOnChromosome(Index - 1, AlleleToUse) _
     Then 'not on chromosome
        NewChromosome.InitialGene Index, AlleleToUse
        Jndex = AlleleListDuplicate(AlleleList, AlleleToUse)
        If Jndex > 0 _
        Then ' on saved chromosome list
           AlleleList.Remove Jndex
        End If
     End If
     If Not AlleleOnChromosome(Index - 1, AlleleToSave) And _
        AlleleListDuplicate(AlleleList, AlleleToSave) = 0 _
     Then ' on neither list
       Set Gene = New Gene
       Gene.Allele = AlleleToSave
       AlleleList.Add Item:=Gene
     End If
   End If
 Next Index
 For Index = 1 To NumberGenes ' final pass
   If NewChromosome.Allele(Index) = "" _
   Then
     NewChromosome.InitialGene Index, AlleleList(1).Allele
     AlleleList.Remove 1
   End If
 Next Index
End Sub
```

Figure 25.15 Genetic algorithm—sub *SpliceChromosome*.

```
Function AlleleOnChromosome(SearchLength As Integer, _
                            AlleleToFind As String) As Boolean
    Dim _
      Index As Integer
    AlleleOnChromosome = False
    For Index = 1 To SearchLength
      If NewChromosome.Allele(Index) = AlleleToFind _
      Then
          AlleleOnChromosome = True
          Exit Function
      End If
    Next Index
End Function
```

Figure 25.16 Genetic algorithm—function *AlleleOnChromosome*.

```
Function AlleleListDuplicate(AlleleList As Collection, _
                             AlleleToFind As String) As Integer
    Dim _
      Index As Integer
    AlleleListDuplicate = 0
    For Index = 1 To AlleleList.Count
      If AlleleList.Item(Index).Allele = AlleleToFind _
      Then
          AlleleListDuplicate = Index
          Exit Function
      End If
    Next Index
End Function
```

Figure 25.17 Genetic algorithm—function *AlleleListDuplicate*.

```
Sub Mutate()
    Rn(1) = Random(12)
    For Index = 1 To Population.Count
      If Rn(1) < Population.Item(Index).CumulativeProbability _
      Then
          Exit For
      End If
    Next Index
    Set MutationParent = Population(Index)
    MutateChromosome ' Parent
End Sub
```

Figure 25.18 Genetic algorithm—sub *Mutate*.

AlleleListDuplicate, in Figure 25.17, is a support function to assist with implementation of the splicing procedure. A zero is returned when no duplicate allele exits; otherwise, the index location of the duplicate is returned.

Mutate, shown in Figure 25.18 and referenced by sub *ProduceOffspring*, uses the probability distribution function to select a chromosome for mutation and then calls sub *MutateChromosome*.

MutateChromosome, in Figure 25.19, implements the previously described mutation procedure.

```
Sub MutateChromosome()
  Dim _
    BegIndex As Integer, _
    EndIndex As Integer, _
    TmpRn As Single
  Set  NewChromosome = New Chromosome
  NewChromosome.GenerationIndex = GenerationIndex
  For Index = 1 To NumberGenes
    NewChromosome.InitialGene Index, MutationParent.Allele(Index)
  Next Index
  Rn(1) = Random(13)
  Rn(2) = Random(14)
  If Rn(2) < Rn(1) _
  Then
    TmpRn = Rn(1)
    Rn(1) = Rn(2)
    Rn(2) = TmpRn
  End If
  BegIndex = CInt(NumberGenes * Rn(1))
  EndIndex = CInt(NumberGenes * Rn(2))
  Do While (1)
    Do While (BegIndex <= NumberGenes)
      If (NewChromosome.Allele(BegIndex) <> "XX") _
      Then
        Exi t Do
      End If
      BegIndex = BegIndex + 1
    Loop
    Do While (EndIndex > 0)
      If (MutationParent.Allele(EndIndex) <> "XX") _
      Then
        Exit Do
      End If
      EndIndex = EndIndex      - 1
    Loop
    If (BegIndex >= EndIndex) _
    Then
      Exit Do
    End If
    NewChromosome.InitialGene BegIndex, MutationParent.Allele(EndIndex)
    NewChromosome.InitialGene EndIndex, MutationParent.Allele(BegIndex)
    BegIndex = BegIndex + 1
    EndIndex = EndIndex      - 1
  Loop
End Sub
```

Figure 25.19   Genetic algorithm—sub *MutateChromosome*.

Differences in data requirements for the traditional simulation model in Section 24.4 and the search algorithm/simulation are minimal. An initial population, as presented in Figure 25.20, is required. The population may be a mixture of randomly generated configurations and configurations that the modeler wants included. The only other difference is that the number of generations to produce must be specified.

The user interface for the search algorithm/simulation model is shown in Figure 25.21. Notice that the options are *Run Search Program* or *Run Simulation*. When *Run Simulation* is used, the modeler must detail the dock configuration as shown in columns I and J. These columns are blank for the search algorithm because, as indicated in Figure 25.5, the configuration is controlled by the search program.

The best configuration recommendations for a small problem are presented in Figure 25.22. This model begins with a very small population, and, as discussed, only five offspring were produced for each generation. As for most genetic algorithms, the model must be tuned to the class of problem. Previously listed design issues must be considered before making conclusions about the quality of recommendations from these solutions.

Configuration 1			Configuration 2			Configuration 3			Configuration 4			Configuration 5		
Strip			Strip			Strip			Strip			Strip		
Door	Dest	Door	Door	Dest	Door	Door	Dest	Door	Door	Dest	Door	Door	Dest	Door
22	BHM1	21	23	MRT1	21	23	MRT1	21	25	MRT1	21	21	BHM1	29
24	CLT1	23	25	NAS1	27	25	BHM1	29	27	NAS1	23	23	CLT1	31
26	GBO1	25	31	BHM1	29	31	CLT1	27	29	BHM1	33	35	GBO1	33
28	KNX1	27	33	CLT1	35	33	GBO1	37	31	CLT1	35	37	KNX1	30
30	KNX2	29	24	GBO1	37	24	KNX1	30	26	GBO1	37	22	KNX2	32
32	KNX3	31	26	KNX3	22	26	KNX2	28	28	KNX3	22	24	KNX3	26
34	KNX4	33	32	KNX4	28	32	KNX3	22	30	KNX4	24	34	KNX4	28
36	MRT1	35	34	KNX1	30	34	KNX4	36	32	KNX1	34	36	MRT1	25
	NAS1	37		KNX2	36		NAS1	35		KNX2	36		NAS1	27

Figure 25.20 Genetic algorithm—initial population.

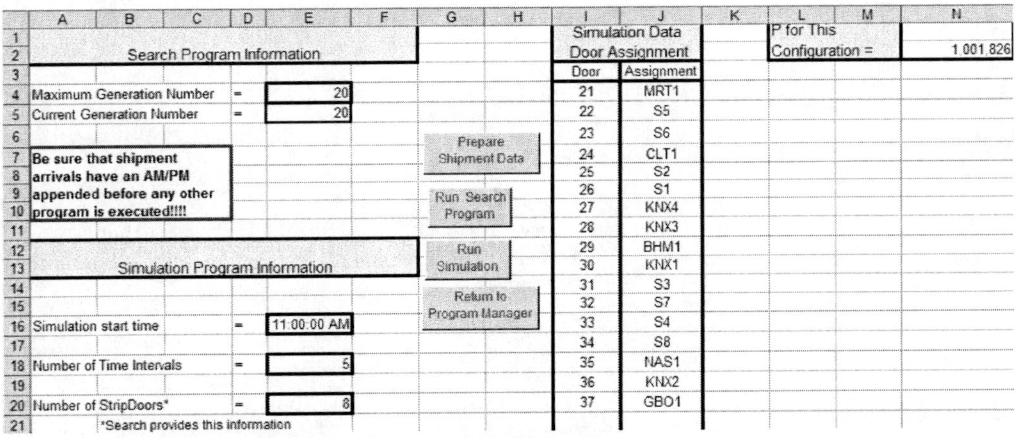

Figure 25.21 Genetic algorithm—user initialed spreadsheet.

Doors/F(x)	Best Dock Configurations 961,604	970,466	971,616	974,222	978,020
8	XX	XX	XX	XX	XX
9	XX	XX	XX	XX	XX
10	XX	XX	XX	XX	XX
11	XX	XX	XX	XX	XX
12	XX	XX	XX	XX	XX
13	XX	XX	XX	XX	XX
14	XX	XX	XX	XX	XX
15	XX	XX	XX	XX	XX
16	XX	XX	XX	XX	XX
17	XX	XX	XX	XX	XX
18	XX	XX	XX	XX	XX
19	XX	XX	XX	XX	XX
20	XX	XX	XX	XX	XX
21	MRT1	MRT1	MRT1	MRT1	MRT1
22	S5	KNX3	KNX3	KNX3	S5
23	KNX4	KNX4	S1	S1	S1
24	S6	S5	S5	S5	S6
25	S2	S2	S2	S2	S2
26	KNX3	S6	S6	S6	KNX3
27	S1	S1	KNX4	NAS1	KNX4
28	CLT1	CLT1	CLT1	CLT1	CLT1
29	BHM1	BHM1	BHM1	BHM1	BHM1
30	KNX1	KNX1	KNX1	KNX1	KNX1
31	S3	S3	S3	S4	S3
32	S7	S7	S7	S7	S7
33	S4	S4	S4	KNX4	S4
34	S8	S8	S8	S8	S8
35	NAS1	NAS1	NAS1	S3	NAS1
36	KNX2	KNX2	KNX2	KNX2	KNX2
37	GBO1	GBO1	GBO1	GBO1	GBO1

Figure 25.22   Genetic algorithm—best solutions.

25.6 Summary

This chapter presented techniques for using discrete event simulation with traditional research tools. A case study demonstrated features of DEEDS that enable the design of experiments to be incorporated into the analysis with minimal incremental effort. We also used a case study to demonstrate the use of search algorithms with DEEDS.

References

1. Law, A. M. et al., *Simulation Modeling and Analysis*, 4th ed. New York: McGraw-Hill Book Company, 2007.
2. Lemis, L. M. and S. K. Park, *Discrete Event Simulation: A First Course*. Upper Saddle River, NJ: Prentice-Hall, Inc., 2006.
3. Mitsuo, G. and R. Cheng, *Genetic Algorithms and Engineering Optimization*. New York: Wiley, 2000.
4. Rardin, R. L., *Optimization in Operations Research*. Upper Saddle River, NJ: Prentice-Hall, Inc., 1998.
5. Taha, H. A., *Operations Research: An Introduction*. Upper Saddle River, NJ: Prentice-Hall, Inc., 2007.

Appendix A:
Excel 2003 Installation

Before starting the remaining installation, **be sure that Excel is closed**. Select the *Browse* option. In the DEEDS CD, choose directory *Setup* as shown. Double-click on the *setup* icon.

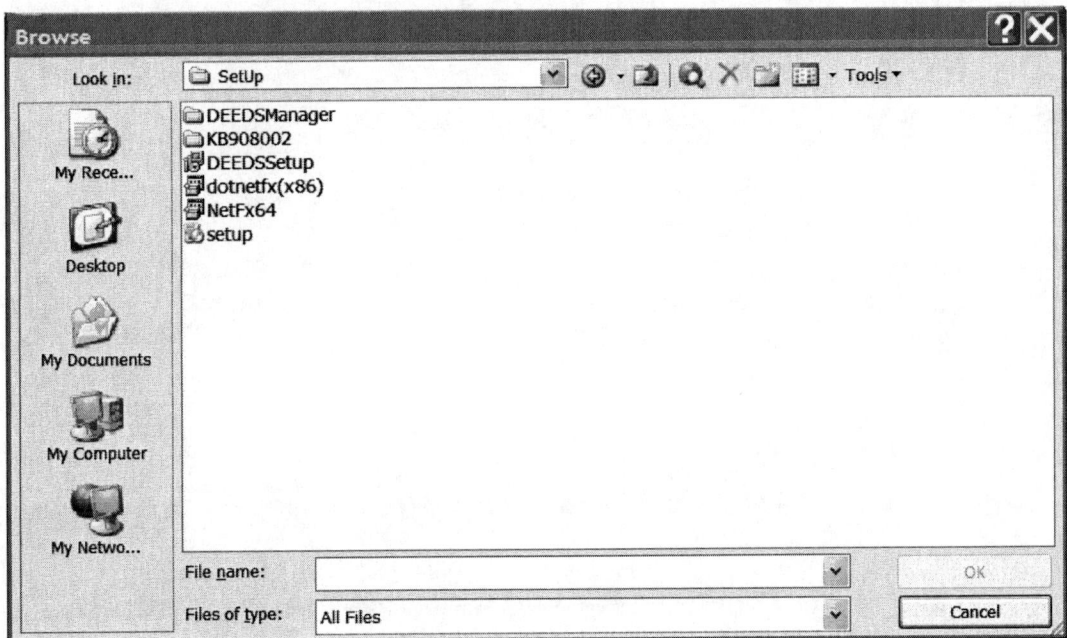

If Framework 2.0 is not on your computer, then, as shown, accept the terms of the licensing agreement.

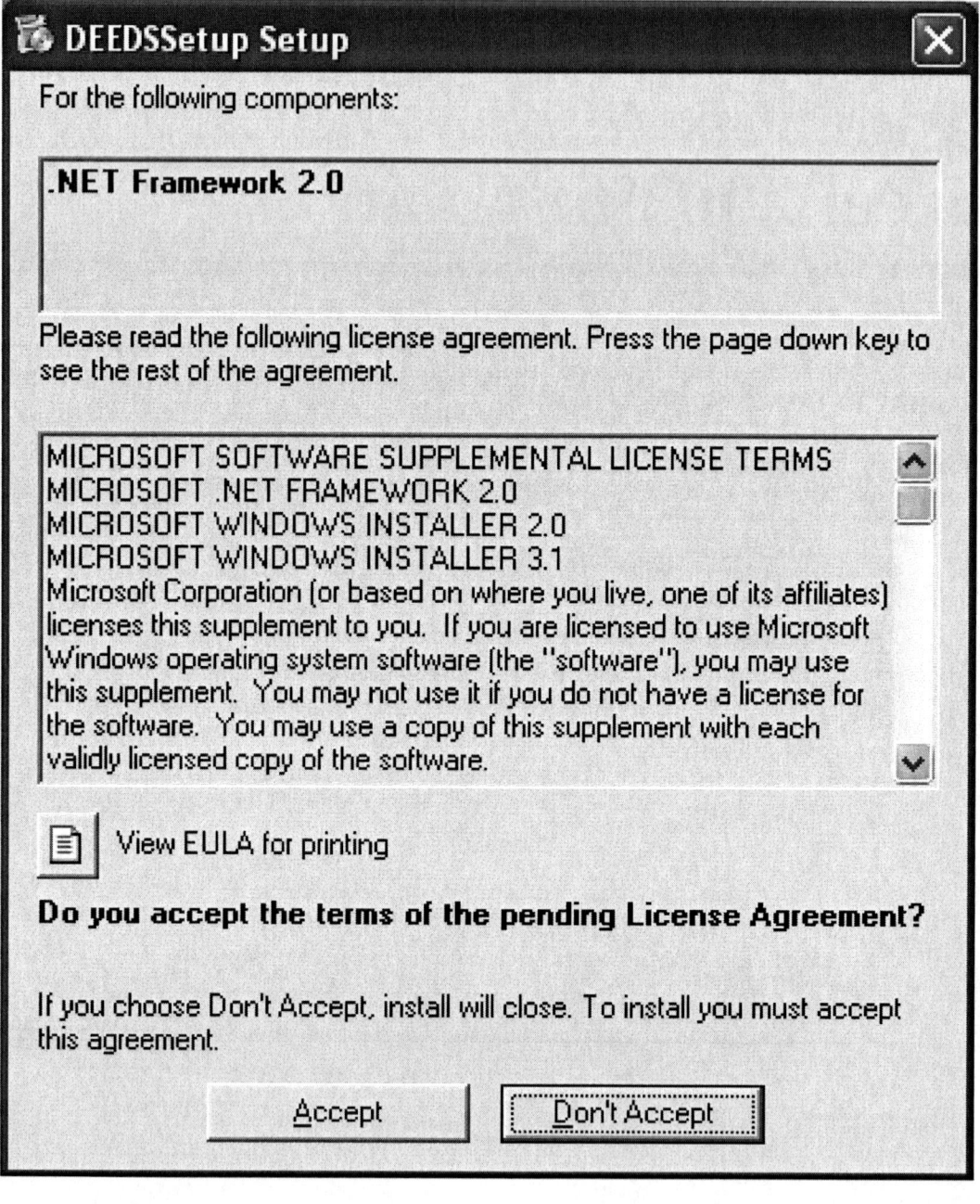

The select *Install* on the following:

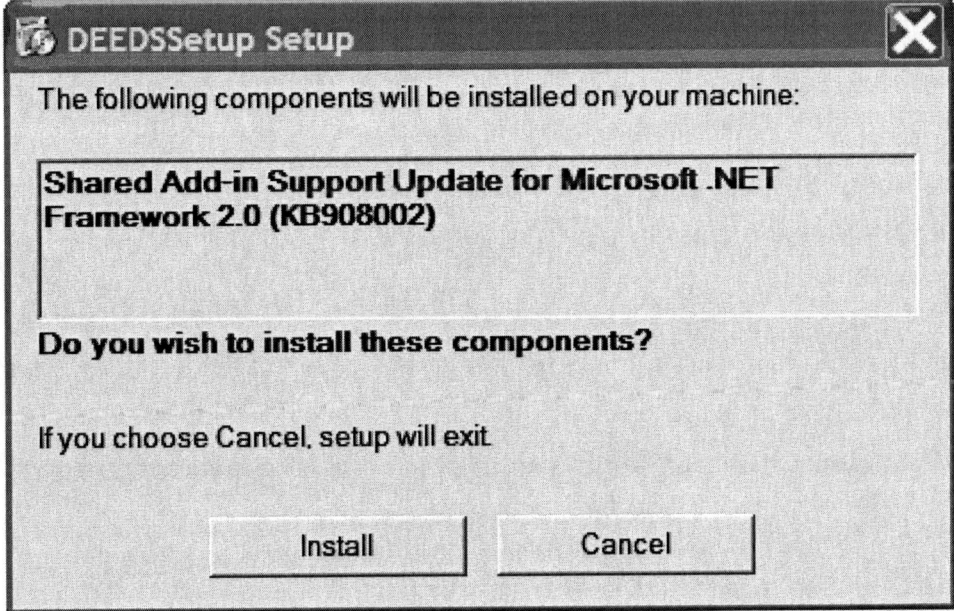

And *Next* on this frame:

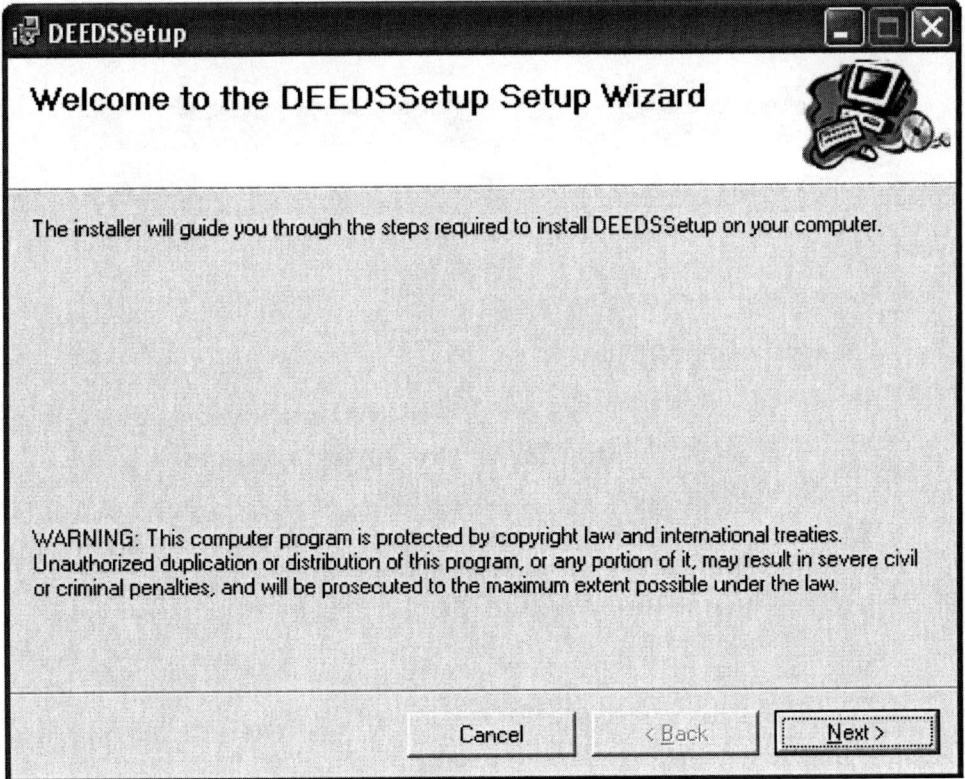

The default install directory may be changed; otherwise, select *Next*.

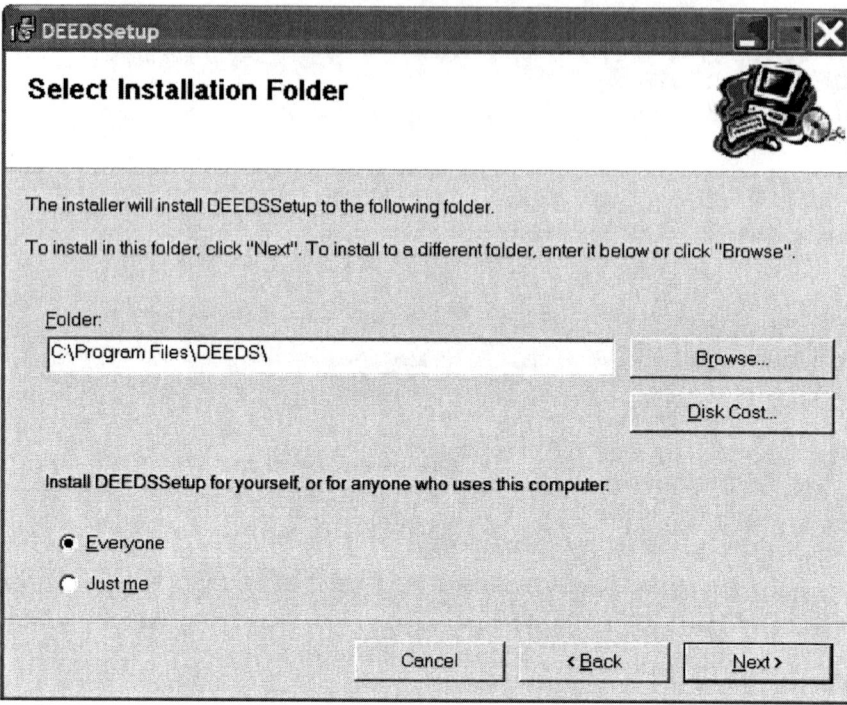

The following image shows the last opportunity to terminate the installation. Select *Next* to continue.

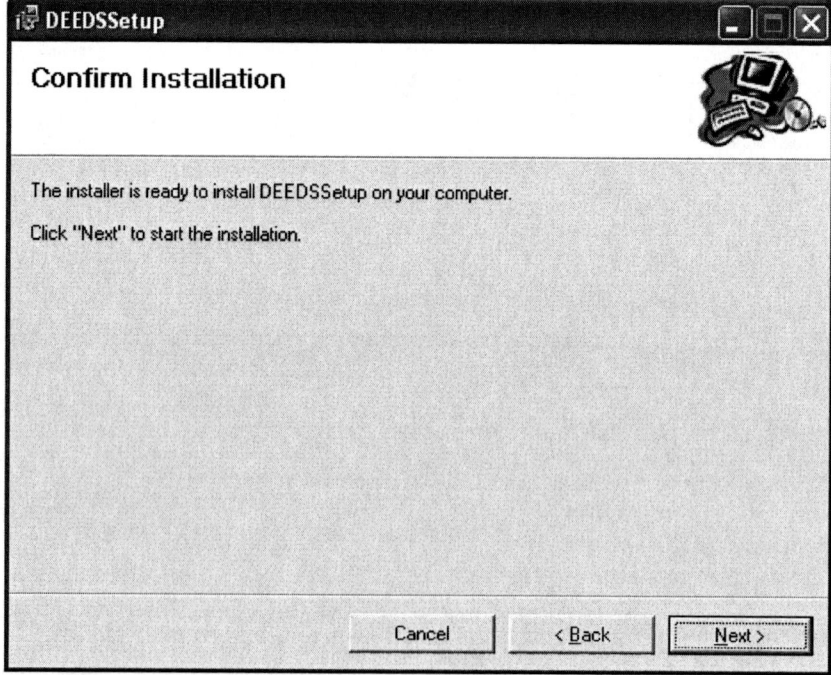

The setup application installs and registers the dll, and so on.

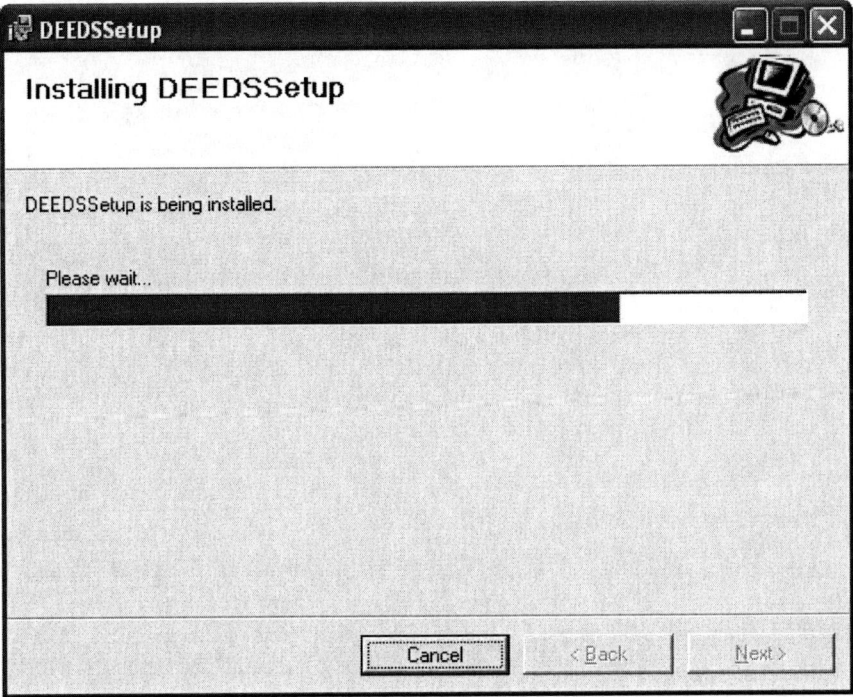

Setup is now complete.

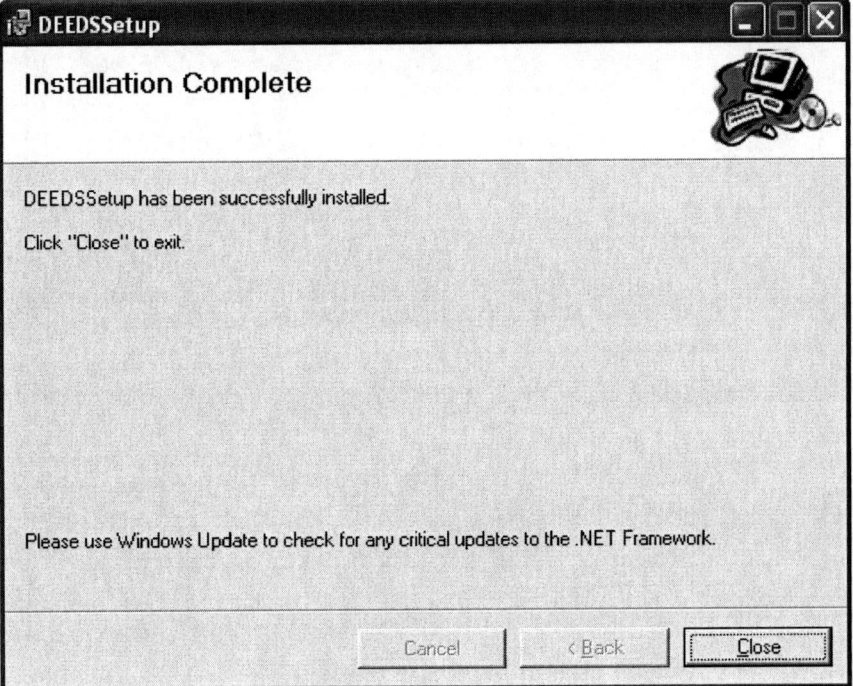

The above installation is performed only one time. Use the *Windows XP* control panel to uninstall DEEDS.

Again, select the *Browse* option. In the DEEDS CD, choose directory *Setup* and then directory *DEEDSManager* to obtain the following:

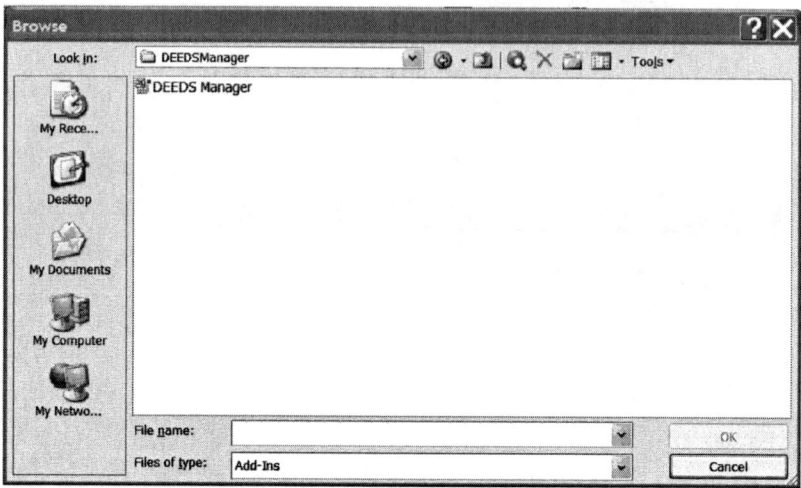

Copy *DEEDS Manager* to directory C:\Program Files\DEEDS. Now, open Excel and enable macros by selecting *Tools → Macros → Security* to obtain the following:

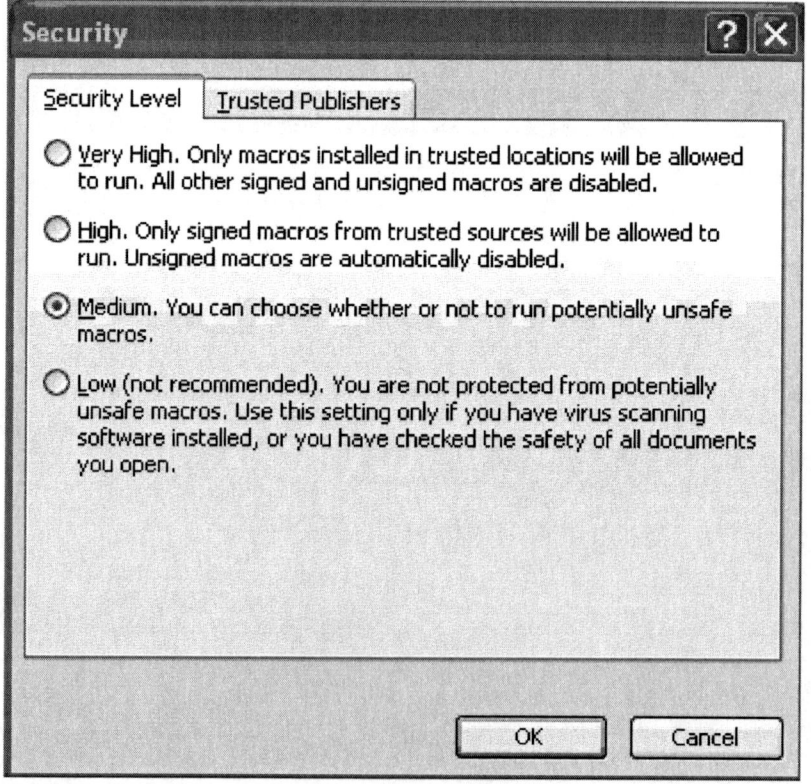

Set the security level to *Medium*.

In *Excel* select *Tools* → *Add-Ins* to obtain the following:

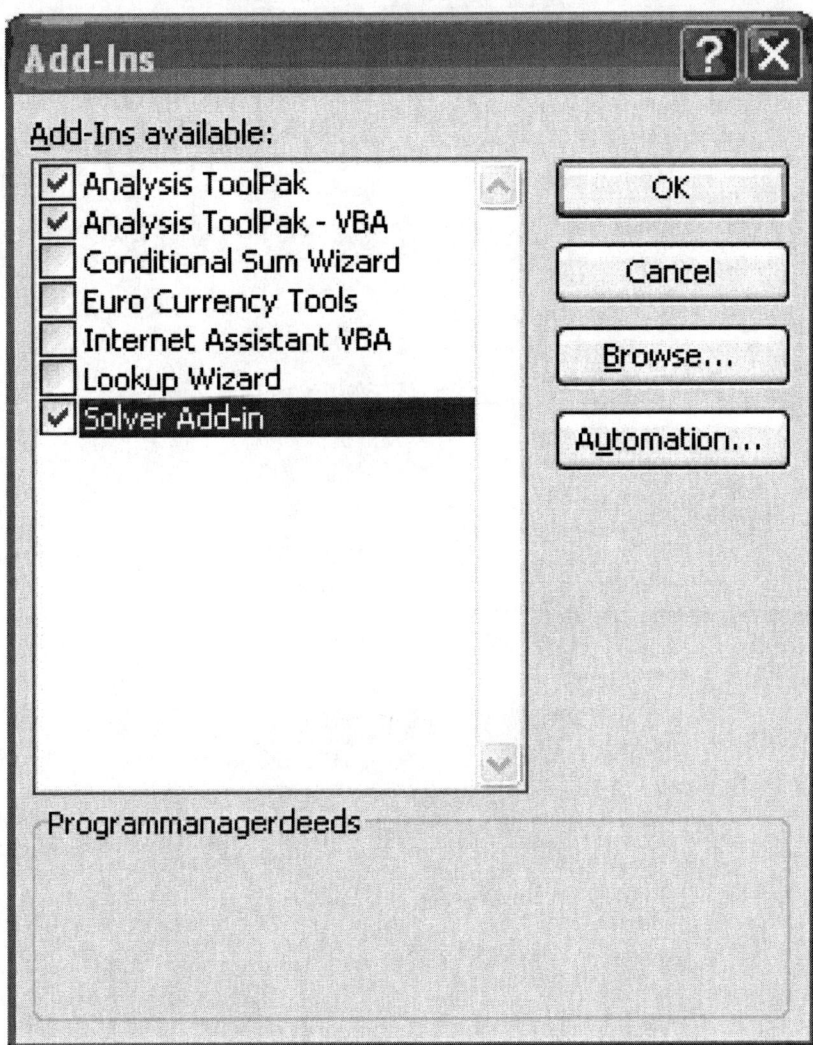

Select the *Browse* option; choose directory *C:\Program Files\DEEDS* to obtain the following:

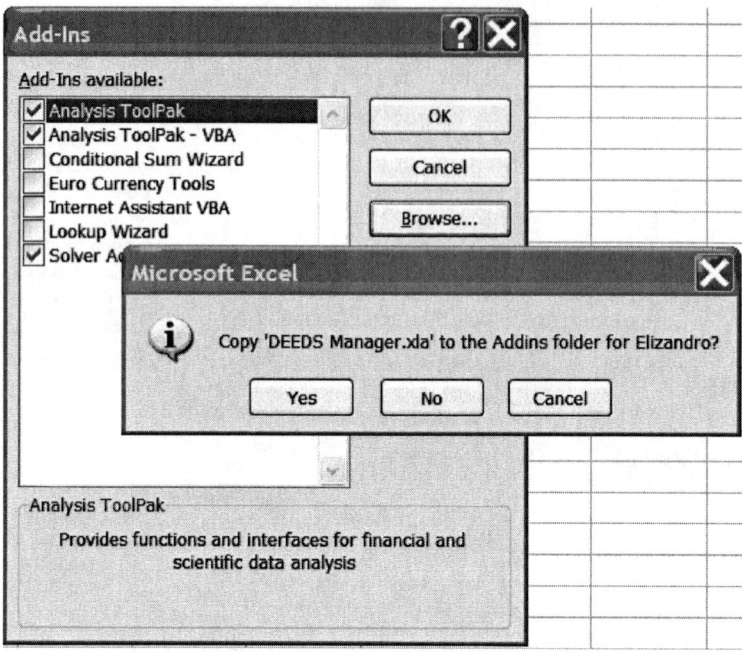

Double-click *No* and the *Add-Ins* will appear as shown.

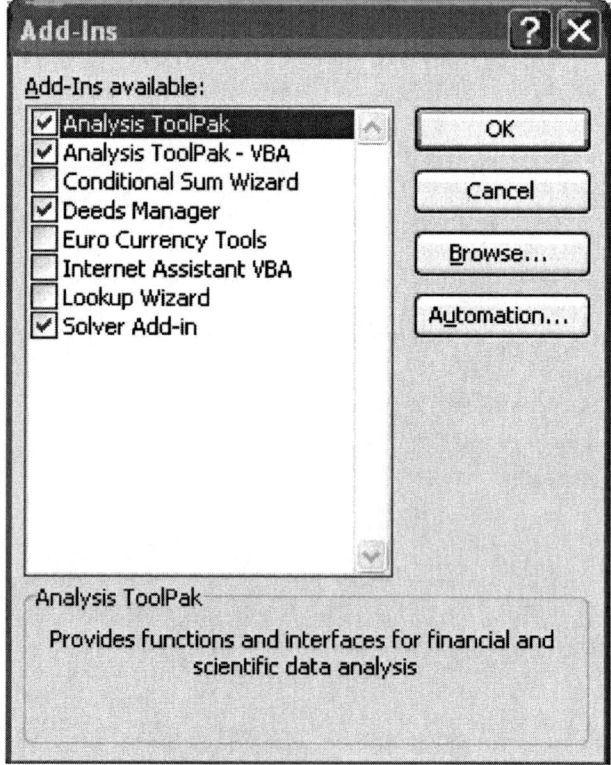

The remaining instructions are for each new DEEDS project. Open Excel and select *Tools → Macro → Macros*.

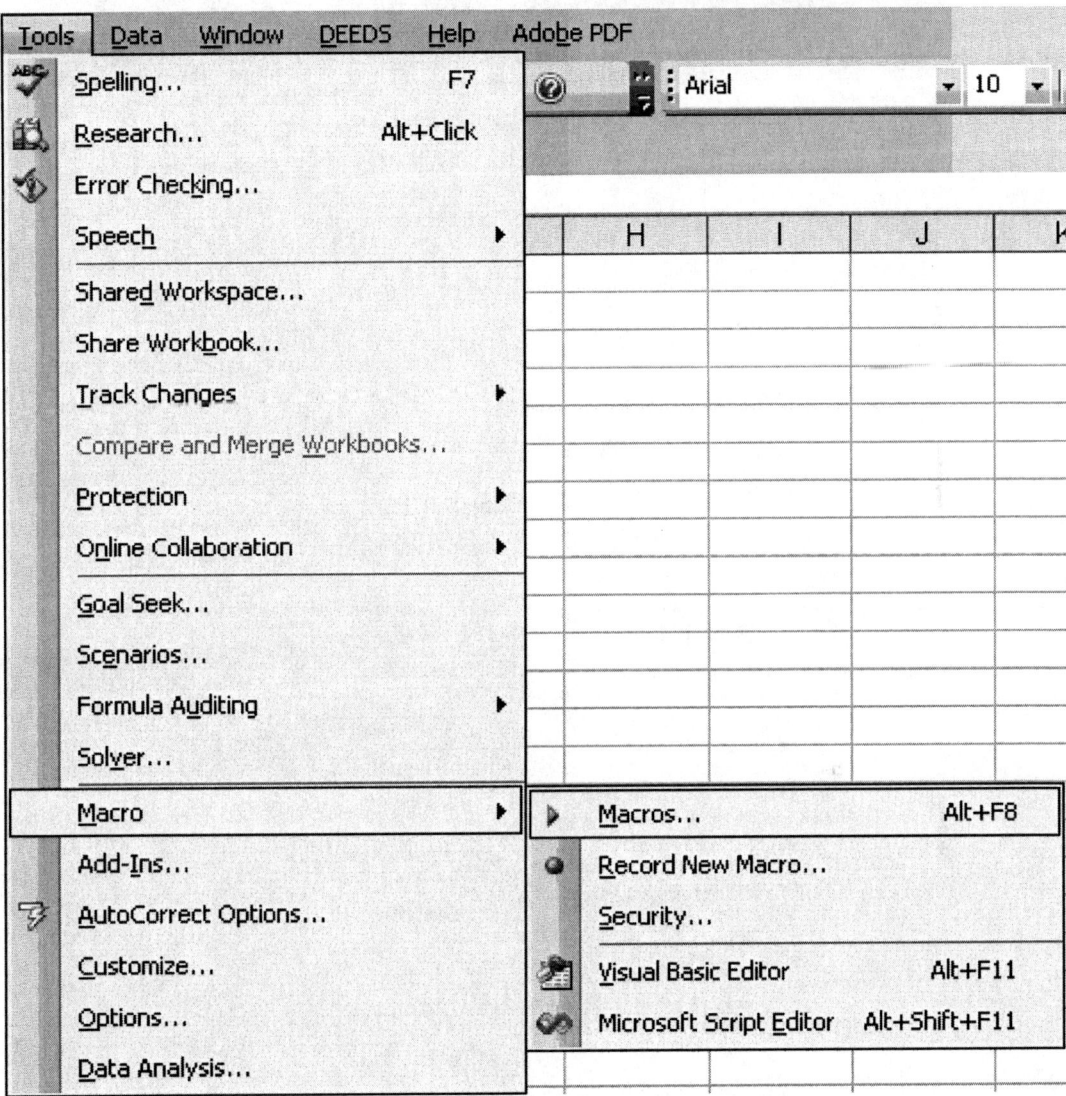

Enter the macro name, *newproject*, and select *Run* to format the spreadsheets.

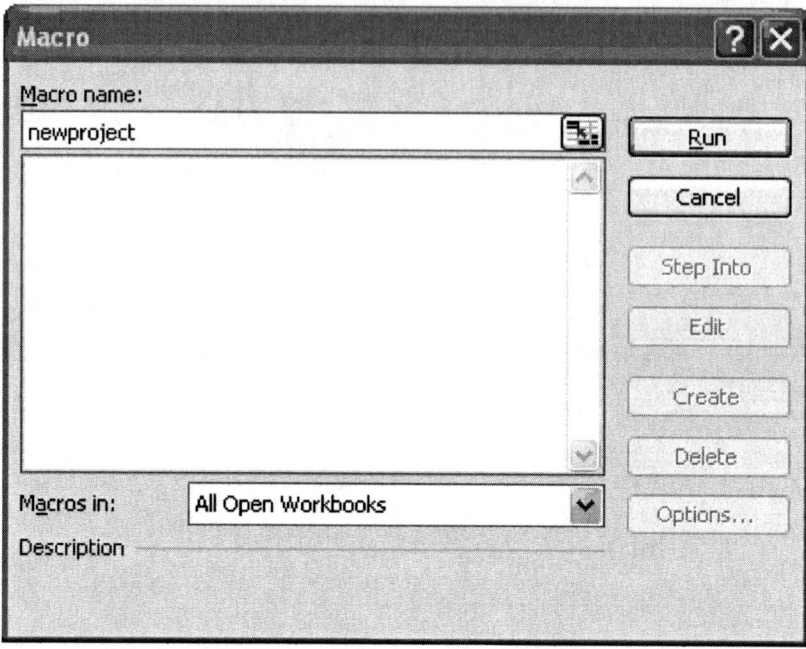

The following program segments are on spreadsheet *Initial*:

```
Option Explicit
' Controller Module
Public Distribution as Distribution
Public Simulator as Simulator
Public Sub StartSimulation()
  Dim _
    PrimaryEvent As Transaction
  Set Simulator = New Simulator
  Set Distribution = New Distribution
  Distribution.InitialPRNG
  Simulator.Initial
  EventsManager.Initial
  Set PrimaryEvent = Simulator.PrimaryEvent
  Do While (1)
     If PrimaryEvent Is Nothing _
    Then
      EventsManager.ShutDown
      Simulator.ShutDown
      Exit Do
    Else
      EventsManager.Parser PrimaryEvent
      Set PrimaryEvent = Simulator.PrimaryEvent
    End If
  Loop
End Sub
```

```
Option Explicit
' This WorkBook Module
Private Sub Workbook_Open()
  Dim _
    HelpMenu As CommandBarControl, _
    NewMenu As CommandBarPopup, _
    MenuItem As CommandBarControl, _
    SubMenuItem As CommandBarButton
  Set HelpMenu = Application.CommandBars(1).FindControl(ID:=30010)
  If HelpMenu Is Nothing Then
      Set NewMenu = Application.CommandBars(1).Controls _
      .Add(Type:=msoControlPopup, temporary:=True)
  Else
      Set NewMenu = Application.CommandBars(1).Controls _
      .Add(Type:=msoControlPopup, Before:=HelpMenu.index, temporary:=True)
  End If
  NewMenu.Caption = "&DEEDS"
  Set MenuItem = NewMenu.Controls.Add(Type:=msoControlButton)
  With MenuItem
   .Caption = "Start Simulation"
   .OnAction = "StartSimulation"
  End With
  Set MenuItem = NewMenu.Controls.Add(Type:=msoControlButton)
  With MenuItem
   .Caption = "Program Manager"
   .OnAction = "OpenProgramManager"
  End With
End Sub
Private Sub Workbook_BeforeClose(Cancel As Boolean)
  CloseWorkbook
End Sub
```

Controller and *EventsManager* modules are now created. From Excel, select *Tools → Macro → Visual Basic Editor* to enter the Visual Basic for Applications (VBA) environment. Select *Insert → Module* twice to create the two modules shown.

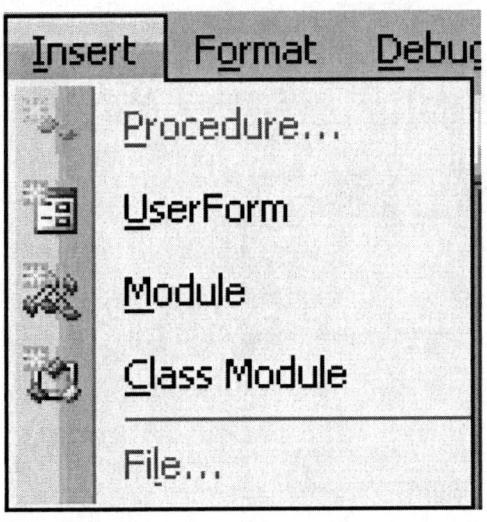
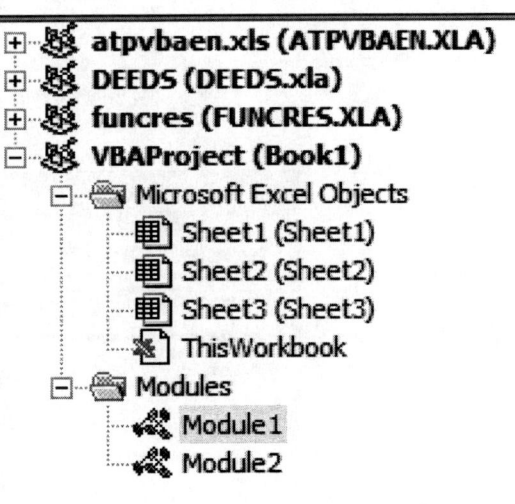

With *Module1* highlighted, select *View → Properties Window* and change *Module1* name to *Controller*. In a similar manner change *Module2* name to *EventsManager*.

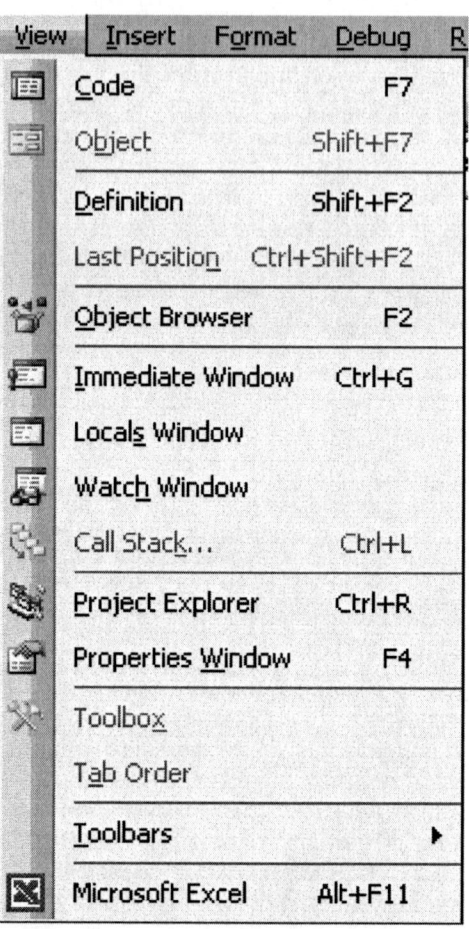

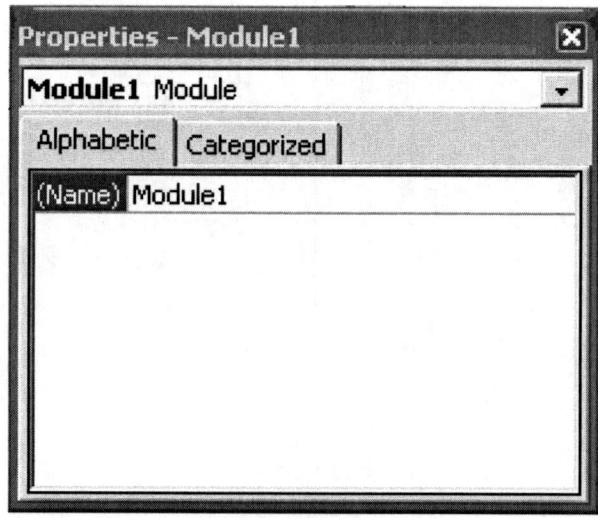

Now copy the program segment *ThisWorkbook* from the *InitialModel* spreadsheet to *ThisWorkbook* shown above the modules and under *VBAProject*, then copy the program segment *Controller Module* from the *InitialModel* spreadsheet to the *Controller* module. Delete these programs from the spreadsheet. The final step before developing the simulation program is to reference *Deeds* and *DeedsManager* from the VBA environment. In VBA select *Tools* → *References* and as shown below, choose the DEEDS option.

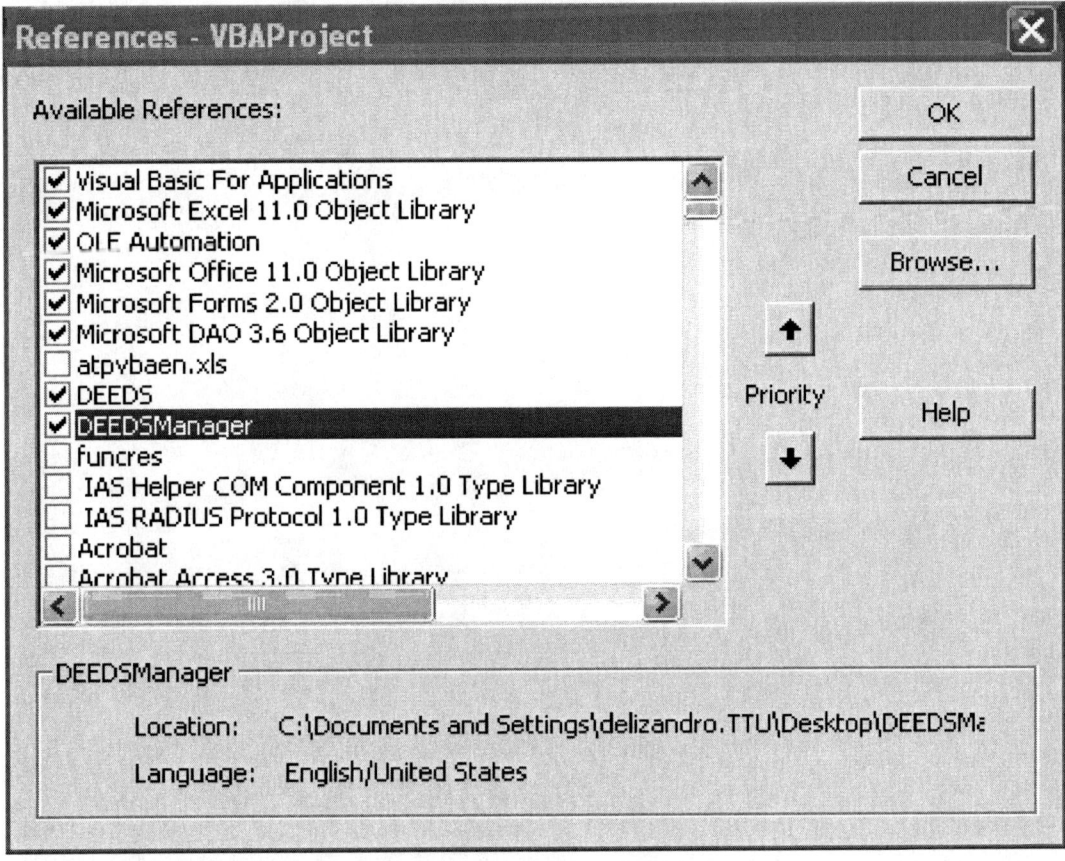

In a similar manner, select *Browse* to find *DeedsManager* in directory *C:\Program Files\DEEDS*. You are now ready to develop a simulation model.

Appendix B:
Excel 2007 Installation

The installation files for *DEEDS* are in the companion CD. In the CD are the following directories and files:

Name	Date modified	Type	Size
DEEDS2003	8/8/2011 3:39 PM	File folder	
DEEDS2007	11/6/2011 10:04 PM	File folder	
NewProject	8/8/2011 3:38 PM	Microsoft Office E...	223 KB
StartSimulation	8/8/2011 3:28 PM	Text Document	1 KB

Open the *DEEDS2007* directory to obtain:

Name	Date modified	Type
Chapter2	12/24/2010 7:56 PM	File folder
Chapter3	12/24/2010 9:53 PM	File folder
Chapter4	12/24/2010 7:45 PM	File folder
Chapter10	12/24/2010 7:45 PM	File folder
Chapter11	12/26/2010 10:59 ...	File folder
Chapter12	12/24/2010 7:45 PM	File folder
Chapter13	12/24/2010 7:45 PM	File folder
Chapter14	12/26/2010 10:26 ...	File folder
Chapter15	12/24/2010 7:45 PM	File folder
Chapter16	12/24/2010 7:45 PM	File folder
Chapter17	12/24/2010 7:45 PM	File folder
Chapter18	12/24/2010 7:45 PM	File folder
Chapter19	12/24/2010 7:45 PM	File folder
Chapter20	12/24/2010 7:45 PM	File folder
Chapter21	12/24/2010 7:45 PM	File folder
Chapter22	12/24/2010 7:45 PM	File folder
Chapter23	12/24/2010 7:45 PM	File folder
SetUp	12/24/2010 7:45 PM	File folder

In the Setup directory, click on *DEEDSSetup*.

Name	Date modified	Type	Size
DEEDSManager	12/26/2010 8:32 PM	File folder	
DEEDSSetup	7/8/2010 2:40 PM	Windows Installer ...	979 KB
dotnetfx(x86)	1/27/2010 10:30 PM	Application	22,960 KB
NetFx64	1/27/2010 10:30 PM	Application	46,290 KB
setup	7/8/2010 2:39 PM	Application	471 KB

Then click on *Next* when the following prompt is available.

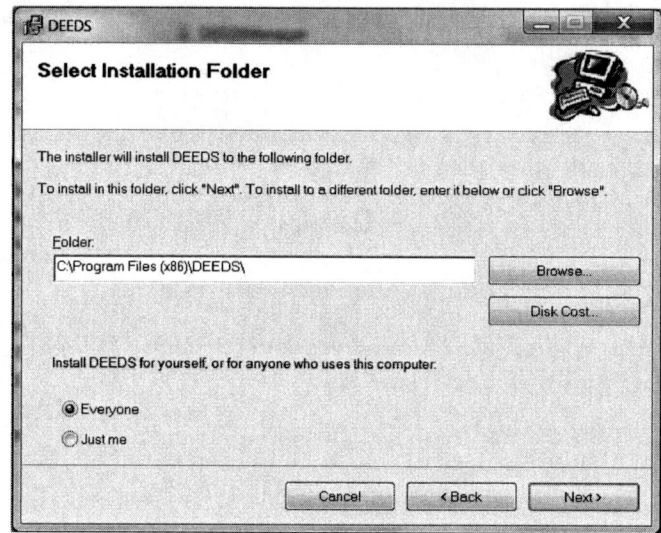

On the following prompt, again click on *Next*.

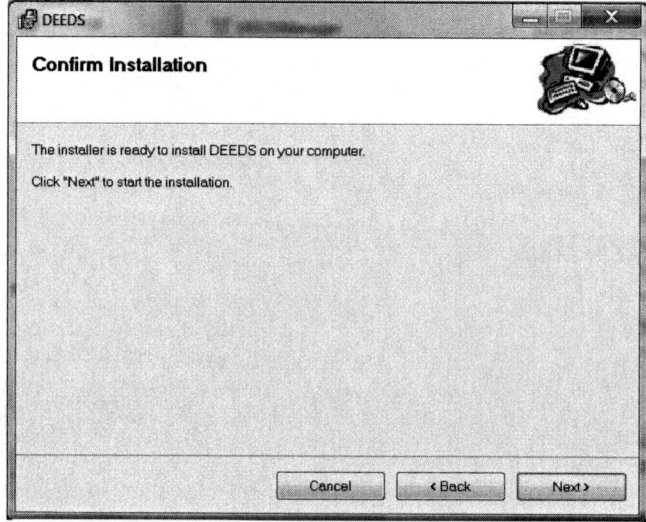

The *DEEDS* installation is now complete. However, the *DeedsManager* file must be copied to the *DEEDS* directory created by the installation program. From the *DEEDSManager* directory in the *SetUp* directory, copy the *DEEDSManager.xlam* file to the *Program Files (x86)\DEEDS* directory. Ensure that the *DEEDSManager* file was correctly copied to the *DEEDS* directory by examining the *DEEDS* directory, as shown.

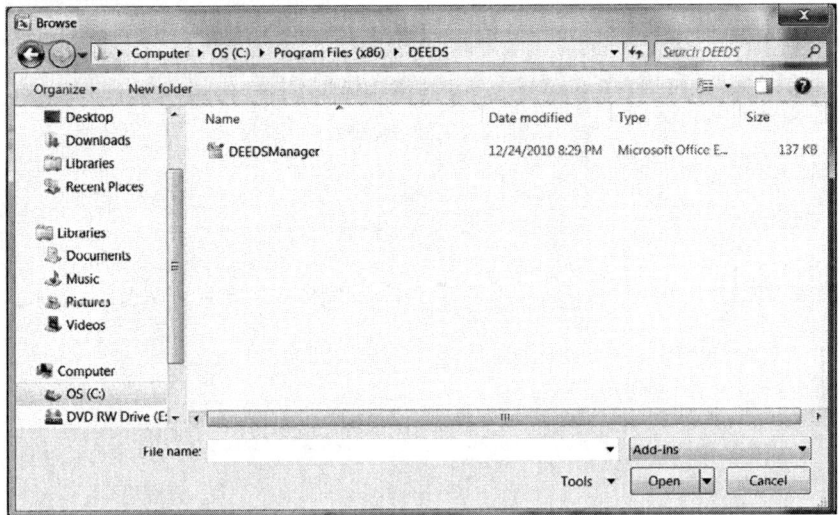

However, *DEEDSManager.xlam*, an *addin* file, must be visible to EXCEL. Click on the *Office Button* in the upper right corner of Excel to obtain the Excel Options shown.

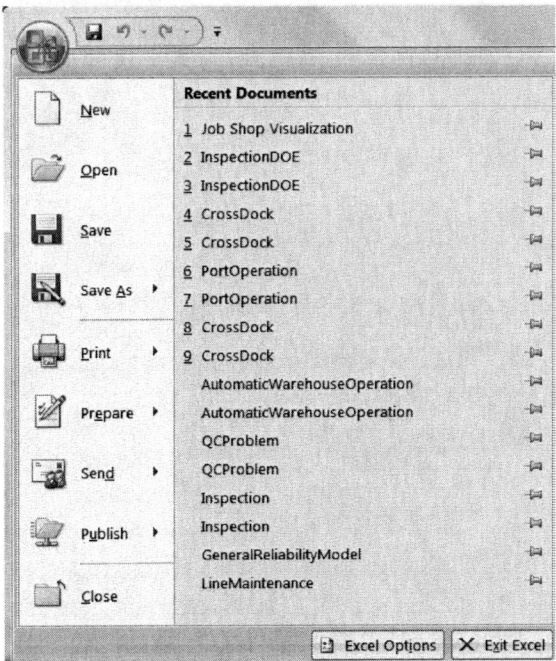

Now click on Excel Options at the bottom of the menu to obtain the following menu. To enable access to the VBA environment, check the *Show Developer tab in the Ribbon* option.

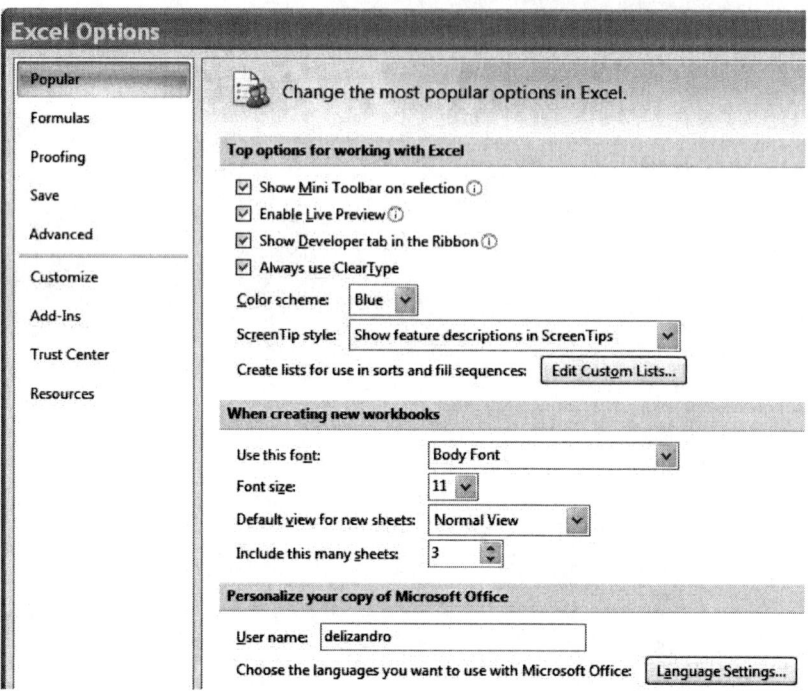

Then, double click the *Add-ins* option (7th item in the left column) of the *Excel Options* menu to obtain the following menu.

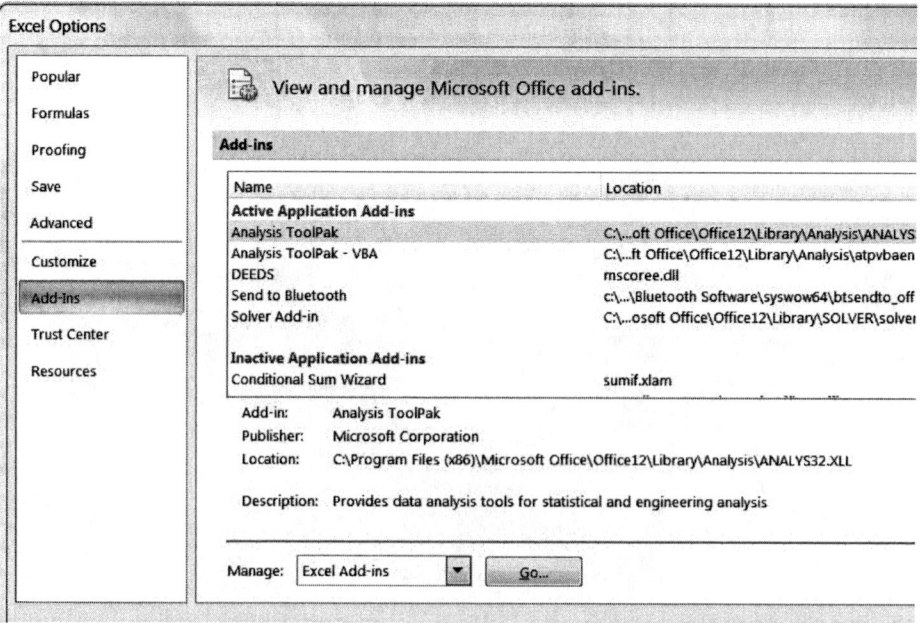

At the bottom of the *View and manage Microsoft Office add-ins* select *Go* on the Manage: Excel Add-ins. Browse in *Program Files\DEEDS* and select *DeedsManager.xlam* and *OK*. Before and after menus are shown below.

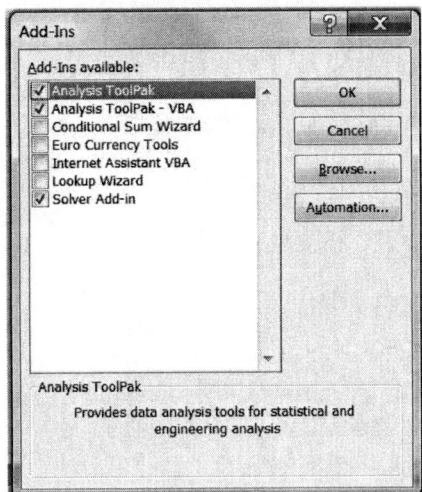 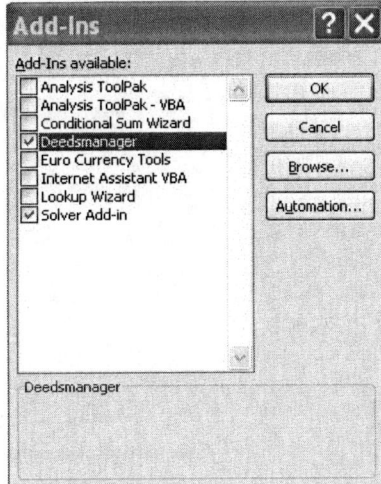

Before launching a simulation program, as described in Chapter 8, ensure that *DEEDS* and *DEEDSManager* are also visible in VBA. In the VBA environment, click on *Tools* and then *References* to obtain the References—VBA Project menu.

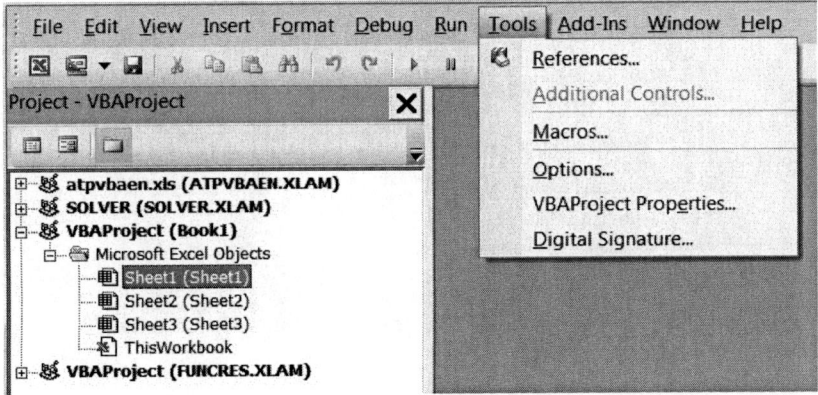

If the *DEEDS* and *DEEDSManager* options are available, as shown on the left, simply check the box and then *OK*.

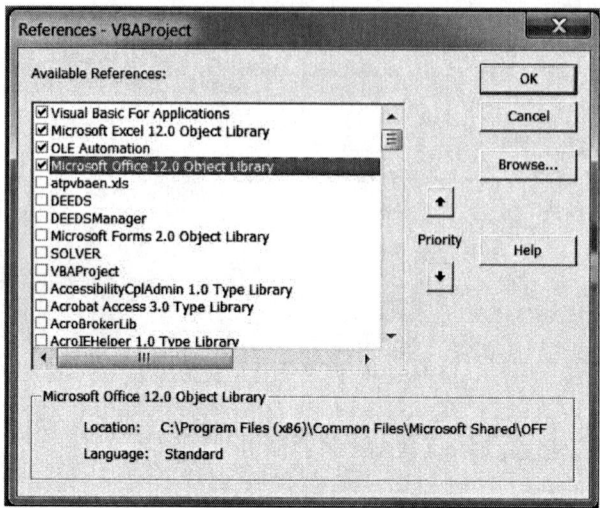

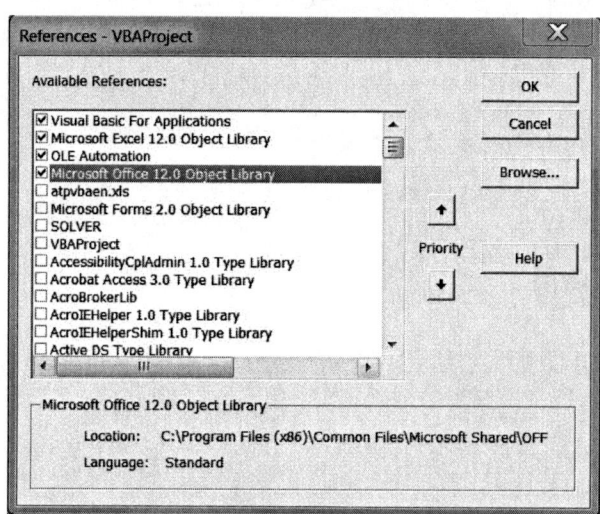

If the menu is similar to the one on the right, then browse the *Program Files (x86)\DEEDS* directory. Set the Files of type option to All Files(*.*) to obtain the following menu. In two steps, and the order unimportant, open the *DEEDS.tlb* and the *DEEDSManger* file.

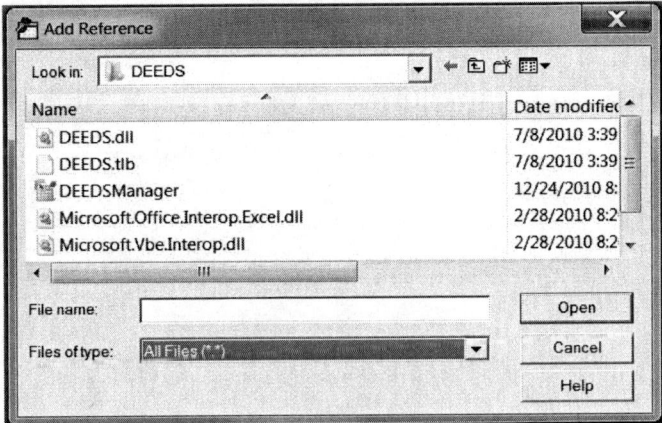

The simulation program now has access to all *DEEDS* files needed to perform the simulation. Each time a simulation workbook is open, the following Security Warning option is presented.

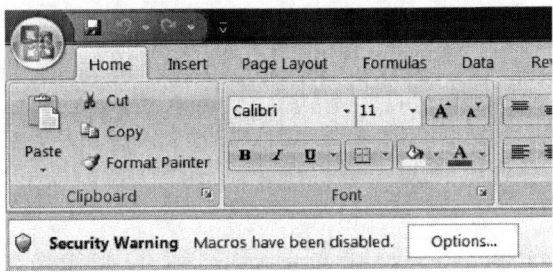

Click on *options* to obtain the following:

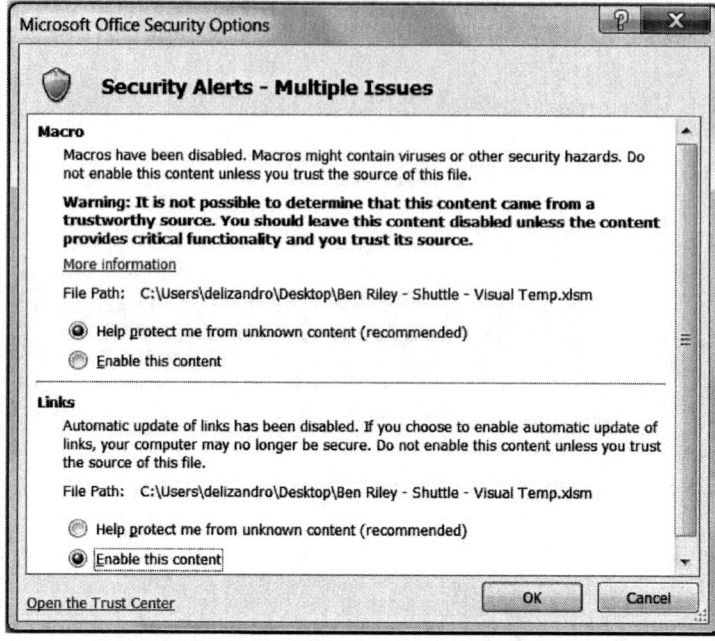

Now click on *Enable this content* in the *Macro* section. The *Add-Ins* option, as shown below, will now be visible on the Excel Ribbon. Clicking on *DEEDS* and then the appropriate option launches the desired feature of *DEEDS*.

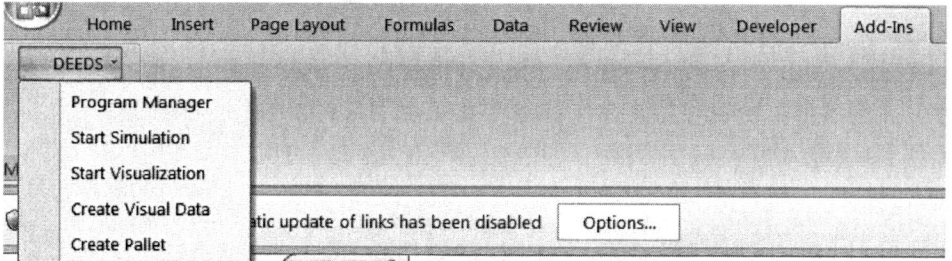

The *NewProject* workbook in the companion CD has all of the references initialized and may be used for developing a new simulation project.

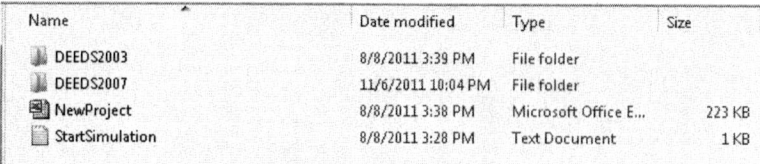

Appendix C: Classes and Procedures

A. Source

Name	T	Parameters	Description	Error Conditions
Name	F		Returns Name As String	
EventType	F		Returns EventType As String	
New-Distribution	S	Distr As String, P1 As Variant, P2 As Variant, P3 As Variant	Next arrival scheduled from Distribution	
Priority	F		Returns Source Priority As Byte	
NewPriority	S	Priority As Integer	Priority for next arrival scheduled	
Schedule-NextArrival	S		Arrival scheduled from Distribution	
Active	F		Returns Source Status As Boolean (True → not suspended)	
Suspend	S		Arrival removed from PEL	
Restart	S			Arrival already scheduled

T represents the type of VB program segment: F ⇒ function and S ⇒ sub.

B. Queue

Name	T	Parameters	Description	Error Conditions
Name	F		Returns Name As String	
Capacity	F		Returns Capacity As Integer	
NewCapacity	S	NewCapacity As Integer	New value for Capacity	
Discipline	F		Returns Discipline As String (LIFO or FIFO)	
NewDiscipline	S	NewDiscipline As String	Changes Discipline	Invalid queue discipline
View	F	ViewIndex As Integer	Returns Transaction reference	ViewIndex exceeds contents
Length	F		Returns number of transactions As Integer	
Full	F		Returns Boolean (Full → True)	
ListLq	S	MaximumObservations As Integer	Writes Maximum Observations of Lq statistic to queue output spreadsheet.	
ListWq	S	MaximumObservations As Integer	Writes Maximum Observations statistic to queue output spreadsheet.	
AverageLq	F		Returns average queue length since last observation As Single	
AverageWq	F		Returns average time in queue since last observation As Single	
Join	S	Joining As Transaction	Transaction queued according to Discipline	Invalid transaction
JoinFront	S	Joining As Transaction	Transaction in front of Queue	Invalid transaction
JoinAfter	S	AfterIndex As Integer, Joining As Transaction	Transaction queued behind indexed transaction	Invalid transaction, invalid AfterIndex

B. Queue (*Continued*)

Name	T	Parameters	Description	Error Conditions
JoinOn HighAttribute	S	AttrIndex As Integer, Joining As Transaction	Transaction queued on ascending attribute value	Invalid transaction, invalid AttrIndex
JoinOn LowAttribute	S	AttrIndex As Integer, Joining As Transaction	Transaction queued on descending attribute value	Invalid transaction, invalid AttrIndex
JoinOnPriority		Joining As Transaction	Transaction queued on priority	Invalid transaction
Remove	S	RemoveIndex As Integer, NumberRemoved As Integer	Delete range of transactions	Invalid remove range
Depart	F		Removes transaction from front and returns transaction reference	Empty Queue
Detach	F	DetachIndex As Integer	Removes transaction from indexed location and returns transaction reference	DetachIndex exceeds contents

T represents the type of VB program segment: F ⇒ function and S ⇒ sub.

C. Facility

Name	T	Parameters	Description	Error Conditions
Name	F		Returns Name As String	
EventType	F		Returns EventType As String	
NewDistribution	S	Distr As String, P1 As Variant, P2 As Variant, P3 As Variant	Next arrival scheduled from Distribution	
NumberOf-Servers	F		Returns Number Of Servers As Integer	
NumberOf-BusyServers	F		Returns Number Of Busy Servers As Integer	
NumberOf-IdleServers	F		Returns Number Of Idle Servers As Integer	

(*Continued*)

C. Facility (*Continued*)

Name	T	Parameters	Description	Error Conditions
AverageNumber OfBusyServers	F		Returns average number of busy servers for current observation	
View	F	ViewIndex As Integer	Returns transaction reference and transaction remains in the Facility	ViewIndex exceeds number in Facility
Length	F		Returns Number of Transactions As Integer	
NewNumber-Servers	S	NumberOf-Servers	New value for NumberOfServers	
Engage	S	Client As Transaction, Servers As Integer	Servers engaged but not on PEL	Insufficient idle servers, invalid transaction
EngageAnd-Service	S	Client As Transaction, Servers As Integer [MyServiceTime As Single]	Servers engaged and scheduled on PEL	Insufficient idle servers, invalid transaction
Service	S	Client As Transaction [MyServiceTime As Single]	Transaction scheduled for service completion	Invalid transaction, transaction must be in Facility
ReceiveServers	S	Client As Transaction, Number-Received As Integer		Invalid transaction, insufficient idle servers
ReturnServers	S	Client As Transaction, Number-Returned As Integer	Servers returned but transaction remains in the Facility	Invalid transaction, transaction on PEL, transaction must have at least one server
Disengage	S	Client As Transaction	Disengages servers and Facility	Invalid transaction, transaction not in Facility
Interrupt	F	InterruptIndex As Integer	Interrupts indexed transaction and returns transaction reference	InterruptIndex exceeds number in Facility

T represents the type of VB program segment: F ⇒ function and S ⇒ sub.

D. Delay

Name	T	Parameters	Description	Error Conditions
Name	F		Returns Name As String	
EventType	F		Returns EventType As String	
NewDistribution	S	Distr As String, P1 As Variant, P2 As Variant, P3 As Variant	Next arrival scheduled from Distribution	
Length	F		Returns number in Delay As Integer	
Start	S	Resident As Transaction [MyDelayTime As Single]	Transaction begins Delay	Invalid transaction
View	F	ViewIndex As Integer	Returns Reference As Transaction	Invalid ViewIndex
Interrupt	F	InterruptIndex As Integer	Remove indexed transaction and returns transaction reference	Invalid Interrupt-Index, transaction not in the Delay

T represents the type of VB program segment: F ⇒ function and S ⇒ sub.

E. Transaction

Name	T	Parameters	Description	Error Conditions
SourceID	F		Returns SourceID As String	
TransID	F		Returns TransactionID As Integer	
SerialNumber	F		Returns Transaction serial number as Long	
CreationTime	F		Returns CreationTime As Single	
Priority	F		Returns Priority As Byte	
DimAttributes	S	NewSize As Integer	Resizes the number of Transaction attributes to NewSize	

(Continued)

E. Transaction (*Continued*)

Name	T	Parameters	Description	Error Conditions
NewPriority	S	Priority As Byte	Changes Priority	
EventType	F		Returns EventType As String	
NewEventType	S	EventType As String	Changes Transaction EventType	
Scheduled	F		Returns Boolean (OnPEL ⇒ True)	
TimeStamp	F		Returns TimeStamp As Single	
NewTimeStamp	S	TimeStamp As Single	Changes Transaction TimeStamp	
Copy	F	Returns Transaction	Returns Copy of Transaction (different SerialNumber and CreationTime)	
Assign	S	AttrIndex As Integer, AttrValue As Variant	Attribute assigned attribute value	Invalid AttrIndex
A	F	AttrIndex As Integer	Returns attribute value As Variant	Invalid AttrIndex
TransitTime	F	AttrIndex As Integer	Returns CurrentTime – Attribute Value(AttrIndex) As Single	Invalid AttrIndex
SystemTime	F		Returns CurrentTime – CreationTime As Single	

T represents the type of VB program segment: F ⇒ function and S ⇒ sub.

F. Statistic

Name	T	Parameters	Description	Error Conditions
Name	F		Returns Name As String	
Average	F		Returns average value since last observation As Single	
Collect	S	Value As Single	Changes time-based statistic or records observation-based statistic	
List	S	MaximumObservation As Integer	List Maximum-Observation number of statistical variable	
TimeBetween-Arrivals	S		Collects time between arrivals	

T represents the type of VB program segment: F ⇒ function and S ⇒ sub.

G. PDF and Table

Name	T	Parameters	Description	Error Conditions
UserPDF RV	F	[RnID As Integer]	Returns PDF value As Single	
UserTables F	F	X As Single	Returns F(X) As Single	

T represents the type of VB program segment: F ⇒ function and S ⇒ sub.

H. Simulator

Name	T	Parameters	Description	Error Conditions
NewTransaction	F		Returns new Transaction (initialized with CreationTime and SerialNumber)	
CurrentTime	F		Returns clock time As Single	
Current-Observation	F		Returns ObservationNumber As Integer	
ViewPEL	F		Returns PEL reference As Collection	
Instance	F	Name As String, SetOfObjects As Collection	Returns Object As reference to instance with Name match	
IndexOf	F	Name As String, SetOfObjects As Collection	Returns Index As Integer to instance with Name match	
InsertInPEL	S	NextEvent As Transaction	Insert NextEvent in PEL	
RemoveFromPEL	S	SerialNumber As Long	Remove transaction from PEL	
Simulator-Message	S	Msg As String, Optional Trans	Displays Msg and time stamp on SimulatorMessages in InitialModel	
TraceReport	S	Msg As String [Trans]	Displays Msg and time stamp on TraceReport in InitialModel	
PrintTransaction	S	Trans As Transaction	Prints Trans information on spreadsheet Collections	
PrintPEL	S		Prints PEL information on spreadsheet Collections	
PrintNodes	S	ParamArray, NodeObject As Variant	Prints node information on spreadsheet Collections	
EndSimulation	S		Collects system data and ends the simulation	

T represents the type of VB program segment: F ⇒ function and S ⇒ sub.

Appendix D: Histograms Using Excel

To facilitate histogram construction, we use the *Histogram* program in Excel to tally frequencies, calculated percentages, and graph the relative frequencies and cumulative relative frequency. *Histogram* is in the *Data Analysis* option of Excel *Tools*. Before looking for *Histogram*, verify that *Add-Ins* includes the *Data Analysis* option. *Add-Ins* are also in the *Tools* option. *Histogram* is used with the following data to construct a continuous and discrete histogram.

	A	B	C	D	E	F	G	H	I	J	K
1	0.13	1.84	4.03	4.22	5.19	7.82	6.83	6.07	5.29	1.58	Bins
2	1.86	1.29	0.86	4.94	2.36	2.32	2.96	0.28	3.72	0.90	3
3	0.18	0.97	9.70	1.19	3.13	1.36	1.39	1.19	0.73	0.72	6
4	1.65	2.88	0.83	3.48	2.67	2.77	0.25	0.14	0.50	0.85	9
5	1.53	2.63	1.02	5.43	6.85	2.24	1.47	0.01	0.87	0.53	12
6	3.11	11.91	1.12	0.83	6.33	0.42	2.53	2.80	2.83	3.55	
7	5.69	0.31	1.50	5.19	0.94	3.40	1.51	6.02	1.70	5.42	
8	0.74	3.94	6.88	0.74	12.08	10.12	6.15	2.22	3.16	1.83	
9	0.51	3.09	2.85	5.77	1.37	0.16	2.11	0.17	1.41	1.80	
10	1.06	12.13	0.04	3.15	2.86	2.29	0.09	3.79	1.13	11.93	

The observations are sampled from a continuous distribution and entered into a spreadsheet, as shown above. The user then chooses histogram bins for class intervals based on the range of observations. Bin values are the upper limit of each class interval. A bin size of 3 with the first upper limit of 3 was chosen.

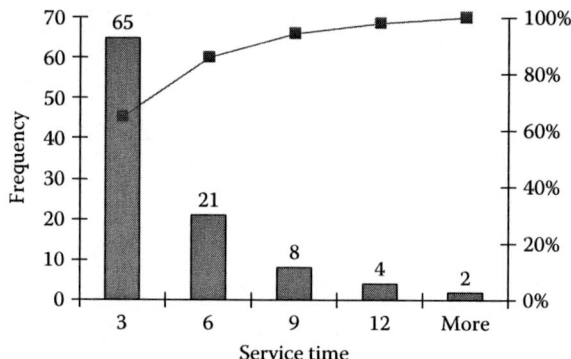

In the above *Histogram* form, the user specifies the spreadsheet range of the data, the range for bins, and the range for the *Histogram* program output. Output options *Chart Output* and *Cumulative Percentage* produce the following histogram and summary table on bins and frequencies within each bin.

By iteratively changing the bin description and invoking *Histogram*, the user is able to produce an "acceptable" histogram with minimal development time. A right click of the mouse when the cursor is on the right vertical axis enables you to format the axis. Setting the *Maximum* to 1.0 and *Major Unit* to 0.25 rescales the right axis.

Bins	Frequency	Cumulative %	Relative f_i
3	65	65.00	0.65
6	21	86.00	0.21
9	8	94.00	0.08
12	4	98.00	0.04
More	2	100.00	0.02

Only the first three columns of above table are *Histogram* output. *Relative* f_i is from a cell formula that computes the difference between current and previous cell values of *Cumulative %*. Notice that the first cell of *Cumulative %* is also the first cell of *Relative* f_i. Below is the relative frequency graph from Excel's *XY Scatter Chart*, to plot *Relative* f_i. The resulting function is an approximation of the continuous distribution.

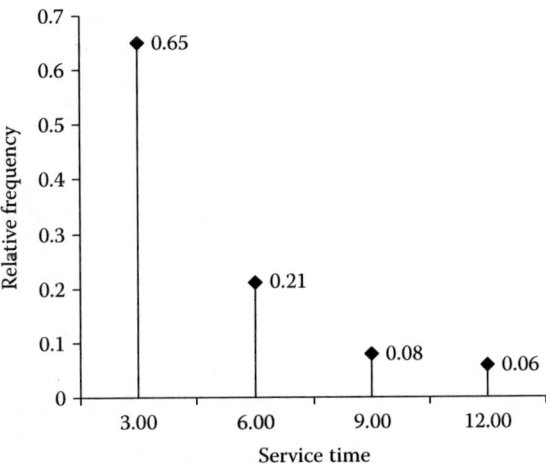

Scatter Chart options to produce the relative frequency graph are shown in the following figures:

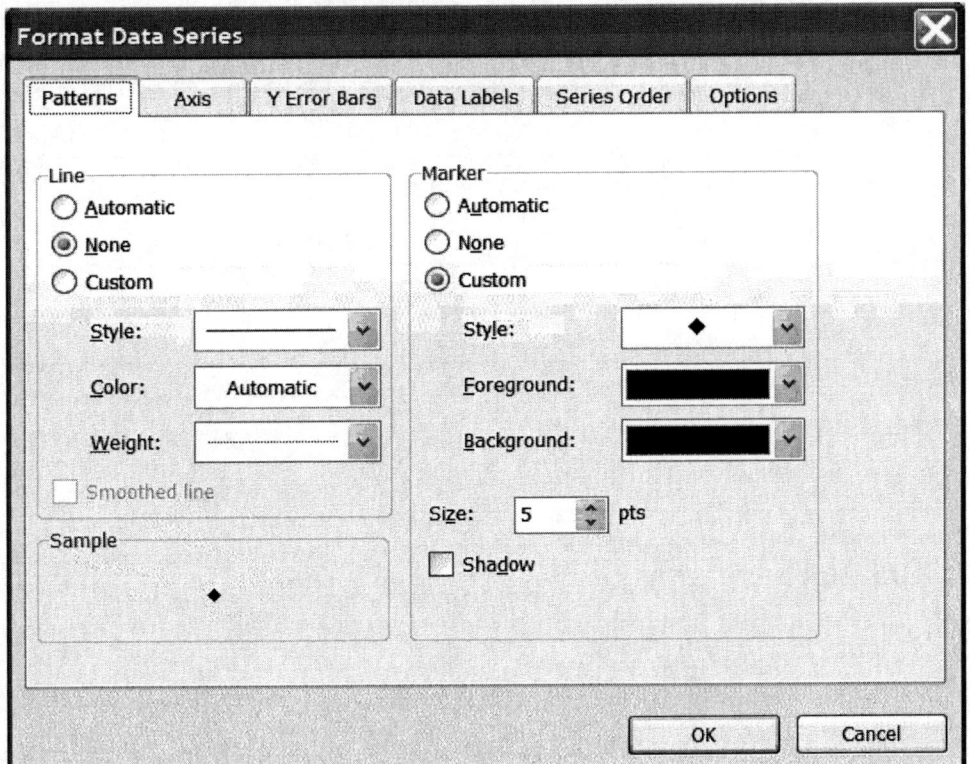

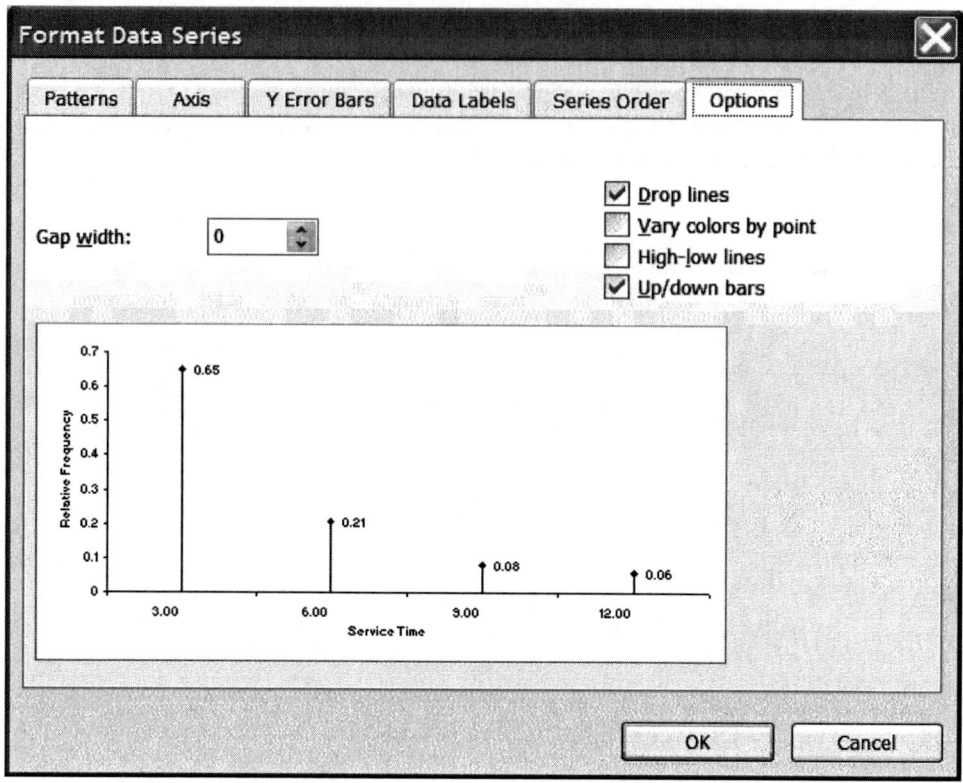

The following table summarizes the computations of the χ^2-test given the hypothesized exponential density function with mean 3.15 minutes.

Cell Number	Cell Boundary	Cell Mid-point	Observed Frequency O_i	Observed Frequency O_i	Ox_i/n_i	$nF_x(a_i)$	E_i	$(O_i - n_i)^2/n_i$
1	[0,3)	1.5	65	65	0.975	61.42	61.42	0.21
2	[3,6)	4.5	21	21	0.945	85.11	23.70	0.31
3	[6,9)	7.5	8	8	0.6	94.26	9.14	0.14
4	[9,12)	10.5	4	6	0.63	97.78	3.53	1.73
5	[12,∞)		2					
				100	3.15		100.00	2.39

Notice that cells 4 and 5 are combined because, as a general rule, it is recommended that the expected frequency E_i in any cell have no less than 5 points. Also, because of the tail of the exponential density function, 2.22% of the cumulative area above the upper interval limit of 12.0 is included in cell 4. As the mean of the exponential is estimated from the observations, the χ^2 will have $4 - 1 - 1 = 2$ degrees of freedom. For a significance level $\alpha = 0.05$, we obtain a critical value from the chi-square tables of $\chi^2_{2,.95} = 7.81$. From data in the above table, the user must calculate theoretical frequency E_i and the χ^2 statistic in the last column and then decide whether the hypothesis about the distribution should be accepted. Because the calculated χ^2 value of 2.39 in the table is less than the critical value of 7.81, we accept the hypothesis that the observed data is from an exponential distribution with mean 3.15 minutes.

This is also an iterative procedure that can be facilitated by using the *CHITEST* function in Excel. As shown in the figure below, *CHITEST* uses the observed and theoretical data to calculate the probability of a χ^2 statistic greater than 2.39. In this example, the probability is 0.495. The rejection criterion is when the calculated probability is less than α.

A potential problem with the χ^2 test for continuous random variables is the lack of definite rules for selecting cell width. In particular, the "grouping" of data and their representation by cell midpoints may lead to loss of useful information. In general, the objective is to select the cell width "sufficiently" small to minimize the adverse effect of midpoint approximation, but large enough to capture information regarding the shape of the distribution.

Index